CISM COURSES AND LECTURES

The series presents lecture notes, monographs, edited works and proceedings in the field of Mechanics, Engineering, Computer Science and Applied Mathematics.
Purpose of the series in to make known in the international scientific and technical community results obtained in some of the activities organized by CISM, the International Centre for Mechanical Sciences.

INTERNATIONAL CENTRE FOR MECHANICAL SCIENCES

COURSES AND LECTURES - No. 336

NON-EQUILIBRIUM THERMODYNAMICS WITH APPLICATION TO SOLIDS

DEDICATED TO THE MEMORY OF PROFESSOR THEODOR LEHMANN

EDITED BY

W. MUSCHIK
TECHNICAL UNIVERSITY, BERLIN

Springer-Verlag Wien GmbH

Le spese di stampa di questo volume sono in parte coperte da
contributi del Consiglio Nazionale delle Ricerche.

This volume contains 33 illustrations.

Originally published by CISM, Udine in 1993.

In order to make this volume available as economically and as rapidly as possible the authors' typescripts have been reproduced in their original forms. This method unfortunately has its typographical limitations but it is hoped that they in no way distract the reader.

ISBN 978-3-211-82453-5 ISBN 978-3-7091-4321-6 (eBook)
DOI 10.1007/978-3-7091-4321-6

PREFACE

This monograph is made up of notes for the Advanced School on "Non-Equilibrium Thermodynamics with Applications to Solids" we gave at the International Centre for Mechanical Sciences in Udine in September 1992. This school was prepared together with our late colleague Theodor Lehmann. Thus we all dedicated this school to the memory of him.

Because non-equilibrium thermodynamics is a still growing discipline in vivid development it was the aim of the school to put emphasis on the basic ideas behind the different approaches and on their application, especially to solids. Therefore starting out with fundamentals of non-equilibrium thermodynamics a general part follows on thermodynamics of solids, including thermoplasticity. Basic ideas, phenomenological as well as microscopic ones, underlying extended thermodynamics and its application are discussed. Non-equilibrium thermodynamics of electromagnetic solids, especially dielectric relaxation and elastic superconductors are treated. The important question of stability in plasticity is investigated with regard to constitutive equations. Obviously it is impossible to present in 4-5 days all aspects of non-equilibrium thermodynamics of solids. But as the detailed discussion on a high level between the lecturers and the participants demonstrates, the school gave way for a new understanding and for further individual studies.

First of all I thank warmly my colleagues P. Haupt, G. Lebon, G.A. Maugin, and H. Petryk for their encouragement in organizing, preparing, and performing the school. Then I have to thank the participants for stimulating questions and vivid discussions, and last not least, it is my pleasant duty to thank the authorities of CISM for inviting us to present this school and for the great hospitality of the Centre. In particular I want to express my gratitude to Professor Sandor Kaliszky for his idea and sponsoring to deliver such a school.

W. Muschik

Dedication to Theodor Lehmann

* 10 August, 1920 † 29 August, 1991

This volume is dedicated to Professor Dr.-Ing. Dr. h.c. Theodor Lehmann. The subject of Theodor Lehmann was engineering science and natural philosophy. He saw the motivation and the origin of engineering mechanics in practical applications as well as in theoretical physics. He studied mechanical engineering in Breslau and Hannover, doctorated in the field of fluid mechanics and then went to an industrial manufactory to work on metal forming processes. There he might have felt that the foundations of thermoplasticity are insufficient: He moved back to the University of Hannover and wrote a Habilitation thesis, entitled "Einige Betrachtungen zu den Grundlagen der Umformtechnik" (i.e ,"Some Remarks to the Foundations of Metal Forming"). For a period of 35 years he worked in general mechanics, continuum mechanics, thermodynamics and plasticity theory.

Lehmann had a widespread experience in quite different branches of his subject and a deep knowledge of its history and literature. He played an active and creative role in the development of modern continuum mechanics. He was appreciated all over the world as a great scientist and quite often as a very good friend. Theodor Lehmann was open minded and helpful to everyone; he was always willing to share his knowledge and experiences with colleagues and especially with young people.

Originally, he promised to participate in this summer school, and we were looking forward to his cooperation and the opportunity to learn from him - not only in the field of science. We are not able to replace him.

CONTENTS

FUNDAMENTALS OF NONEQUILIBRIUM THERMODYNAMICS

W. Muschik
Technical University of Berlin, Berlin, Germany

Abstract
Starting out with a survey on different thermodynamical theories concepts of nonclassical thermodynamics such as state space, process, projection, and nonequilibrium contact quantity are discussed. Using the dissipation inequality for discrete systems the existence of a non-negative entropy production is investigated. Internal variables are defined by introducing concepts concerning their properties. Then field formulation of thermodynamics is achieved by use of nonequilibrium contact quantities introduced above. Extended thermodynamics is shortly discussed in comparison with the non-extended one. Material axioms and how to exploit the dissipation inequality are items of the last two chapters.

1 Survey

Thermodynamics is concerned with the general structure of Schottky systems [1]. By definition these systems exchange heat, work and material with their environment. Their states are described by state variables which are elements of a suitable state space. It is useful to distinguish between considering only one Schottky system in an equilibrium environment, a so-called discrete system, and between dealing with a set of sufficiently small Schottky systems, each denoted by its position and time, exchanging heat, work and material with its adjacent Schottky systems. In doing so we get a so-called field formulation of thermodynamics which needs different mathematical tools with regard to the description of discrete systems. But the general thermodynamical fundamentals are the same in both cases, although they may appear in a different form. Traditionally these fundamentals of thermodynamics are formulated by "Laws" which are enumerated from zero to three. Here these laws are discussed in the framework of different thermodynamical theories of which a survey is given as an introduction.

1.1 Probabilistic and Deterministic Theories

Thermodynamical theories can be divided into two classes, the probabilistic and the deterministic theories (Fig. 1.1). The probabilistic theories themselves are decomposed into stochastic, statistical, and transporttheoretical branches which are operating with totally different concepts. The deterministic theories are split up into those which describe discrete systems and others which deal with continuumtheoretical concepts.

Stochastic thermodynamics [2] [3] is characterized by a measure space $[\Omega, \mathcal{A}, \mathcal{P}]$ - a Kolmogorov probability algebra - which is defined on the state space of a microdomain. These microdomains form mesodomains to which the probabilistic description of the microdomains is transferred by introducing suitable mean values. The site of the mesodomain is identified with the material coordinate so getting the connection between stochastic and deterministic thermodynamics.

Statistical theories are marked by a distribution function or by a density operator which may depend on a relevant set of observables (so-called Beobachtungsebene) [4]. We use such a description for the foundation of nonequilibrium contact quantities, such as contact temperature or nonequilibrium chemical potentials [5].

Transporttheoretical methods also use a distribution function f(p,q,t) which is defined in contrast to the distribution functions of the statistical methods on the (2f+1)-dimensional μ-space of the single molecule having f degrees of freedom [6].

In contrast to the probabilistic theories the deterministic or phenomenological theories do not take into consideration the molecular structure of materials, whereas probabilistic theories even embrace this structure by microscopic models using master equations, molecular dynamics or density operators. The basic concept of phenomenological theories is that of the macroscopic variable. These quantities describe the state of the system which can be retraced immediately to measuring quantities of the system. Examples are volume, pressure, temperature, mass density, charge density, magnetization, pressure tensor, internal energy, etc.

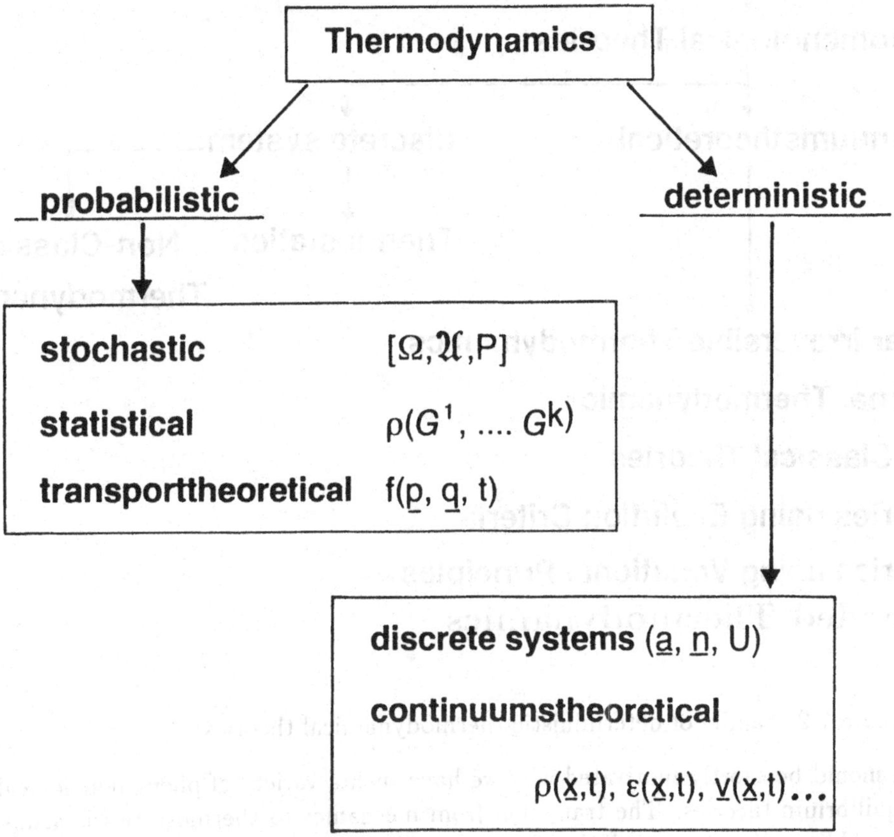

Figure 1.1: Diagram showing distinct classes of thermodynamical theories.

As mentioned above deterministic theories are divided into those which describe discrete systems, these are Schottky systems, and those which apply continuumtheoretical methods. As Fig. 1.2 shows there exist a lot of similar, but different deterministic thermodynamical theories.

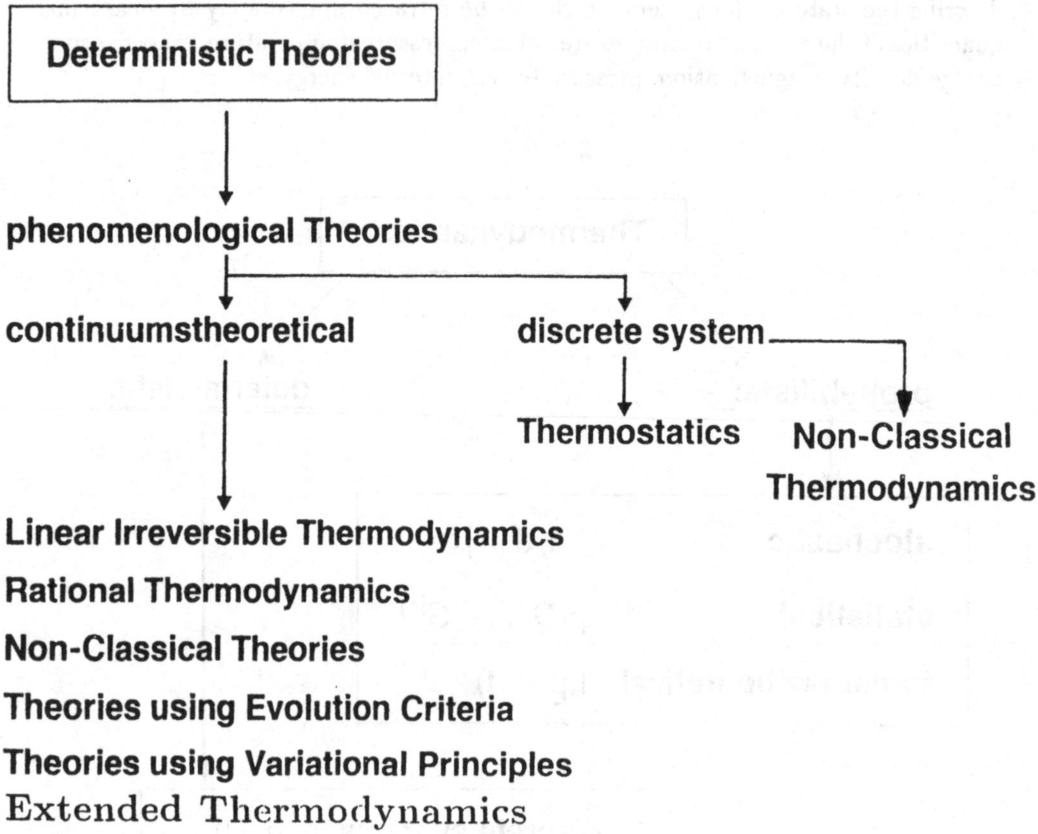

Figure 1.2: Family of deterministic thermodynamical theories.

It should be shortly motivated why we have such a variety of phenomenological nonequilibrium theories: The transition from mechanics to thermostatics is achieved by adding thermodynamical quantities to the mechanical ones. Besides other quantities especially temperature and entropy are added. Because both these quantities are defined by measuring rules in equilibrium, the transition from mechanics to thermostatics is possible without any problem. Now the question arises how to define temperature and entropy in nonequilibrium? In principle this question can be answered differently, and therefore no natural extension of thermostatics to thermodynamics exists [7]. Either temperature and entropy will redefined for nonequilibrium, or they are taken for primitive concepts, i.e. their mathematical existence is presupposed and first of all a physical verification remains open. Here we deal mainly with a non-classical approach to thermodynamics starting out with discrete systems and transferring the results to continuum thermodynamics. Non-classical thermodynamics is characterized by a dynamical nonequilibrium concept of temperature, whereas other theories use the hypothesis of local equilibrium (irreversible thermodynamics) or introduce temperature as a primitive concept (rational thermodynamics).

1.2 Discrete Systems and Field Formulation

A system $\mathcal{G}$ which is separated by a partition $\partial\mathcal{G}$ from its surroundings $\bar{\mathcal{G}}$ is called a discrete system or a Schottky system [1], if the interaction between $\mathcal{G}$ and $\bar{\mathcal{G}}$ can be described by the heat exchange $\dot{Q}$, the power exchange $\dot{W}$, and the material exchange $\dot{\boldsymbol{n}}^e$ (Fig. 1.3).

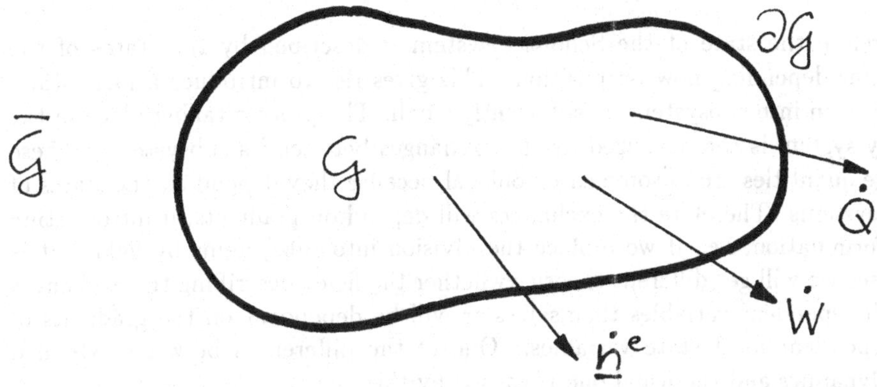

Figure 1.3 : A Schottky system $\mathcal{G}$ exchanges heat, power, and material through $\partial\mathcal{G}$ with its surroundings $\bar{\mathcal{G}}$.

The description of a thermodynamical system by a discrete one is restricted because the system is taken for a black box exchanging quantities with its surroundings. As we will see below (sect. 1.2.) the state variables of the discrete system belong to the system as a whole. The exchange quantities between the discrete system and its vicinity depend also on the state of the vicinity. Because the concept of a discrete system is so easy, we can use it for general considerations as introducing states, processes, exchange and contact quantities, and for formulating the laws of thermodynamics. Because no gradients appear in the description of a discrete system, this concept is often used in thermostatics. But also complex machines in engineering sciences are described by discrete systems, if only the exchange quantities are of interest. Thus for fundamental considerations and for a reduced description of complex situations, the concept of a discrete system is a very useful one.

For getting more information the Schottky system can be divided into discrete subsystems (Fig. 1.4) which exchange heat, power, and material with the adjacent discrete systems.

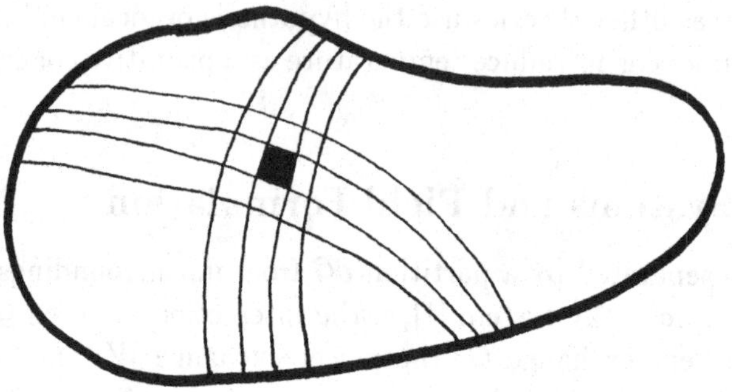

Figure 1.4: A Schottky system divided into discrete subsystems.

Therefore the state of the Schottky system is described by the states of the subsystems depending now on position. This gives rise to introduce a local state, if the division into subsystems is sufficiently small. The process taking place in the Schottky system is characterized by the exchanges between its subsystems. These exchange quantities are in some case nonlocal because they depend on the states of two subsystems. Therefore the exchanges will depend on gradients, if introducting a field formulation, i.e. if we replace the division into subsystems by fields. It is easy to see we will get different theories whether the fields describing the exchanges will be independent variables themselves or will be dependent on the gradients of the independent local state variables. One of the differences between extended thermodynamics and the usual one is caused by this fact.

As we will see below (sect. 1.6.) the transfer from discrete systems to field formulation gives some insight in how to introduce the fields of temperature and entropy. Both quantities are strictly defined only in equilibrium, but in field formulation they should be applicable to nonequilibrium. How to define them in this case is a question of non-classical theories. But most of the theories in field formulation use temperature and entropy as primitive concepts, as we will see in the next section.

1.3 Rational Thermodynamics

Historically Rational Thermodynamics [8] is the thermal extension of Rational Mechanics. This extension is simply performed by adding the fields of temperature, entropy, and entropy flux density as primitive concepts, i.e. these fields are not defined physically. Although manifold, some representative prototypes of Rational Thermodynamics can be specified [9]. We distinguish between (Fig. 1.5) Clausius-Duhem theories which all use an in time local dissipation inequality as an analytical expression describing the Second Law, and between other theories which use dissipations inequalities being global in time or which use evolution criteria [10], [11]. Here we will not treat the latter ones and also not those Clausius-Duhem theories which are global in position [12].

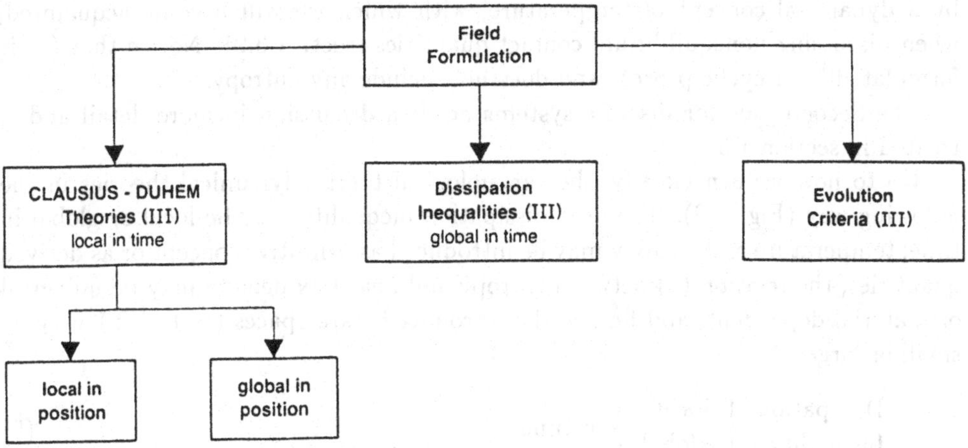

Figure 1.5: Different theories in field formulation.

Clausius-Duhem theories presuppose the existence of the in time local *Clausius-Duhem dissipation inequality* from which they got their name:

$$\rho\dot{s} + \nabla \cdot \boldsymbol{\Phi} - \gamma = \sigma \geq 0. \tag{1}$$

Here the entropy density s and the entropy flux density

$$\boldsymbol{\Phi} = \boldsymbol{q}/\Theta + \boldsymbol{k} \tag{2}$$

are considered as constitutive equations [13] ($\boldsymbol{k} \neq \mathbf{o}$) or as being determined by heat flux over temperature [14] ($\boldsymbol{k} = \mathbf{o}$) ($\rho$ is the mass density,γ is the entropy supply,σ is the entropy production density).

The in time local dissipation inequality (1) represents only a sufficient formulation of the second law because this is an in time global statement (see sect. 1.4.2). Field formulations taking this into account are given by [15], [16]

$$\int_A^B [\dot{s}^{eq} + (1/\rho)\nabla \cdot (\boldsymbol{q}/\Theta)]\, dt \geq 0 \tag{3}$$

and

$$\oint [(1/\rho)\nabla \cdot (\boldsymbol{q}/\Theta) - (r/\Theta)]\, dt \geq 0. \tag{4}$$

In both inequalities an undefined nonequilibrium temperature Θ appears, which is taken for a primitive concept. In (3) an equilibrium entropy s^{eq} is used, which can be interpreted as the entropy of an accompanying process [17], [18] (see sect. 1.5.3). This concept originates from non-classical thermodynamics which is characterized by a dynamical concept of temperature, with which we will become acquainted, when discussing nonequilibrium contact quantities (sect. 1.3.2). Notice that (4) is formulated for a cyclic process and does not include any entropy.

The Second Law for discrete systems needs a discussion in more detail and is treated in section 1.5.3.

Up to now we can classify phenomenological thermodynamical theories in the following way (Fig. 1.6): The used dissipation inequality may be local or global in time; temperature and entropy may be introduced as primitive concept or as derived quantities; the relation (2) between entropy and heat flux density may be universal or material dependent; and finally, the introduced state spaces (sect. 1.2.) may be small or large.

$$\begin{matrix}\text{Dissipation}\\ \text{Inequality}\end{matrix} \left\{ \begin{matrix}\text{local}\\ \text{global}\end{matrix} \right\} \text{in time} \tag{5}$$

$$\begin{matrix}\text{Temperature}\\ \text{Entropy}\end{matrix} \left\{ \begin{matrix}\text{primitive concept}\\ \text{derived quantity}\end{matrix} \right. \tag{6}$$

$$\text{relation } \boldsymbol{q} \leftrightarrow \boldsymbol{\Phi} \left\{ \begin{matrix}\text{universal}\\ \text{material dependent}\end{matrix} \right. \tag{7}$$

$$\text{State Space} \left\{ \begin{matrix}\text{small}\\ \text{large}\end{matrix} \right. \tag{8}$$

Figure 1.6: Categories for classifying phenomenological thermodynamical theories.

1.4 Different Formulations of the Second Law

Formulations of the second law for a discrete system are historically the oldest ones [19]. First of all, we have to distinguish between formulations which use work criteria, the so-called Sears-Kestin statement (Fig. 1.7) [20]. This states that, for adiabatic processes which are cyclic in the work variables, the work performed on the system ist not negative. At first sight it seems that the Sears-Kestin statement - because it deals especially with adiabatic processes - is only necessary but not sufficient for describing the physical content of the Second Law [21] which is expressed by the *Clausius inequality*

$$\oint [\dot{Q}(t)/T(t) - s(t) \cdot \dot{n}^e(t)]dt \leq 0 \tag{9}$$

(T thermostatic temperature of the system's vicinity and its s molar entropy). But by adding two further axioms relate to reversible processes and to compound systems - general items not affecting or affected by the Second Law and often presupposed tacitly - Sears-Kestin statement becomes equivalent to Clausius inequality [22]. Here we will not refer to formulations of the Second Law which use work criteria.

Although there is a huge flood of papers concerning the second law, two streams can easily be distinguished (Fig. 1.7): one stems from Clausius and Lord Kelvin [23], the other from Carathéodory [24] and Born [25].

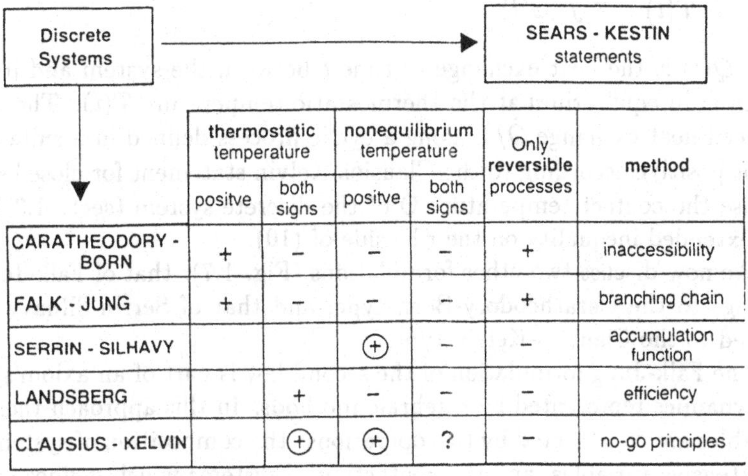

	thermostatic temperature		nonequilibrium temperature		Only reversible processes	method
	positve	both signs	positve	both signs		
CARATHEODORY - BORN	+	−	−	−	+	inaccessibility
FALK - JUNG	+	−	−	−	+	branching chain
SERRIN - SILHAVY			⊕	−	−	accumulation function
LANDSBERG		+	−	−	−	efficiency
CLAUSIUS - KELVIN		⊕	⊕	?	−	no-go principles

Figure 1.7: Classification for formulations of the Second Law for discrete systems.

Although both formulations are equivalent in the field of thermodynamics, they are, in general, different [26].

Those in the Carathéodory-Born stream [27] make use of a statement of the adiabatic inaccessibility of states in the vicinity of each state [28]. But this statement is only valid for reversible processes, and yields a foliation of the state space into surfaces of constant entropy which is, of course, the thermostatic entropy. Also the existence of the thermostatic temperature can be proved, but only for positive absolute temperatures.

Those in the Clausius-Kelvin stream state that selected irreversible processes operating between two heat reservoirs, which need not have only positive absolute temperatures, are existent or non-existent [29]. Therefore, the Clausius-Kelvin formulation, representing a no-go principle for certain processes, is not restricted to reversible processes and positive absolute temperatures. The temperatures which appear in this formulation of the second law are thermostatic temperatures of the heat reservoirs controlling the process. Also in the case of negative absolute temperatures [30] the processes are presupposed to be so fast - and therefore irreversible - that the nonequilibrium systems of negative absolute temperature can be taken for reservoirs.

The question arises: Is it possible to extend the Clausius-Kelvin formulation to nonequilibrium analogues of the thermostatic temperature [31]? This can indeed be proved by the use of an additional axiom of the simulation of processes [32]. We get for closed systems

$$\oint \frac{\dot{Q}(t)}{T(t)}dt \leq \oint \frac{\dot{Q}(t)}{\Theta(t)}dt \leq 0 \tag{10}$$

Here $\dot{Q}(t)$ is the heat exchange at time t between the system and its environment which is in equilibrium at the thermostatic temperature $T(t)$. The integral of the reduced heat exchange $\dot{Q}/T$ along a cyclic process defined in a suitable state space is not positive according to the Clausius-Kelvin statement for closed systems (9). If we use the contact temperature Θ of the discrete system (sect. 1.3.2), we will get the extended inequality on the r.h. side of (10).

We now discuss two other formulations (Fig. 1.7): that of Falk-Jung [33], which belongs to the Carathéodory-Born type, and that of Serrin-Silhavy [34], which is related to the Clausius-Kelvin type.

The Falk-Jung formulation of the second law is part of an axiom system of thermodynamics represented by algebraic methods. In this approach thermodynamical variables are constructed by two operations: the composition of systems which leads to extensive variables, and the contacting of systems which defines intensive variables. Processes are described by a transition relation, by means of which chains of transitions can be defined. The adiabatic isolation of a system has the structure of a chain of transitions, with no branching, by means of which the empirical entropy is defined. This procedure which uses the adiabatical isolation to determine the empirical entropy, reduces the Falk- Jung formulation to reversible processes. Therefore with this technique, only the first part of the second law can be formulated.

The Serrin-Silhavy formulation uses temperature, work and heat as primitive concepts. It is usual for work and heat to be primitive concepts in thermodynamical theories, but - as we discussed above - temperature should not be a primitive concept without any physical interpretation. So Serrin introduces a "thermal manifold consisting of the set of hotness levels". Of course, the hotness level represents an empirical temperature, but nothing is said about what empirical temperature represents; it could be that of the discrete nonequilibrium system or that of its environment. The latter is, of course, the correct interpretation. After this, Serrin introduces the so-called accumulation function celebrated by Man [35] which in usual terminology is given by

$$Q(\mathcal{P},\tau) := \mathcal{P}\oint \dot{Q}(t)\Delta[\tau - T(t)]dt, \tag{11}$$

with the step function

$$\Delta(x) = \begin{cases} 1, & x > 0 \\ 0, & x \leq 0. \end{cases} \tag{12}$$

The second law can now be formulated as

$$Q(\mathcal{P},\cdot) \geq 0 \rightarrow Q(\mathcal{P},\cdot) = 0, \tag{13}$$

which is only valid for positive absolute temperatures [36].

As we saw above, all formulations of the second law for discrete systems are global in time (Clausius-Kelvin type) or presuppose topological properties of the state space (Carathéodory-Born type) which are only given in equilibrium. Thus in thermodynamics of discrete systems no in time local formulation of the second law exists. Consequently in time local formulations as the dissipation inequality (1) need additional assumptions beyond the second law.

2 State Spaces, Processes, and Projections

For describing the state of a discrete system we need a state space. The chemical composition of the system is determined by the mole numbers $\boldsymbol{n}$, mechanical and geometrical properties are described by the work variables $\boldsymbol{a}$. We get a general state by adding thermodynamical variables $\boldsymbol{z}$

$$Z := (\boldsymbol{a}, \boldsymbol{n}, \boldsymbol{z}) \in \mathcal{Z} \tag{14}$$

$\mathcal{Z}$ is called (nonequilibrium)*state space of a discrete system* [17]. The power exchange of the discrete system $\mathcal{G}$ with its vicinity $\bar{\mathcal{G}}$ is homogeneous in the time rates of the work variables

$$\dot{W}(t) := \boldsymbol{F}(t) \cdot \dot{\boldsymbol{a}}(t) \tag{15}$$

($\boldsymbol{F}(t)$ are the generalized forces belonging to $\mathcal{G}$). The mole number of the mass exchange is defined by

$$\dot{\boldsymbol{n}}^e := \dot{\boldsymbol{n}} - \dot{\boldsymbol{n}}^i, \qquad \dot{\boldsymbol{n}}^i := \dot{\boldsymbol{\nu}}^+ \cdot \dot{\boldsymbol{\xi}}(t) \tag{16}$$

($\dot{n}^i$ are the time rates of the mole numbers due to chemical reactions, $\dot{\xi}$ are the reaction speeds, $\dot{\nu}^+$ is the transposed matrix of the stoichiometric equations

$$\dot{\nu} \cdot M = o, \tag{17}$$

(M are the mole masses). The heat exchange $\dot{Q}(t)$ is a quantity which can be measured by calorimetry and which is introduced here as a primitive concept.

The thermodynamical variables z include the internal energy U of $\mathcal{G}$, the contact temperature Θ which is independent of U [17], and if necessary, the internal variables α. Contact temperature and internal variables will be treated in more detail in sect.1.3.2. and in 1.5.

We presuppose that we know what isolating, adiabatic, power-insulating, and material- insulating partitions are,and how to define them (for more details see [37], sect. 1.1.1). Then the equilibrium conditions for discrete systems are easy to specify:

Def.: Time-independent states of isolated systems are called *states of equilibrium.*

Especially simple systems are those which do not contain adiabatic partitions:

Def.: A system is called *thermal homogeneous*, if it does not contain any adiabatic partitions in its interior.

Obviously for describing a state of equilibrium we do not need as many variables as in nonequilibrium. Therefore the question arises how many variables span the equilibrium subspace? The answer is given by the *Zeroth Law* which bases on experience and which is here formulated as an "empirem" (axiom whose validity is founded on experience).

Empirem: (0th Law): The state space of thermal homogeneous system in equilibrium is

$$Z^* := (a, n, *; \alpha_m(a, n, *)). \tag{18}$$

$*$ is exact one additional thermodynamical variable, and α_m is one of the k possible equilibrium values of the internal variables α which are determined by $(a, n, *)$.

As we will see below, this thermodynamical variable $*$ may be the internal energy U of the system or its thermostatical temperature T.

If the state of a system changes in time, we say the system undergoes a process:

Def.: A path in state space is called a *process*

$$Z(\cdot) := (a, n, z)(\cdot) \tag{19}$$

especially

$$Z(t) = (a, n, z)(t), \qquad t \in [t_1, t_2]. \tag{20}$$

Experience shows that equal processes in different systems can induce different values of other variables not included in the state space. A very simple example are two gaseous systems having equal volume and temperature but different pressure. Of course each of the two systems consists of another gas. Different materials differ from each other by different material properties which are described by a map $\mathcal{M}$ called the *constitutive map*. The domain of $\mathcal{M}$ depends on the chosen state space: There are state spaces so that $\mathcal{M}$ is local in time as in the theory of thermoelastic materials or in theories using internal variables. There are other state spaces on which $\mathcal{M}$ is not local in time as in materials showing after-effects or hysteresis. Therefore we classify [38]:

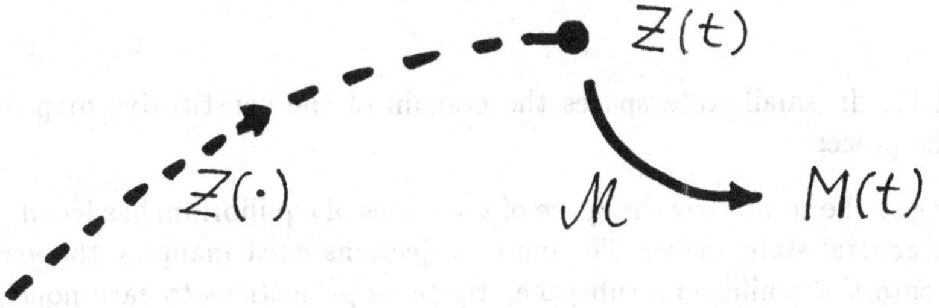

Figure 1.8: The material property $M(t)$ is only determined by the state $Z(t)$.

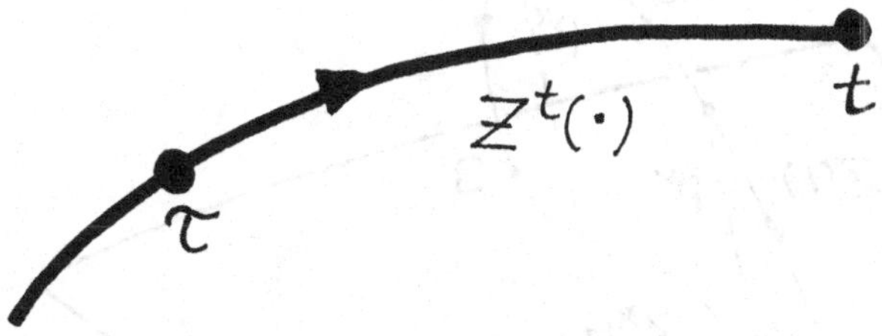

Figure 1.9: The history $Z^t(\cdot)$ of the process $Z(\cdot)$ is defined by the path of $Z(\cdot)$ up to time t. Def.: A *state space* is called *large*, if material properties M are defined

by maps local in time (Fig. 1.8)

$$\mathcal{M} : Z(t) \to M(t), \qquad \text{for all } t. \tag{21}$$

For defining after-effects we need the concept of the history:

Def.: For a fixed time t and real $s \geq 0$

$$Z^t(s) := Z(t - s), \qquad s \in [0, \tau], \tag{22}$$

is called the *history of the process* $Z(\cdot)$ between $t - \tau$ and t (Fig. 1.9).

Def.: A *state space* is called *small*, if material properties M are defined by maps on process histories (Fig. 1.10)

$$\mathcal{M} : \ Z^t(\cdot) \to M(t), \qquad \text{for all } t. \tag{23}$$

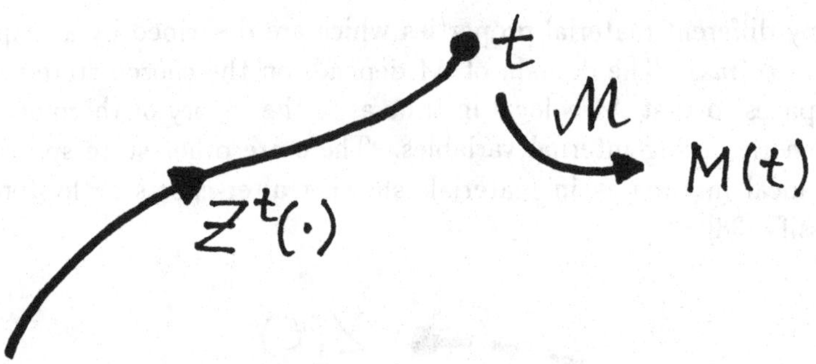

Figure 1.10: In small state spaces the domain of the constitutive map is the history of the process.

According to the zeroth law the space of the states of equilibrium has less dimensions as the general state space. Therefore projections exist mapping the general state space onto the equilibrium subspace. By these projections to each nonequilibrium process an equilibrium "process" or better a trajectory in equilibrium subspace is attached (Fig. 1.11).

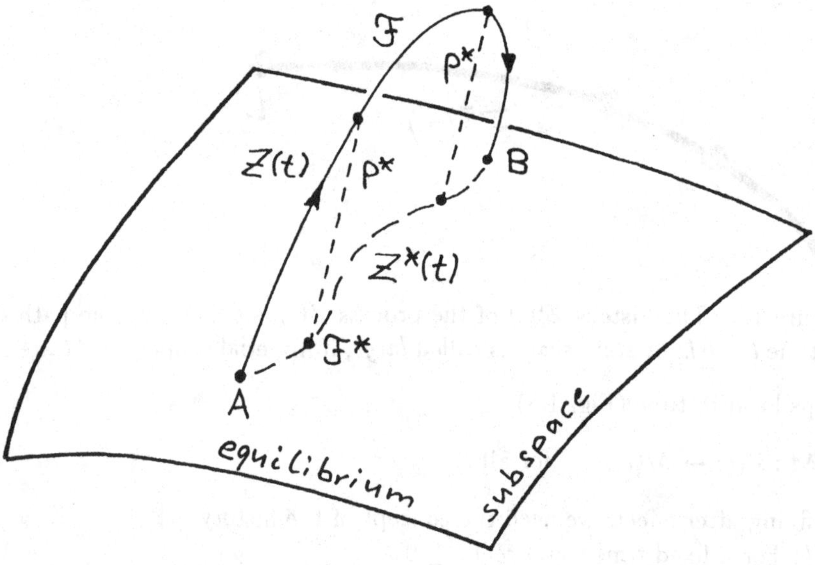

Figure 1.11: Equilibrium subspace represented as hypersurface in state space. The projection P^* maps $Z(t)$ into $Z^*(t)$ so inducing the accompanying process $\mathcal{F}^*$ from the real process $\mathcal{F}$.

Def.: A trajectory $Z^*(\cdot)$ in the equilibrium subspace induced by a projection P^*, $(P^*)^2 = P^*$, of a process $Z(\cdot)$

$$P^* Z(t) = Z^*(t) \tag{24}$$

$$P^*(\boldsymbol{a}, \boldsymbol{n}, \boldsymbol{z})(\cdot) = (\boldsymbol{a}, \boldsymbol{n}, *; \boldsymbol{\alpha}_m)(\cdot) \tag{25}$$

is called an *accompanying process* [39].

Because in equilibrium material properties do not depend on process histories (that is a demand for constructing the state space)

$$M^*(t) = \mathcal{M}(Z^*(t)) \tag{26}$$

we get by the definition of M^* due to (23)

$$M^*(t) = \mathcal{M}(Z^{*t}(\cdot)) \tag{27}$$

the following property of the constitutive map

$$\mathcal{M}(Z^{*t}(\cdot)) = \mathcal{M}(Z^*(t)). \tag{28}$$

Additionally the equilibrium values M^* of $\mathcal{M}$ have to be compatible with its nonequilibrium values M by satisfying the

Embedding Axiom:

$$\mathcal{F}\int_A^B \dot{M}(t)dt = M_B^{eq} - M_A^{eq} = \mathcal{F}^* \int_A^B \dot{M}^*(t)dt \tag{29}$$

$$A, B \in \text{equil. subsp.}$$

If we especially apply this embedding axiom to a quantity which is defined on equilibrium subspace, and we look for an extension of this equilibrium quantity to nonequilibrium, this extension has to satisfy the embedding axiom, because otherwise it would not be compatible with the earlier defined equilibrium quantity: i.e. a nonequilibrium entropy - in whatever way defined - has to obey the embedding axiom in order to be in agreement with the well known concept of equilibrium entropy.

By experience we know that in adiabatic isolated system the work exchange is independent of the process, if the initial and the final state are the same in all processes under consideration: If

$$\text{ad: } 1 \to 3 \to 2, \quad \text{ad: } 1 \to 4 \to 2 \tag{30}$$

are two adiabatic processes, then the work exchange

$$W_{13} + W_{32} = W_{14} + W_{42} \tag{31}$$

is independent of the path in the state space between 1 and 2. This can be formulated as

Empirem: (1st Law): A function of state$U(Z)$ exists, so that for adiabatic processes α

$$\text{ad: } 1 \rightarrow 2$$

the work W_{12}^{α} in a system at rest is represented by

$$W_{12}^{\alpha} = U_2 - U_1, \quad U_j := U(Z_j), \tag{32}$$
$$\text{ad: } \dot{W} = \dot{U}, \text{ in a restsystem.} \tag{33}$$

If the process constraint of being adiabatical is cancelled, we define:

Def.: In closed restsystems the heat exchange is defined by

$$\dot{Q} := \dot{U} - \dot{W}, \tag{34}$$
$$\text{closed: } Q_{12}^{\alpha} = U_2 - U_1 - W_{12}^{\alpha}, \text{ in a restsystem.} \tag{35}$$

Proposition: If E is the total energy $E := U + E_{kin} + Y$, Y = potential energy, and L' the power of the forces without a potential, the first law for closed systems writes

$$\text{closed: } Q_{12}^{\alpha} + L'^{\alpha}_{12} = E_2 - E_1 \tag{36}$$

By (32) we defined by mechanics the internal energy of all states Z_2 which are adiabatically connected with Z_1. But there may be states which are not adiabatically connected with Z_1, and first of all for those states no internal energy can be defined by adiabatical work. It can be proved [40] that also in this case a unique gauge of the internal energy exists. Thus $U(Z)$ is defined for all equilibrium and nonequilibrium states Z.

According to (36) the first law for closed systems is

$$0 = \dot{U} - \dot{Q} - \dot{W}. \tag{37}$$

If the partition $\partial\mathcal{G}$ is now permeable for material, the left hand side of (37) is no longer zero, but homogeneous in the external rates of the mole numbers $\dot{\boldsymbol{n}}^e$

$$\boldsymbol{X} \cdot \dot{\boldsymbol{n}}^e := \dot{U} - \dot{Q} - \dot{W} \tag{38}$$

For elucidating the meaning of $\boldsymbol{X}$ we especially consider a 1-component equilibrium system $\mathcal{G}$ surrounded by a vicinity $\partial\mathcal{G}$ of equal composition, equal thermostatic temperature, and equal pressure. If now the partition $\partial\mathcal{G}$ is removed, volume and mole number of $\mathcal{G}$ changes by constant pressure, temperature, and vanishing heat exchange:

$$\dot{p} = 0, \quad \dot{Q} = 0. \tag{39}$$

We get

$$\dot{W} := -p\dot{V} = -(pV)^{\cdot}, \tag{40}$$

and

$$\dot{U} - \dot{W} = \boldsymbol{X} \cdot \dot{\boldsymbol{n}}^e = (U + pV)^{\cdot}. \tag{41}$$

This yields

$$\dot{Q} = 0, \ \dot{p} = 0 : \quad \boldsymbol{X} \cdot \dot{\boldsymbol{n}}^e = \dot{H} := (U + pV)^{\cdot}. \tag{42}$$

Because $\dot{\boldsymbol{n}}^e$ is here the only variable we have

$$X \equiv \frac{\partial H}{\partial n}, \tag{43}$$

and the general case we get from (41) the first law of open discrete restsystems

$$\dot{U} = \dot{Q} + \dot{W} + \boldsymbol{h} \cdot \dot{\boldsymbol{n}}^e, \tag{44}$$

$$\boldsymbol{h} := \frac{\partial H}{\partial \boldsymbol{n}} \equiv \boldsymbol{X}. \tag{45}$$

3 Nonequilibrium Contact Quantities

3.1 Preliminary Considerations: The "Dynamic" Pressure

We consider an isolated system consisting of two subsystems (Fig.1.12). Both subsystems are separated by a movable adiabatic piston which is impervious to material. At the beginning one subsystem is in equilibrium, the other is in an arbitrary state. Introducing the mean pressure $\bar{P}$ by

$$\int_{\text{piston}} P(\boldsymbol{x}, t)\dot{\boldsymbol{x}} \cdot d\boldsymbol{f} = \bar{P} \int_{\text{piston}} \dot{\boldsymbol{x}} \cdot d\boldsymbol{f} = \bar{P}\dot{V}, \tag{46}$$

we get for the initial time the inequality

$$\left(\bar{P} - P^{eq}\right) \dot{V} \geq 0. \tag{47}$$

Of course P and V are functions of time. If we contact the nonequilibrium system at time t with a suitable equilibrium system of the pressure $P^{eq}(t)$, (46) is also valid for all contact times

$$\left[\bar{P}(t) - P^{eq}(t)\right] \dot{V}(t) \geq 0. \tag{48}$$

We interpret this inequality as follows: The "dynamic" pressure $\bar{P}(t)$ of the nonequilibrium system at time t can be measured by the pressure $P^{eq}(t)$ of an equilibrium system being contacted at time t with the nonequilibrium system by a partition only permeable to work.

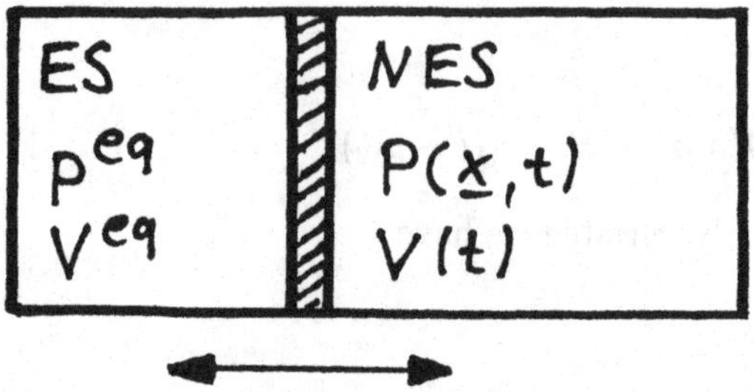

Figure 1.12 : An isolated system consisting of two subsystems separated by a partition which is only permeable to work.

Both pressures are equal

$$\bar{P}(t) = P^{eq}(t), \tag{49}$$

if

$$\dot{V}(t) = 0. \tag{50}$$

Therefore $\dot{V} = 0$ is the indicator for the equality of the dynamic pressure and the pressure of the equilibrium system. By this procedure the nonequilibrium pressure can be defined by the equilibrium pressure.

Generalizing pressure and volume to the generalized forces and the work variables (sect. 1.1)

$$P \to A, \quad \dot{V} \to -\dot{\boldsymbol{a}} \tag{51}$$

we get the defining inequality for the generalized dynamic forces $A_k(t)$

$$[A_k^{eq}(t) - A_k(t)]\,\dot{a}_k(t) \geq 0, \qquad k = 1, 2, ..., w. \tag{52}$$

Here A_k^{eq} is the equilibrium quantity for gauging, $\dot{a}_k$ the indicator, and A_k is the defined nonequilibrium quantity.

A very simple example which can be treated exactly is:

Proposition: A linear damped oscillator with elongation x is affected by a constant force

$$\ddot{x} + 2\gamma\dot{x} + \omega_0^2 x = P = \text{const.} \tag{53}$$

The following inequalities are valid:

$$\dot{x}(t)\left[(P/\omega_0^2) - x(0)\right] \geq 0, \text{ for } t \to +0, \tag{54}$$

$$\dot{x}(0)\left[(P/\omega_0^2) - x(0)\right] \geq 0, \tag{55}$$

if the following initial conditions are satisfied:

$$x(0) = (P/\omega_0^2) + \epsilon, \quad \dot{x}(0) = 0. \tag{56}$$

Here $\dot{x}(t \to +0)$ is the indicator, $x(0)$ the deflection of the measuring device gauged in equilibrium, and P the nonequilibrium quantity which can be measured by the zero of
$\dot{x}(t \to +0)$. (55) and (52) are analogous to each other.

3.2 Contact Temperature

We consider two subsystems of an isolated system which are separated by a partition being only permeable for heat exchange. As in sect. 1.3.1 one subsystem is in equilibrium at the beginning of the contact at time t, and the other subsystem is an arbitrary state. Because of the chosen partition, here the heat exchange between both subsystems is the indicator, and the defining inequality for the "dynamic" temperature of the nonequilibrium system writes

$$\left[\frac{1}{\Theta}(t) - \frac{1}{T}(t)\right] \dot{Q}(t) \geq 0 \tag{57}$$

T is the thermostatic temperature of the contacting equilibrium system at time t. $\Theta(t)$ is called *contact temperature of the thermal contact* between equilibrium and nonequilibrium system [41]. The contact temperature depends on individual properties of this thermal contact. In equilibrium $\Theta = T$ holds independently of its special nature.

As easy to see Θ and U are independent of each other: Two subsystems of an isolated system exchange only heat (Fig.1.13). One subsystem is controlled by different other subsystems which are always in equilibrium. We now choose the thermostatic temperature of the controlling reservoirs so that $\dot{Q}(t) = 0$. This special choice is achieved by $T(t) = \Theta(t)$. Because $\dot{Q} = 0$ we get for this partition $\dot{U} = 0$ according to the first law, but $\Theta(t)$ depends on t because NES is not in equilibrium.

Consequently U and Θ are independent of each other.

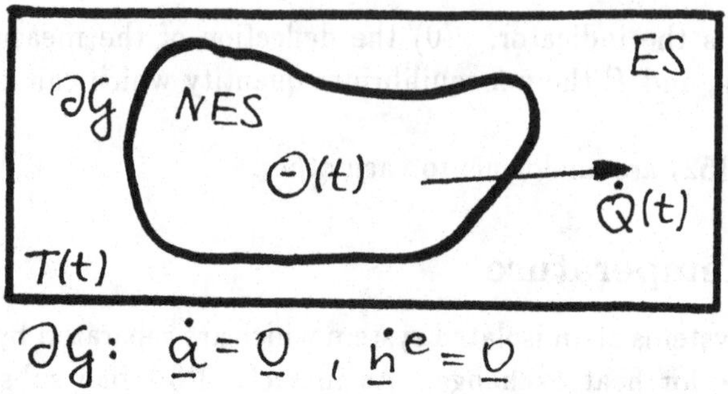

Figure 1.13 : A nonequilibrium system (NES) of the contact temperature $\Theta(t)$ is contacted with a set of equilibrium systems (ES) of the thermostatic temperature $T(t)$. The partition is permeable only for heat.

We now introduce the minimal (nonequilibrium) state space [42]

$$Z = (\boldsymbol{a}, \boldsymbol{n}, U, \Theta). \tag{58}$$

In equilibrium we get

$$Z_0 = (\boldsymbol{a}, \boldsymbol{n}, U, T), \quad T = C(\boldsymbol{a}, \boldsymbol{n}, U), \tag{59}$$

where C is the caloric equation of state in equilibrium. Because in equilibrium the contact temperature becomes dependent on the variables of the equilibrium subspace its dimensions is in case of the minimal state space

$$\dim\{Z_0\} = \dim\{Z\} - 1. \tag{60}$$

The contact temperature introduced by the inequality (57) is a nonequilibrium analogue of the thermostatic temperature with which it coincides in equilibrium. Here the contact surface $\partial\mathcal{G}$ was not divided into subsurfaces so that Θ belongs to $\partial\mathcal{G}$ as a whole. But (57) is also valid for such a subsurface $\partial\mathcal{G}^\alpha$:

$$\partial\mathcal{G} = \bigcup_\alpha \partial\mathcal{G}^\alpha, \quad \left[\frac{1}{\Theta^\alpha} - \frac{1}{T^\alpha}\right] \dot{Q}^\alpha \geq 0 \tag{61}$$

Here Θ^α is the contact temperature belonging to $\partial\mathcal{G}^\alpha$, and T^α the thermostatic temperature of the same contact surface $\partial\mathcal{G}^\alpha$ of the controlling reservoir. The division of $\partial\mathcal{G}$ into subsurfaces is the first step towards a field formulation. Here two questions arise: How to translate (61) into field formulation, and what about T^α, if the contacting vicinity is also in nonequilibrium? Both questions are answered in sect. 1. .

Two different accompanying processes (25) can be introduced onto minimal state space: The U-projection P^U defined as

$$P^U(\boldsymbol{a}, \boldsymbol{n}, U, \Theta) = (\boldsymbol{a}, \boldsymbol{n}, U), \tag{62}$$

and the Θ-projection P^Θ

$$P^\Theta(\boldsymbol{a}, \boldsymbol{n}, U, \Theta) = (\boldsymbol{a}, \boldsymbol{n}, U^\Theta), \tag{63}$$

where U^Θ is defined by the caloric equation of state

$$\Theta = C(\boldsymbol{a}, \boldsymbol{n}, U^\Theta). \tag{64}$$

The two projections coincides, if the original process is reversible. Therefore we can get a measure of irreversibility by the "distance" of both projections.

3.3 Chemical Potentials

We now consider two subsystems of an isolated system, one in equilibrium, the other in an arbitrary state. The thermostatic temperature of the equilibrium system is equal to the contact temperature of the other system, and also the generalized forces have the same value:

$$\Theta(t) = T(t) \rightarrow \dot{Q}(t) = 0, \tag{65}$$

$$\boldsymbol{A}(t) = \boldsymbol{A}^{eq}(t) \rightarrow \dot{\boldsymbol{a}}(t) = \mathbf{o}. \tag{66}$$

Here the indicators according to sect. 1.3.1 are the external rates of the mole numbers. The defining inequality of the "dynamical" chemical potential is

$$[\mu_k^{eq}(t) - \mu_k(t)]\, \dot{n}_k^e \geq 0 \qquad k = 1, 2, ..., K, \tag{67}$$

with K as the number of components. A formalization and a strict treatment of the contact quantities here introduced is possible [43].

As demonstrated nonequilibrium contact quantities belong to the exchange quantities of discrete systems: The pressure or - as we will see below - the pressure tensor belongs to the power exchange, the contact temperature to the heat exchange, and the dynamical chemical potentials to the material exchange.

4 Dissipation Inequality

In sect. 1.1.4 we gave a survey on different formulations of the second law. The main result was that formulations of the second law can be divided into classes: Formulations which are global in time like Clausius' inequality (time integrals) and those which are local in time like the formulation: Entropy production is always non-negative. Of course both the formulations are not equivalent because from formulations local in time those being global in time can be derived, but not vice versa.

Here we will mention five versions of the second law:

I. Clausius' inequality for open systems:

$$\oint \left\{ \frac{\dot{Q}(t)}{T(t)} - \boldsymbol{s}(t) \cdot \dot{\boldsymbol{n}}^e \right\} dt \leq 0 \tag{68}$$

Here $T(t)$ is the thermostatic temperature of the equilibrium vicinity and $\boldsymbol{s}(t)$ its molar entropies.

II. Entropy does not decrease in isolated systems. By release of constraints a process

$$\mathcal{F}: \ A(eq) \stackrel{isol.}{\rightarrow} B(eq) \tag{69}$$

undergoes starting out from a state of equilibrium $A(eq)$ to another state of equilibrium $B(eq)$. Then we have

$$S_B^{eq} \geq S_A^{eq}. \tag{70}$$

Another interpretation of II may be starting out from a state of nonequilibrium C:

III.

$$\mathcal{F}: \ C \stackrel{isol.}{\rightarrow} B(eq) \tag{71}$$

$$S_B^{eq} \geq S_C. \tag{72}$$

Here the question arises, what the definition of the nonequilibrium entropy S_C is.

Another understanding of II gives a statement about the time rate of entropy

IV: $\mathcal{F}$ isolated : $\dot{S}(t) \geq 0, \qquad (73)$

in which also a nonequilibrium entropy appears.

If we use a field formulation for describing the system, we can formulate the second law:

V. There exists a dissipation inequality which is local in position and time:

$$\frac{\partial}{\partial t}\sum_{\alpha}\rho_\alpha\hat{s}_\alpha + \nabla\cdot\left(\frac{\boldsymbol{q}}{\Theta} + \boldsymbol{k} + \sum_{\alpha}\rho_\alpha\hat{s}_\alpha\boldsymbol{v}_\alpha\right) \geq 0. \tag{74}$$

Here we have the following fields at $(\boldsymbol{x}, t)$ of the component α:
mass density: ρ_α, velocity: $\boldsymbol{v}_\alpha$, specific entropy: $\hat{s}_\alpha$, heat flux density: $\boldsymbol{q}$, entropy flux density: $\boldsymbol{\phi}$, with the definition:

$$\boldsymbol{k} := \boldsymbol{\phi} - \boldsymbol{q}/\Theta, \tag{75}$$

where Θ is a temperature whose definition is at first as undetermined as the definition of a nonequilibrium entropy.

It is obvious that I to V are *not equivalent* to each other. Here we deal with the following question:

If we start out with the basic assumptions

i) Thermostatics is presupposed to be known,

ii) Clausius' inequality holds,

can we answer the question:
Do there exist dissipation inequalities local in time and position?

4.1 First Law and Entropies

We consider a discrete system $\mathcal{G}$ in its equilibrium surroundings $\mathcal{G}^*$ from which it is separated by the surface $\partial\mathcal{G}$ [32]. We introduce a small state space (23)

$$Z := (\boldsymbol{a}, \boldsymbol{n}, U, , \ldots; T^*, \boldsymbol{A}^*, \boldsymbol{\mu}^*) \tag{76}$$

Here $\boldsymbol{a}$ are the work variables, $\boldsymbol{n}$ the mole numbers, U the internal energy, Θ the contact temperature (57), T^* the thermostatic temperature of $\mathcal{G}^*$, $\boldsymbol{A}^*$ the generalized forces of $\mathcal{G}^*$, and $\boldsymbol{\mu}^*$ the chemical potentials of $\mathcal{G}^*$. Somewhat different from (14) we here introduce the intensive variables of the surroundings as parameters.

The equilibrium caloric equation of state defined on the equilibrium subspace

$$T = C(\boldsymbol{a}, \boldsymbol{n}, U), \quad \frac{\partial C}{\partial U} > 0, \tag{77}$$

gives a 1-1 map between the internal energy and the thermostatic temperature T of the considered system, if it is in equilibrium. The quantities U^* and U^Θ are defined by

$$T^* = C(\boldsymbol{a}, \boldsymbol{n}, U^*), \tag{78}$$

$$\Theta = C(\boldsymbol{a}, \boldsymbol{n}, U^\Theta). \tag{79}$$

By projections P^U, P^Θ , and P^* onto the equilibrium subspace (18) we introduce accompanying processes (25):

$$P^U\, Z(t) = (\boldsymbol{a}, \boldsymbol{n}, U)(t) = Z_0^U(t), \tag{80}$$
$$P^\Theta\, Z(t) = (\boldsymbol{a}, \boldsymbol{n}, U^\Theta)(t) = Z_0^\Theta(t), \tag{81}$$
$$P^*\, Z(t) = (\boldsymbol{a}, \boldsymbol{n}, U^*)(t) = Z_0^*(t), \tag{82}$$
$$P^U\, P^U = P^U, \text{ etc.} \tag{83}$$

The first law of an open system runs as follows (38)

$$\dot{U}(t) = \dot{Q}(t) + \boldsymbol{A}(t) \cdot \dot{\boldsymbol{a}}(t) + \boldsymbol{h}(t) \cdot \dot{\boldsymbol{n}}^e \tag{84}$$

Applying the projection P^* we get along the accompanying process $P^* Z(t)$:

$$\dot{U}^*(t) \;=\; \dot{Q}^*(t) + \boldsymbol{A}^*(t) \cdot \dot{\boldsymbol{a}}(t) + \boldsymbol{h}^*(t) \cdot \dot{\boldsymbol{n}}^e, \tag{85}$$
$$\boldsymbol{h}^* \;=\; \frac{\partial}{\partial \boldsymbol{n}}(U^* - \boldsymbol{A}^* \cdot \boldsymbol{a}) = \boldsymbol{h}^*(\boldsymbol{A}^*, \boldsymbol{n}, T^*). \tag{86}$$

The equilibrium entropy along the U-projection $P^U Z(t)$ is

$$S^{eq}(t) = S^{eq}(\boldsymbol{a}, \boldsymbol{n}, U)(t), \tag{87}$$

and we get Gibbs' fundamental equation as usual:

$$T\dot{S}^{eq} = \dot{U} - \boldsymbol{A}^U \cdot \dot{\boldsymbol{a}} - \boldsymbol{\mu}^U \cdot \dot{\boldsymbol{n}}^e. \tag{88}$$

$\boldsymbol{n}$ is the total rate of mole numbers of $\mathcal{G}$. By other projections we get further Gibbs' equations:

$$P^\Theta: \quad \Theta\dot{S}^{eq} = \dot{U}^\Theta - \boldsymbol{A}^\Theta \cdot \dot{\boldsymbol{a}} - \boldsymbol{\mu}^\Theta \cdot \dot{\boldsymbol{n}}^e \tag{89}$$
$$P^*: \quad T^*\dot{S}^{eq} = \dot{U}^* - \boldsymbol{A}^* \cdot \dot{\boldsymbol{a}} - \boldsymbol{\mu}^* \cdot \dot{\boldsymbol{n}}^e \tag{90}$$

These equations are projections onto equilibrium subspace of a nonequilibrium trajectory along which we have to define a nonequilibrium entropy. How can such a definition look like? Of course we have to use the contact quantities, contact temperature Θ, dynamic generalized forces $\boldsymbol{A}$, and the dynamic chemical potentials $\boldsymbol{\mu}$:

$$\Theta\dot{S} := \dot{U} - \boldsymbol{A} \cdot \dot{\boldsymbol{a}} - \boldsymbol{\mu} \cdot \dot{\boldsymbol{n}} + \Theta\Sigma. \tag{91}$$

The entropy production Σ fixes the definition of S and determines it. Because we use small state spaces (23) the constitutive equations are functionals of the process history (22):

$$S(t) = \mathcal{S}(Z^t(\cdot)) \tag{92}$$
$$\boldsymbol{A}(t) = \mathcal{A}(Z^t(\cdot)) \tag{93}$$
$$\boldsymbol{\mu}(t) = \mathcal{M}(Z^t(\cdot)) \tag{94}$$

Proposition: Using the definition

$$\Theta \boldsymbol{s} := \boldsymbol{h} - \boldsymbol{\mu} \tag{95}$$

we get by inserting the first law into the defining equations of the time rate of entropy:

$$\Theta \dot{S} := \dot{Q} + \Theta \boldsymbol{s} \cdot \dot{\boldsymbol{n}}^e - \boldsymbol{\mu} \cdot \dot{\boldsymbol{n}}^i + \Theta \Sigma \tag{96}$$

If we consider an isolated system, we see that the last two terms of the right-hand side of (96) are the irreversible parts of $\Theta \dot{S}$. The different projections of (96) are

$$P^* : \quad T^* \dot{S}^{eq} = \dot{Q}^* - T^* \boldsymbol{s}^* \cdot \dot{\boldsymbol{n}}^e, \tag{97}$$

$$P^U : \quad T \dot{S}^{eq} = \dot{Q}^* - T \boldsymbol{s}^U \cdot \dot{\boldsymbol{n}}^e, \tag{98}$$

$$P^\Theta : \quad \Theta \dot{S}^{eq} = \dot{Q}^\Theta - \Theta \boldsymbol{s}^\Theta \cdot \dot{\boldsymbol{n}}^e. \tag{99}$$

Of course the definition of a nonequilibrium entropy is *not at all unique.* Another possibility is

$$T^* \dot{S}' := \dot{Q} + T^* \boldsymbol{s}^* \cdot \dot{\boldsymbol{n}}^e - \boldsymbol{\mu}^\Theta \cdot \dot{\boldsymbol{n}}^i + \Theta \Sigma'. \tag{100}$$

The definitions (96) and (100) and other possiblities cannot be arbitrary. They have to be compatible with Clausius' inequality and the embedding axiom being introduced in the next section.

4.2 Clausius' Inequality and Embedding Axiom

Clausius' inequality refers to a nonequilibrium trajectory which becomes cyclic by closing it using a projection (Fig.1.14). This kind of cyclic process which contains at least one equilibrium state is marked by +. Clausius' inequality for open systems then writes

$$\oint_+ \left[\frac{\dot{Q}(t)}{T^*(t)} + \boldsymbol{s}^*(t) \cdot \dot{\boldsymbol{n}}^e(t) \right] dt \leq 0 \tag{101}$$

Notice that T^* and $\boldsymbol{s}^*$ belong to the system's surroundings $\mathcal{G}^*$. Integrating $\dot{S}$ (96) and $\dot{S}'$ (100) along the considered cyclic process we get

$$\oint_+ \dot{S}' \, dt = \oint_+ \left[\frac{\dot{Q}}{T^*} + \boldsymbol{s}^* \cdot \dot{\boldsymbol{n}}^e \right] dt + \oint_+ \left[\frac{\Theta}{T^*} \Sigma' - \frac{\boldsymbol{\mu}^\Theta}{T^*} \cdot \dot{\boldsymbol{n}}^i \right] dt \, , \tag{102}$$

$$\oint_+ \dot{S} \, dt = \oint_+ \left[\frac{\dot{Q}}{\Theta} + \boldsymbol{s} \cdot \dot{\boldsymbol{n}}^e \right] dt + \oint_+ \left[\Sigma - \frac{\boldsymbol{\mu}}{\Theta} \cdot \dot{\boldsymbol{n}}^i \right] dt \, . \tag{103}$$

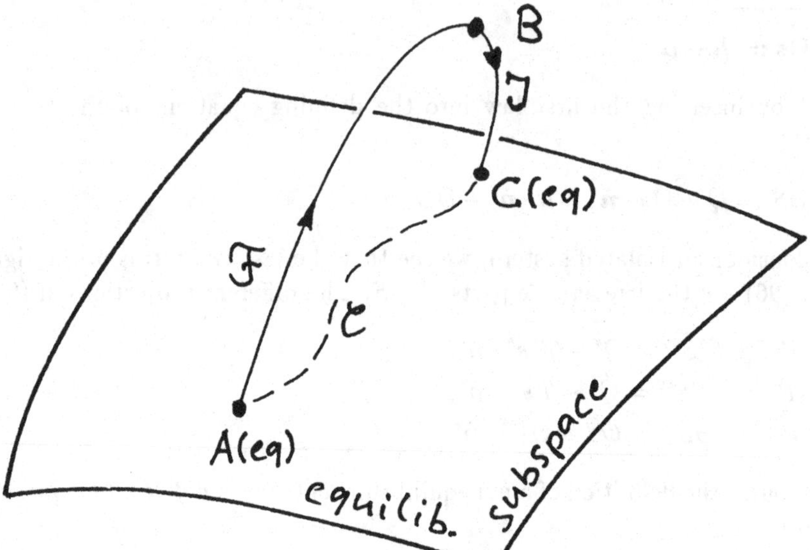

Figure 1.14 : Cyclic process consisting of an arbitrary part $\mathcal{F}$ starting out from equilibrium A(eq), of an isolated part $\mathcal{J}$, and of an accompanying process $\mathcal{C}$.

Because definitions of nonequilibrium entropies have to be compatible with the well-known equilibrium entropy we formulate the

Embedding Axiom:

$$\oint_+ \dot{S}\,dt = \oint_+ \dot{S}'\,dt = 0 \tag{104}$$

Notice that in general for arbitrary cyclic processes

$$\oint \dot{S}\,dt \neq 0 \tag{105}$$

is valid, because in small state spaces the nonequilibrium entropy is as a constitutive equation a functional of the process history and not a state function. By use of (101) we get from (102) a dissipation inequality

$$\oint_+ \left[\frac{\Theta}{T^*}\Sigma' - \frac{\boldsymbol{\mu}^\Theta}{T^*}\cdot\dot{\boldsymbol{n}}^i\right] dt \geq 0\,. \tag{106}$$

The analogous inequality following from (103)

$$\oint_+ \left[\Sigma - \frac{\boldsymbol{\mu}}{\Theta}\cdot\dot{\boldsymbol{n}}^i\right] dt \geq 0\,. \tag{107}$$

does not follow from Clausius' inequality (101) because we need its extended formulation [44]:

$$\oint_{+} \left[\frac{\dot{Q}}{\Theta} + \boldsymbol{s} \cdot \dot{\boldsymbol{n}}^{e} \right] dt \leq 0 \,. \tag{108}$$

In the special case of closed systems we can easily see that (108) is an extension of (101) because for closed systems (108) yields

$$\oint_{+} \frac{\dot{Q}}{\Theta} dt \geq \oint \frac{\dot{Q}}{T^{*}} dt \,. \tag{109}$$

4.3 Non-negative Entropy Production

Consider an arbitrary process starting out from equilibrium (Fig.1.14):

$$\mathcal{F}: \; A(eq) \rightarrow B \,. \tag{110}$$

In B the system will be isolated and will go to equilibrium:

$$\mathcal{J}: \; B \rightarrow C(eq), \quad \dot{Q} = 0, \quad \dot{\boldsymbol{a}} = \mathbf{o}, \quad \dot{\boldsymbol{n}}^{e} = \mathbf{o} \,. \tag{111}$$

The process is closed by an equilibrium trajectory

$$\mathcal{C}: \; C(eq) \rightarrow A(eq), \quad \Sigma = 0, \quad \dot{\boldsymbol{n}}^{i} = \mathbf{o} \,. \tag{112}$$

Proposition: The dissipation inequalities (100) and (91) yield for the cyclic process $\mathcal{F} \vee \mathcal{J} \vee \mathcal{C}$

$$\mathcal{F} \int_{A}^{B} \frac{1}{T^{*}} \left[\Theta \Sigma' - \boldsymbol{\mu}^{\Theta} \cdot \dot{\boldsymbol{n}}^{i} \right] dt \geq S'_{B} - S_{C}^{eq} \,, \tag{113}$$

$$\mathcal{F} \int_{A}^{B} \left[\Sigma - \frac{1}{\Theta} \boldsymbol{\mu} \cdot \dot{\boldsymbol{n}}^{i} \right] dt \geq S_{B} - S_{C}^{eq} \,. \tag{114}$$

The change of entropy $S_B - S_C^{eq}$ cannot be estimated because we do not know the sign of the left-hand integrals. But in the case $B \rightarrow A(eq)$ (Fig.1.17), i.e. $\mathcal{F}$ connects $A(eq)$ and $B(eq)$ and the system is isolated, the left-hand integrals in (113) and (114) vanish. We get

$$A(eq) \overset{isol.}{\rightarrow} C(eq) \qquad S_C^{eq} \geq S_A^{eq} \,. \tag{115}$$

We now integrate the definition of $\dot{S}$ (91) and of $\dot{S}'$ (100) along $\mathcal{J}$. By use of (111) this yields:

$$\mathcal{J} \int_{B}^{C} \left[\Sigma - \frac{1}{\Theta} \boldsymbol{\mu} \cdot \dot{\boldsymbol{n}}^{i} \right] dt = S_C^{eq} - S_B \,, \tag{116}$$

$$\mathcal{J} \int_{B}^{C} \frac{1}{T^{*}} \left[\Theta \Sigma' - \boldsymbol{\mu}^{\Theta} \cdot \dot{\boldsymbol{n}}^{i} \right] dt = S_C^{eq} - S'_{B} \,. \tag{117}$$

In contrast to (114) and (113) these are *equations* for the differences of entropy. Because $\mathcal{J}$ is a process in an isolated system we know some more about the left-hand integrals in (116) and (117) than about those in (113) and (114): Because we are in small state spaces these integrals are functionals of the process history up to the state

$$Z^B = (\boldsymbol{a}^B, \boldsymbol{n}^B, U^B, \Theta^B, \dots; T_B^*, \boldsymbol{A}_B^*, \boldsymbol{\mu}_B^*) \ . \tag{118}$$

Because of the isolation along $\mathcal{J}$ after passing B the history influences the value of the integrals only until the time t_B from which the isolation begins. Along $\mathcal{J}$ the integral depends only on the differences between B and C of those variables which change despite the isolation. These variables are the contact temperature Θ, or according to (64) U^Θ , and the mole numbers due to chemical reactions. Therefore we get introducing the abbreviations

$$U^\Theta(t) - U_C =: \xi(t), \quad \boldsymbol{n}(t) - \boldsymbol{n}_C =: \boldsymbol{\eta}(t), \tag{119}$$

$$\mathcal{J}\int_B^C \left[\Sigma - \frac{1}{\Theta}\boldsymbol{\mu}\cdot\dot{\boldsymbol{n}}^i\right] dt = -F(\xi(t_B), \boldsymbol{\eta}(t_B), \dots, Z^{t_B}(\cdot)) \ . \tag{120}$$

Here F is a functional of the process history up to time t_B. Because $Z^{t_B}(\cdot)$ is the only parameter along $\mathcal{J}$ the total differential of F along $\mathcal{J}$ writes:

$$DF = \frac{\partial F}{\partial \xi}(\xi, \boldsymbol{\eta}, \dots, Z^{t_B}(\cdot))\dot{\xi} + \frac{\partial F}{\partial \boldsymbol{\eta}}(\xi, \boldsymbol{\eta}, \dots, Z^{t_B}(\cdot))\dot{\boldsymbol{\eta}} + \dots \ . \tag{121}$$

Choosing along $\mathcal{J}$

$$\left(\Sigma - \frac{1}{\Theta}\boldsymbol{\mu}\cdot\dot{\boldsymbol{n}}^i\right)(t) = DF(\xi(t), \boldsymbol{\eta}(t), \dots, Z^{t_B}(\cdot)) \tag{122}$$

we get by integrating along $\mathcal{J}$

$$\mathcal{J}\int_B^C \left(\Sigma - \frac{1}{\Theta}\boldsymbol{\mu}\cdot\dot{\boldsymbol{n}}^i\right) dt = F(0, \mathbf{o}, \dots, Z^{t_B}(\cdot)) - F(\xi(t_B), \boldsymbol{\eta}(t_B), \dots, Z^{t_B}(\cdot)) \tag{123}$$

Up to here the functional F is arbitrary. Now we demand

$$F(0, \mathbf{o}, \dots, Z^{t_B}(\cdot)) \equiv 0 \ , \tag{124}$$

and consequently (120) is satisfied.

The "process speed" along $\mathcal{J}$ is determined by constitutive equations

$$\dot{\xi}(t) = X(\xi(t), \boldsymbol{\eta}(t), \dots, Z^{t_B}(\cdot)), \tag{125}$$
$$\dot{\boldsymbol{\eta}}(t) = E(\xi(t), \boldsymbol{\eta}(t), \dots, Z^{t_B}(\cdot)). \tag{126}$$

In another diction (125) and (126) are relaxation rate equations in an isolated system of a process starting out from B at time t_B. For enforcing non-negative entropy production we choose

$$DF \geq 0, \tag{127}$$

which can be achieved by

$$\operatorname{sgn}\frac{\partial F}{\partial \xi} \stackrel{!}{=} \operatorname{sgn}\dot{\xi}, \quad \operatorname{sgn}\frac{\partial F}{\partial \eta} \stackrel{!}{=} \operatorname{sgn}\dot{\eta}\,. \tag{128}$$

This is a very weak condition to F because it only concerns signs. In contrast to F the functionals X and E are determined by the material. A possible dependence of F from is ξ sketched in Fig.1.15.

Finally we put together the results. According to (122) and (127) we have

$$\left(\Sigma - \frac{1}{\Theta}\boldsymbol{\mu}\cdot\dot{\boldsymbol{n}}^{i}\right)(t) \geq 0, \text{ along } \mathcal{J}. \tag{129}$$

Consequently according to (116) and (117) we get

$$\begin{aligned} &B \stackrel{isol}{\to} C(eq)\\ &S_C^{eq} \geq S_B\\ &S_C^{eq} \geq S'_B \end{aligned} \tag{130}$$

and beyond that also

$$\begin{aligned} &\dot{S}(t) \geq 0\\ &\dot{S}'(t) \geq 0 \end{aligned} \tag{131}$$

along $\mathcal{J}$.

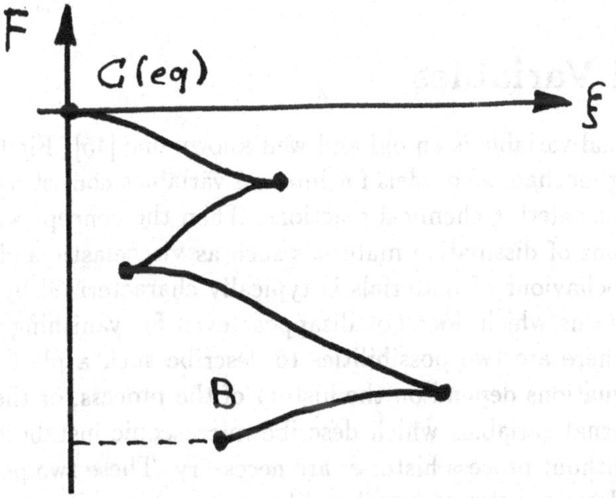

Figure 1.15 : Graph showing the behavior of F taking into consideration (127) and (128).

According to (122) the entropy production Σ and the chemical potentials $\boldsymbol{\mu}$ are also functionals of the process history

$$\Sigma(t) = \Sigma(Z^t(\cdot)) \,, \tag{132}$$
$$\boldsymbol{\mu}(t) = \boldsymbol{\mu}(Z^t(\cdot)) \,. \tag{133}$$

These equations show that the entropy production and the chemical potentials do not depend on the time derivatives of the state space variables and consequently they are independent of the process direction in state space. But along $\mathcal{J}$ DF is definite, and because of its independence of the process direction it is definite in every direction. So we get:

$$\left(\Sigma - \frac{1}{\Theta}\boldsymbol{\mu} \cdot \dot{\boldsymbol{n}}^i\right)(t) \geq 0 \qquad \text{holds generally.} \tag{134}$$

(96) and (100) yield

$$\dot{S} \geq \frac{\dot{Q}}{\Theta} + \boldsymbol{s} \cdot \dot{\boldsymbol{n}}^e \,, \tag{135}$$
$$\dot{S}' \geq \frac{\dot{Q}}{T^*} + \boldsymbol{s}^* \cdot \dot{\boldsymbol{n}}^e \,. \tag{136}$$

Consequently we proved a statement which is used in continuum thermodynamics as a formulation of the second law without any remark about its connection to Clausius' inequality [32]: It is always possible, but not uniquely, to give a material dependent definition of entropy production so that it will be definite.

5 Internal Variables

The notion of internal variable is an old and well-known one [45]. First of all it was used by considering mechanical models for internal variables consisting of dashpots and springs or by considering chemical reactions. Then the concept was applied to constitutive equations of dissipative materials such as viscoelastic and viscoplastic ones [46]. Plastic behaviour of materials is typically characterized by a hysteresis in their stress diagrams which does not disappear even for vanishing stretch velocity. In principle there are two possibilities to describe such a plastic behaviour: The constitutive equations depend on the history of the process, or the state space is extended by internal variables which describe microscopic instabilities inducing plastic behaviour without process histories are necessary. These two possibilities do characterize the difference between small and large state spaces (sect.1.2). Here we introduce internal variables i.e. large state spaces for avoiding process histories in the domain of the constitutive maps. This demands the formulation of the Zeroth Law which states the equilibrium subspace in thermal homogeneous systems is spanned by the work variables, the mole numbers and by one additional thermodynamical variable, such as the internal energy (18). In plastic materials equilibrium states include besides the variables mentioned above the equilibrium values of the internal variables which depend on the work variables and on the internal energy [47]. But this dependence of the internal variables in equilibrium on the work variables and on the internal energy may be not unique in special classes of materials: More than one set of values of the internal variables belongs to given work variables and given internal energy. As discussed below this is the reason for getting hysteresis phenomena in cyclic processes. From a more theoretical point of view it is important that in general the equilibrium subspace is many-valued.

Whereas in equilibrium the internal variables are (not uniquely) dependent on the variables spanning the equilibrium subspace, in nonequilibrium the internal variables are independent variables obeying a special rate equation. This rate equation cannot directly be influenced by process controlling or other external manipulations. From this fact the name internal variable originates.

5.1 Nonequilibrium State Space

According to (58) we now introduce the nonequilibrium state space of the discrete system under consideration

$$Z = (\boldsymbol{a}, \boldsymbol{n}, U, \Theta, \boldsymbol{\alpha}; T^*, \boldsymbol{\mu}^*, \boldsymbol{F}^*) \tag{137}$$

(T^* is the thermostatic temperature of $\mathcal{G}^*$, $\boldsymbol{\mu}^*$ its chemical potentials, and $\boldsymbol{F}^*$ its generalized forces). According to the 0-th law (18) - as we will see below - the equilibrium subspace of $\mathcal{G}$ is given by

$$Z^0 = (\boldsymbol{a}, \boldsymbol{n}, U, \boldsymbol{\alpha}_m(\boldsymbol{a}, \boldsymbol{n}, U)). \tag{138}$$

Here m is the multiplicity of equilibrium values of $\boldsymbol{\alpha}$ belonging to $(\boldsymbol{a}, \boldsymbol{n}, U)$.

A process $Z(t)$ in $\mathcal{G}$ is represented by a trajectory $\mathcal{T}$ in the nonequilibrium state space. Along $\mathcal{T}$ the time rate of a nonequilibrium entropy has the following form

$$\Theta\dot{S} = \dot{U} - \boldsymbol{F} \cdot \dot{\boldsymbol{a}} - \boldsymbol{\mu} \cdot \dot{\boldsymbol{n}} + \boldsymbol{A} \cdot \dot{\boldsymbol{\alpha}} + \Sigma, \tag{139}$$

($\boldsymbol{\mu}$ are the nonequilibrium chemical potentials of the components of $\mathcal{G}$ [32],$\boldsymbol{A}$ are the affinities,Σ is the entropy production in $\mathcal{G}$ beyond chemical reactions and relaxation processes by internal variables). Introducing the chemical affinities

$$\mathcal{A} := -\boldsymbol{\mu} \cdot \boldsymbol{\nu}^+, \tag{140}$$

here $\boldsymbol{\nu}^+$ is the transposed matrix of the stoichiometric equations

$$\boldsymbol{\nu} \cdot \boldsymbol{M} = \mathbf{o} \tag{141}$$

($\boldsymbol{M}$ are the mole masses), and introducing the time rates of mole numbers due to chemical reactions $\dot{\boldsymbol{n}}^i$ and the reaction speeds

$$\dot{\boldsymbol{n}}^i = \boldsymbol{\nu}^+ \cdot \boldsymbol{\xi}, \tag{142}$$

we get by (140) and (142)

$$-\boldsymbol{\mu} \cdot \dot{\boldsymbol{n}} = \boldsymbol{\mathcal{A}} \cdot \dot{\boldsymbol{\xi}} - \boldsymbol{\mu} \cdot \dot{\boldsymbol{n}}^e. \tag{143}$$

Pay attention to the fact that $\dot{\boldsymbol{\xi}}$ and $\dot{\boldsymbol{n}}^e$ are no time rates of state variables. Inserting (143) into (139) and remembering the First Law for open discrete systems [46],

$$\dot{U} = \dot{Q} + \boldsymbol{F} \cdot \dot{\boldsymbol{a}} + \boldsymbol{h} \cdot \dot{\boldsymbol{n}}^e, \tag{144}$$

($\boldsymbol{h}$ are the molar enthalpies) we get

$$\Theta \dot{S} = \dot{Q} + (\boldsymbol{h} - \boldsymbol{\mu}) \cdot \dot{\boldsymbol{n}}^e + \boldsymbol{\mathcal{A}} \cdot \dot{\boldsymbol{\xi}} + \boldsymbol{A} \cdot \dot{\boldsymbol{\alpha}} + \Sigma. \tag{145}$$

In (145) constitutive quantities appear which are represented by maps (abbreviated by M)

$$(S, \boldsymbol{F}, \boldsymbol{\mu}, \boldsymbol{\mathcal{A}}, \boldsymbol{A}, \Sigma) \equiv M \tag{146}$$

which can be defined on small or large state spaces (23) or (21).

5.2 Concepts

Up to here the internal variables are not characterized yet. A definition of internal variables which is suitable for all situations is hard to write down. Therefore we want to formulate some concepts which are obligatory for all possible definitions of internal variables.

Concept I: The introduction of internal variables $\boldsymbol{\alpha}$ makes possible the use of large state spaces, i.e. the constitutive maps (146) are defined on the state space itself and not on process histories:

$$\boldsymbol{F} = \boldsymbol{F}(Z(t)), \qquad S(t) = S(Z(t)), ... \tag{147}$$

Besides the balance equations (139) and (144) for U and S and the constitutive maps (146) we additionally need rate equations for the internal variables because of (139) or (145). Therefore we formulate the

Concept II: Rate equations for internal variables do not belong to the constitutive equations (146). They are additional material-dependent equations for the time rates of the internal variables which join the balance equations and the constitutive equations.

Time rates of the internal variables as the reaction speeds of chemical reactions are independent of the state of the system's vicinity $\mathcal{G}^*$. At most state variables of the system itself can influence the internal variables. Therefore we formulate the

Concept III: An isolation of the system does not influence the internal variables instantaneously.

Because isolation of a discrete system is defined by

$$\dot{U} = 0, \qquad \dot{\boldsymbol{\alpha}} = \mathbf{o}, \qquad \dot{Q} = 0, \qquad \dot{\boldsymbol{n}}^e = \mathbf{o} \tag{148}$$

the time rate of entropy (139) becomes by use of (143)

$$\Theta\dot{S}^{isol} = \mathcal{A}\cdot\dot{\xi} + A\cdot\dot{\alpha} + \Sigma \geq 0. \tag{149}$$

This inequality expresses the second law (73). Beyond (149) we here presuppose - because of $\dot{\xi}$, $\dot{\alpha}$, and Σ being independent of each other - that each single term in (149) satisfies

$$\mathcal{A}\cdot\dot{\xi} \geq 0, \qquad A\cdot\dot{\alpha} \geq 0, \qquad \Sigma \geq 0, \tag{150}$$

or even more special that inequalities

$$\mathcal{A}_j\dot{\xi}_j \geq 0, \qquad A_k\dot{\alpha}_k \geq 0 \tag{151}$$

hold for all components j and k.

In (143) we have replaced the time rates of the mole numbers $\dot{n}$ by the reaction speeds $\dot{\xi}$ and the time rates of the mole numbers due to external mass exchange $\dot{n}^e$. Both time rates are no time derivatives of state variables in contrast to $\dot{n}$ and $\dot{\alpha}$. Because $\dot{\xi}$ are the analogues to $\dot{\alpha}$ in case of chemical reactions we formulate the

Concept IV: Internal variables may be included in the state space (as α) or may be not (no ξ exist as state variables).

A special example for the rate equations mentioned in concept II are those for the reaction speeds [48]

$$\dot{\xi}_j(t) = a_j(Z(t)[1 - \exp(b_j(t)\mathcal{A}_j(Z(t))] \tag{152}$$

and those for the internal variables

$$\dot{\alpha}(t) = C(Z(t); A(Z(t)). \tag{153}$$

From (153) and (150) we get for all A the inequality

$$A\cdot C(Z(t); A(Z(t)) \geq 0. \tag{154}$$

Proposition: If C is continuous at $A = o$

$$\lim_{\lambda\to 0} C(Z(t); \lambda A(Z(t)) = C(Z(t); o), \tag{155}$$

then (154) induces

$$C(Z(t); o) = o. \tag{156}$$

Proof [49]: From (154) we get for all $\lambda \geq 0$

$$\lambda A\cdot C(Z(t); \lambda A(Z(t)) \geq 0, \tag{157}$$

and therefore

$$A\cdot C(Z(t); \lambda A(Z(t)) \geq 0. \tag{158}$$

Because of (154) this yields in the limit $\lambda \to +0$

$$\boldsymbol{A} \cdot \boldsymbol{C}(Z(t); \mathbf{o}) \geq 0 \tag{159}$$

for all $\boldsymbol{A}$. Therefore we get (156). □

According to (150), (153) and (156) we have

$$\mathcal{A} = \mathbf{o} \to \dot{\boldsymbol{\xi}} = \mathbf{o} \tag{160}$$
$$\text{and } \boldsymbol{A} = \mathbf{o} \to \dot{\boldsymbol{\alpha}} = \mathbf{o}, \tag{161}$$

so we can see that disappering affinities characterize equilibrium:

Definition: The equilibrium conditions are

$$\Theta(U) = T, \qquad \Sigma = 0, \tag{162}$$

$$\mathcal{A} = \mathbf{o} \to \dot{\boldsymbol{\xi}} = \mathbf{o}, \tag{163}$$

$$\boldsymbol{A} = \mathbf{o} \to \dot{\boldsymbol{\alpha}} = \mathbf{o}. \tag{164}$$

From (137), (162), and (164) we get in equilibrium

$$\boldsymbol{A}(\boldsymbol{a}, \boldsymbol{n}, U, \boldsymbol{\alpha}) = \mathbf{o}. \tag{165}$$

which results in

$$\boldsymbol{\alpha} = \boldsymbol{f}(\boldsymbol{a}, \boldsymbol{n}, U), \qquad \dot{\boldsymbol{\alpha}} = \mathbf{o}. \tag{166}$$

Consequently we can formulate the

Concept V: In equilibrium the internal variables become dependent on the variables of equilibrium subspace (138).

In equilibrium the value of the contact temperature Θ is equal to the thermostatic temperature according to (162) [31], whereas the equilibrium value of $\boldsymbol{\alpha}$ in (166) is not unique and has the multiplicity m as mentioned above. This induces a many-valued equilibrium subspace.

In equilibrium subspace (139) and (145) become

$$T\dot{S}^{eq} = \dot{U} - \boldsymbol{F}^{eq} \cdot \dot{\boldsymbol{a}} - \boldsymbol{\mu}^{eq} \cdot \dot{\boldsymbol{n}}^{e}, \tag{167}$$

$$T\dot{S}^{eq} = \dot{Q}^{eq} + (\boldsymbol{h}^{eq} - \boldsymbol{\mu}^{eq}) \cdot \dot{\boldsymbol{n}}^{e} \tag{168}$$

(T is the thermostatic temperature). The superscript "eq" denotes the equilibrium quantities which belong to the reversible process in the equilibrium subspace (138).

Besides the so-called unconstrained equilibrium characterized by (162) to (164) there are other equilibria:

Definition: Conditions of constrained equilibrium are:

$$\Theta(U) = T, \qquad \Sigma = 0, \tag{169}$$

$$\dot{\boldsymbol{\xi}} = \mathbf{o} \wedge \boldsymbol{\mathcal{A}} \neq \mathbf{o}, \tag{170}$$

$$\dot{\boldsymbol{\alpha}} = \mathbf{o} \wedge \boldsymbol{A} \neq \mathbf{o}. \tag{171}$$

According to (170) and (171), (152) and (153) there exist states of constrained equilibrium Z^C with

$$a_j(Z^C) = 0, \qquad \boldsymbol{C}(Z^C; \boldsymbol{A}(Z^C)) = \mathbf{o}. \tag{172}$$

Consequently we get the

Concept VI: There exist constrained equilibria characterized by (172), whereas in equilibrium the affinities disappear.

According to Concept II we need additional rate equations govering the internal variables. Equation (152) is an example for the reaction speeds in case of chemical reactions, whereas (153) together with (154) represents the general form of such a rate equation. A special case exists, if the time rate of the internal variable is proportional to the internal variable itself.

Definition: Internal variables which satisfy

$$\dot{\alpha}_j = -(1/\tau_j)A_j\alpha_j, \qquad \tau_j = const_j > 0, \tag{173}$$

are called relaxation variables.τ_j is the relaxation time of variable α_j. According to (151) we get by (173)

$$sgn\ \dot{\alpha}_j = sgn\ A_j, \qquad \alpha_j \leq 0. \tag{174}$$

5.3 Projectors and Accompanying Processes

To each (nonequilibrium) process there exist different projections onto the equilibrium subspace. Each of these projections generate a trajectory in the equilibrium subspace which is parametrized by the time, because the projection maps point by point, so that the time of the real process is transferred to the quasi-process in the equilibrium subspace (Fig.1.16).

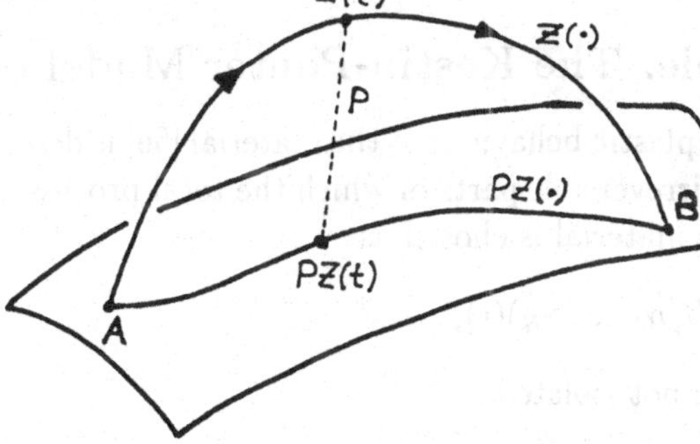

Figure 1.16: The projection P maps the process $Z(\cdot)$ point by point onto the

equilibrium subspace represented as a hypersurface in state space. Consequently the accompanying process $PZ(\cdot)$ is parametrized by t.

We now introduce such projections in the case for which internal variables exist. We consider three special kinds of relaxation variables with different relaxation times:

$$\tau_k < \tau_m, \qquad k < m, \qquad \tau_1 \ll t_0, \qquad \tau_3 \gg t_0. \tag{175}$$

Here t_0 is the duration of the observation. From (173) and (175) we get the time rates of the relaxation variables

$$\dot{\alpha}_1 \to \pm\infty, \qquad \dot{\alpha}_3 \to 0. \tag{176}$$

Therefore α_1 is a very fast and α_3 a very slow relaxation variable, whereas $\dot{\alpha}_2$ has a value between $\dot{\alpha}_1$ and $\dot{\alpha}_2$. We now consider an isolated system which undergoes a process according to (137)

$$Z(t) = (\boldsymbol{a}, \boldsymbol{n}, U, \Theta, \alpha_1^\infty(\boldsymbol{a}, \boldsymbol{n}, U), \alpha_2, \alpha_3^0)(t). \tag{177}$$

Here α_1^∞ is the not necessarily unique value of the fast relaxation variable which according to (166) depends on the equilibrium variables $(\boldsymbol{a}, \boldsymbol{n}, U)$ whereas the value α_3^0 of the slow relaxation variable α_3 is nearly constant during the observation time: α_3 is frozen in. An accompanying process of (177) is defined by a projection of (177) onto the equilibrium subspace (138) taking into consideration the constant relaxation variables. Of course there exist different projections:

$$P^0 Z(t) := (\boldsymbol{a}, \boldsymbol{n}, U, \alpha_3^0), \tag{178}$$

$$P^\infty Z(t) := (\boldsymbol{a}, \boldsymbol{n}, U, \alpha_1^\infty), \qquad \text{or} \tag{179}$$

$$PZ(t) := (\boldsymbol{a}, \boldsymbol{n}, U, \alpha_3^0; \alpha_1^\infty). \tag{180}$$

Using these projections we can characterize plastic behaviour. We refer to

5.4 Example: The Kestin-Ponter Model of Plasticity

In this model the plastic behaviour of the material [50] is described by a rapid change of reversible and irreversible parts of which the total process consists [51]. The state space of a plastic material is chosen as

$$Z(t) = (\boldsymbol{a}, U, \alpha_1, ..., \alpha_N)(t), \tag{181}$$

and the system is not isolated:

$$\boldsymbol{a} = \boldsymbol{a}(t), \qquad U = U(t). \tag{182}$$

All N internal variables are sufficiently fast compared with the time rates of $\boldsymbol{a}$ and U. The number N of the internal variables is very high. Consequently at a fixed time t a part of them is just undergoing a relaxation process, so that their values are not determined by $\boldsymbol{a}$ and U according to (166). Therefore we get by (179)

$$P^{\infty}Z(t) = (\boldsymbol{a}, U, \alpha_{k1}^{\infty}, ..., \alpha_{kL}^{\infty})(t), \quad kJ \in (1,2,...,N), \quad L \leq N, \tag{183}$$

Here the $\alpha_{kJ}^{\infty}$are solution of

$$\boldsymbol{A}(\boldsymbol{a}, U, \alpha_1^{\infty}, ..., \alpha_N^{\infty}) = \mathbf{o} \tag{184}$$

which are not uniquely determined by $\boldsymbol{a}$ and U

$$\alpha_{kJ}^{\infty} = g_{kJ}(\boldsymbol{a}, U), \qquad j = 1, ..., m(kJ), \tag{185}$$

(m is the multiplicity of the value of α_{kJ}^{∞}). The other internal variables $\alpha_{kL+1}, ..., \alpha_{kN}$ are not constant and have values which do not depend on $\boldsymbol{a}$ and U. They are just sufficiently fast relaxing to an equilibrium value which is compatible with (185). The affinity does not vanish

$$\boldsymbol{A}(\boldsymbol{a}, U, \alpha_{k1}^{\infty}, ..., \alpha_{kL}^{\infty}, \alpha_{kL}, ..., \alpha_{kN}) \neq \mathbf{o}. \tag{186}$$

Because $\boldsymbol{a}$ and U are time-dependent, the α_{kJ}^{∞} changes in time according to (185), but if α_{kJ}^{∞} takes a value which is not compatible with (184) a relaxation process will begin.

For demonstration we will assume, there is only one internal variable α [51]. Then (184) and (186) become

$$A(\boldsymbol{a}, U, \alpha^{\infty}) = 0 \qquad \text{or} \qquad A(\boldsymbol{a}, U, \alpha) \neq 0 \tag{187}$$

with multiplicity m (Fig.1.17).

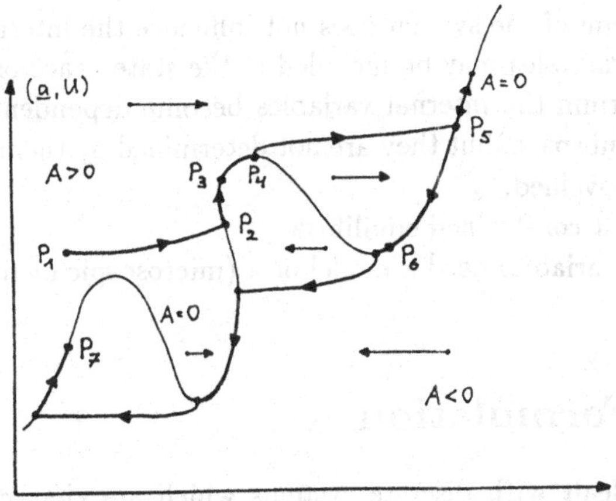

Figure 1.17 : Process P_1 to P_7 in the state space spanned by $(\boldsymbol{a}, U, \alpha)$. P_1 to $P_2(P_4$ to $P_5)$ is a fast relaxation process being irreversible due to the internal variable

α. Along P_2 to P_4 (P_5 to P_6) α is determined by $(\boldsymbol{a}, U)$, and the entropy production due to α disappears. Three $(m = 3)$ states $(\boldsymbol{a}, U, \alpha^{\infty})$ belong to P_1.

The $(\boldsymbol{a}, U) - \alpha -$ *plane* is divided by the graph of (187) into two regions $A > 0$ and $A < 0$. According to (151) $\dot{\alpha}$ is definite in each of the regions. The arrows in Fig. 1.17 represent amount and direction of the "vector field" $\dot{\alpha}$. A process starting from P_1 arrives very fast, $\mid (\boldsymbol{a}, U)^{\cdot} \mid \ll \mid \dot{\alpha} \mid$ at P_2 on the graph of $A = 0$. If now $(\boldsymbol{a}, U)^{\cdot} \neq 0$ the process will arrive at P_3 on a trajectory which is very near to the graph of $A = 0$, because the relaxation time is so short. In P_4 the vector field is determined by $(\boldsymbol{a}, U)^{\cdot}$. Therefore the process leaves the graph of $A = 0$ with a perpendicular tangent going fast to P_5. The way back from P_5 is evidently different from the way to P_5. So P_6 has the same value of $(\boldsymbol{a}, U)$ as P_1, and P_7 the same value of α as P_1.

Using this model the material has the following properties [51]: After loading the stress-free state has a non-zero stretch; there exists a hysteresis in the stress-strain diagram and hardening.

One can imagine the concept of internal variables is flexible enough to describe a lot of phenomena easily. Therefore not to abuse the notion of internal variables they need a physical background. Therefore we formulate

Concept VII: Internal variables need a model or an interpretation.

Kestin and Ponter [50] determine their internal variables by modes of operation of Frank-Read sources. A detailed explanation is given in [51].

Finally we summarize the seven concepts for introducing internal variables into nonequilibrium thermodynamics:

I: The introduction of internal variables makes possible the use of large state spaces,

II: Additional rate equations for internal variables join the balance equations and constitutive equations,

III: An isolation of the system does not influence the internal variables,

IV: Internal variables may be included in the state space or may be not,

V: In equilibrium the internal variables become dependent on the variables of the equilibrium subspace, but they are not determined by them, i.e. the equilibrium subspace is many-valued,

VI: There exist constrained equilibria,

VII: Internal variables need a model or a (microscopic or molecular) interpretation.

6 Field Formulation

Up to here we dealt with discrete systems which are characterized by exchange quantities and a (non-equilibrium) state space. If we divide a system into sufficiently small subsystems, each of them described as a discrete system, we get the so-called field formulation of thermodynamics which uses fields of thermodynamical quantities. Most of these fields contain no difficulty with regard to their definition and interpretation. But the fields of specific entropy and temperature have to be considered in more detail, because up to now contact temperature is only defined for discrete systems.

We will consider three examples to see the translation from discrete systems to field formulation: the balance equations of mass, momentum, and entropy.

6.1 Partial Mass Balance

We consider a discrete open system $\mathcal{G}$ containing K components reacting with each other[52]. The mass balance of the α-th component is

$$^{\alpha}\dot{M}(\mathcal{G}(t)) = {}^{\alpha}\dot{M}_{chem} + {}^{\alpha}\dot{M}(\partial\mathcal{G}(t)). \tag{188}$$

The right-hand terms are the rate of mass of the α-th component due to chemical reactions and due to exchange through the partition $\partial\mathcal{G}$ between $\mathcal{G}$ and its environment. Summing up the components we get

$$\dot{M}(\partial\mathcal{G}(t)) = \sum_{\alpha} {}^{\alpha}\dot{M}(\partial\mathcal{G}(t)) \tag{189}$$

$$0 = \sum_{\alpha}^{\alpha} \dot{M}_{chem}. \tag{190}$$

Equation (190) represents mass conservation.

Dividing $\partial\mathcal{G}$ into subsurfaces we get a surface field formulation on $\partial\mathcal{G}$:

$$^{\alpha}\dot{M}(\mathcal{G}(t)) \to -\oint_{(\partial\mathcal{G}(t))} \dot{\boldsymbol{m}}^{\alpha}(\boldsymbol{x},t)\cdot d\boldsymbol{f}. \tag{191}$$

Here $\dot{\boldsymbol{m}}^{\alpha}$ is the *partial mass flux density* of the α-th component. Dividing $\partial\mathcal{G}$ into subsystems

$$\mathcal{G} = \bigcup_i \mathcal{G}_i, \qquad \mathcal{G}_i \cap \mathcal{G}_k = \emptyset, \qquad i \neq k, \qquad V_i = vol(\mathcal{G}_i), \tag{192}$$

we get

$$\frac{^{\alpha}\dot{M}(\partial\mathcal{G}_i(t))}{V_i} = \frac{^{\alpha}\dot{M}_{chem}}{V_i} - \frac{1}{V_i}\oint_{\partial\mathcal{G}_i} \dot{\boldsymbol{m}}^{\alpha}(\boldsymbol{x},t)\cdot d\boldsymbol{f}. \tag{193}$$

The left-hand side writes:

$$\frac{^{\alpha}\dot{M}(\mathcal{G}_i(t))}{V_i} = \frac{d}{dt}\frac{^{\alpha}M(\mathcal{G}_i)}{V_i} + \frac{^{\alpha}M(\mathcal{G}_i)}{V_i}\frac{\dot{V}_i}{V_i}. \tag{194}$$

By use of Reynolds' transport theorem

$$\frac{d}{dt}\int_{(\mathcal{G}(t))}\Phi dV = \int_{(\mathcal{G}(t))}(\dot{\Phi} + \Phi\nabla\cdot\boldsymbol{w})dV, \tag{195}$$

$$\frac{d}{dt} = \frac{\partial}{\partial t} + \boldsymbol{w}\cdot\frac{\partial}{\partial\boldsymbol{x}}, \tag{196}$$

we get

$$\dot{V}_i = \frac{d}{dt}\int_{(\mathcal{G}_i(t))} dV = \oint_{\partial\mathcal{G}_i}\boldsymbol{w}\cdot d\boldsymbol{f}. \tag{197}$$

Introducing arbitrary small subsystems we get the *partial mass density*

$$\lim_{\mathcal{G}_i\to 0}\frac{{}^{\alpha}\dot{M}(\mathcal{G}_i(t))}{V_i} =: \rho^{\alpha}(\boldsymbol{x},t), \tag{198}$$

the *partial mass production*

$$\lim_{\mathcal{G}_i\to 0}\frac{{}^{\alpha}\dot{M}_{chem}}{V_i} = \tau^{\alpha}(\boldsymbol{x},t) \tag{199}$$

the divergence of the *velocity of the map* $\mathcal{G}(t_0)\to\mathcal{G}(t)$

$$\lim_{\mathcal{G}_i\to 0}\frac{1}{V_i}\oint_{\partial\mathcal{G}_i}\boldsymbol{w}\cdot d\boldsymbol{f} =: \nabla\cdot\boldsymbol{w}, \tag{200}$$

the *partial mass flux density*

$$\lim_{\mathcal{G}_i\to 0}\frac{1}{V_i}\oint_{\partial\mathcal{G}_i}\dot{\boldsymbol{m}}^{\alpha}\cdot d\boldsymbol{f} =: \nabla\cdot\dot{\boldsymbol{m}}^{\alpha}. \tag{201}$$

Introducing the quantities into (193) we have

$$\dot{\rho}^{\alpha} + \rho^{\alpha}\nabla\cdot\boldsymbol{w} = \tau^{\alpha} - \nabla\cdot\dot{\boldsymbol{m}}^{\alpha}. \tag{202}$$

The *partial material velocity* is defined by

$$\rho^{\alpha}\boldsymbol{v}^{\alpha} := \dot{\boldsymbol{m}}^{\alpha} + \rho^{\alpha}\boldsymbol{w}, \tag{203}$$

and (202) yields the *partial mass balance*:

$$\frac{\partial}{\partial t}\rho^{\alpha} = -\nabla\cdot(\dot{\boldsymbol{m}}^{\alpha} + \rho^{\alpha}\boldsymbol{w}) + \tau^{\alpha}. \tag{204}$$

Summing up the components and defining the *mass density*

$$\rho := \sum_{\alpha}\rho^{\alpha}, \tag{205}$$

and the *material velocity*

$$\rho \boldsymbol{v} := \sum_{\alpha} \rho^{\alpha} \boldsymbol{v}^{\alpha} \tag{206}$$

we get the *mass balance equation,*

$$\frac{\partial}{\partial t}\rho = -\nabla \cdot \rho \boldsymbol{v}, \tag{207}$$

and we can formulate the

Axiom of Mixtures: Summing up the components of a partial balance equations the sum looks like a balance of an 1-component system.

6.2 Partial Momentum Balance

The partial momentum of a discrete system is defined by

$$^{\alpha}\boldsymbol{P}(\mathcal{G}(t)) := \int_{(\mathcal{G}(t))} \rho^{\alpha}(\boldsymbol{x},t)\boldsymbol{v}^{\alpha}(\boldsymbol{x},t)dV. \tag{208}$$

The balance of momentum writes

$$^{\alpha}\dot{\boldsymbol{P}}(\mathcal{G}) = {}^{\alpha}\boldsymbol{K} + {}^{\alpha}\dot{\boldsymbol{P}}(\partial\mathcal{G}) + {}^{\alpha}\boldsymbol{M}(\mathcal{G}). \tag{209}$$

Here $^{\alpha}\boldsymbol{K}$ are the forces, $^{\alpha}\dot{P}(\partial\mathcal{G})$ the supply due to the opening of the system, and $^{\alpha}\boldsymbol{M}(\mathcal{G})$ the momentum interaction. The forces can be split into volume forces $^{\alpha}\boldsymbol{F}$ and surface forces $^{\alpha}\boldsymbol{t}$

$$^{\alpha}\boldsymbol{K} = {}^{\alpha}\boldsymbol{F} + {}^{\alpha}\boldsymbol{t}. \tag{210}$$

The division procedure into subsystems yields the *partial acceleration*

$$^{\alpha}\boldsymbol{F}(\mathcal{G}) \rightarrow \rho^{\alpha}(\boldsymbol{x},t)\boldsymbol{f}^{\alpha}(\boldsymbol{x},t), \tag{211}$$

and the volume force

$$\boldsymbol{F}^{\alpha}(\mathcal{G}) = \int_{(\mathcal{G}(t))} \rho^{\alpha}\boldsymbol{f}^{\alpha}dV. \tag{212}$$

The surface forces $^{\alpha}\boldsymbol{t}$ are contact quantities as the generalized forces in (52). Their defining inequality is

$$({}^{\alpha}\boldsymbol{t}^{eq} - {}^{\alpha}\boldsymbol{t}) \cdot \dot{\boldsymbol{g}}^{\alpha} \geq 0, \tag{213}$$

and the indicator $\dot{\boldsymbol{g}}^{\alpha}$ is the time rate of the elongation of a spring fastened at an α-impermeable piston on the surface $\partial\mathcal{G}$ (Fig.1.18). By this procedure we get the surface function of the contact force $^{\alpha}\boldsymbol{t}(\boldsymbol{x},\boldsymbol{n},t)$. The total surface force is

$$^{\alpha}\boldsymbol{t}(\partial\mathcal{G}(t)) = \oint_{(\partial\mathcal{G}(t))} {}^{\alpha}\boldsymbol{t}(\boldsymbol{x},\boldsymbol{n},t)df = -\oint_{(\partial\mathcal{G}(t))} \mathbf{P}^{\alpha}(\boldsymbol{x},t)\cdot d\boldsymbol{f}. \tag{214}$$

The second equation is obtained by the usual reasoning using Cauchy's tetrahedron argumentation. $\mathbf{P}^\alpha$ is the *partial pressure tensor* field.

The supply in (209) is defined by

$$^\alpha\dot{\boldsymbol{P}}(\partial\mathcal{G}(t)) := -\oint_{(\partial\mathcal{G}(t))} \boldsymbol{v}\dot{m}^\alpha \cdot d\boldsymbol{f}. \tag{215}$$

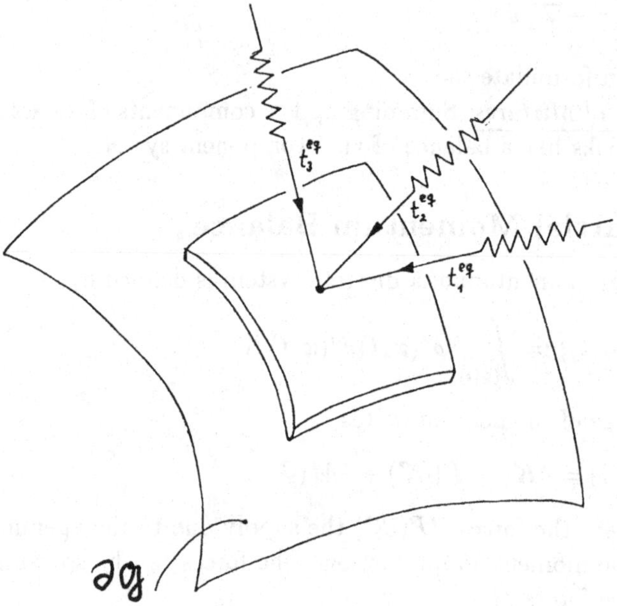

Figure 1.18 : An α-impermeable piston on $\partial\mathcal{G}$ is in relative rest to $\partial\mathcal{G}$. The springs indicate the surface contact force.

The momentum conservation is given by

$$\sum_\alpha {}^\alpha\boldsymbol{M}(\mathcal{G}(t)) = 0. \tag{216}$$

Inserting (208), (214), (215), and (216) into (209) we get the *global partial momentum balance* equation:

$$\frac{d}{dt}\int_{(\mathcal{G}(t))} \rho^\alpha \boldsymbol{v}^\alpha dV = \int_{(\mathcal{G}(t))} \rho^\alpha \boldsymbol{f}^\alpha dV - \oint_{(\partial\mathcal{G}(t))} \mathbf{P}^\alpha \cdot d\boldsymbol{f} - \tag{217}$$

$$- \oint_{\partial\mathcal{G}} \rho^\alpha \boldsymbol{v}(\boldsymbol{v}^\alpha - \boldsymbol{w}) \cdot d\boldsymbol{f} + {}^\alpha\boldsymbol{M}(\mathcal{G}(t)).$$

<u>*Problem*</u>: Using Reynolds' transport theorem we get two equivalent formulations of the *local partial momentum balance* equation:

$$\rho^\alpha \dot{\boldsymbol{v}}^\alpha + \rho^\alpha \boldsymbol{v}^\alpha \nabla \cdot \boldsymbol{w} + \nabla \cdot \mathbf{P}^\alpha = \rho^\alpha \boldsymbol{f}^\alpha - -\nabla \cdot \boldsymbol{m}^\alpha \boldsymbol{v}^\alpha + \boldsymbol{m}^\alpha \Leftrightarrow \tag{218}$$

$$\frac{\partial}{\partial t}(\rho^\alpha \boldsymbol{v}^\alpha) + \nabla \cdot (\rho^\alpha \boldsymbol{v}^\alpha \boldsymbol{v}^\alpha + \mathbf{P}^\alpha) = \rho^\alpha \boldsymbol{f}^\alpha + \boldsymbol{m}^\alpha. \tag{219}$$

Problem: Using the axiom of mixtures the partial quantities and the quantities belonging to the mixtures as a whole are related by:

$$\boldsymbol{P}(\mathcal{G}(t)) = \sum_\alpha {}^\alpha\boldsymbol{P}(\mathcal{G}(t)) \tag{220}$$

$$\dot{\boldsymbol{P}}(\mathcal{G}(t)) = \sum_\alpha {}^\alpha\dot{\boldsymbol{P}}(\mathcal{G}(t)) - \oint_{(\partial\mathcal{G}(t))} \sum_\alpha \rho^\alpha \boldsymbol{v}^\alpha (\boldsymbol{v} - \boldsymbol{v}^\alpha) \cdot d\boldsymbol{f}, \tag{221}$$

$$\boldsymbol{t} = \sum_\alpha {}^\alpha\boldsymbol{t} - [\dot{\boldsymbol{P}}(\mathcal{G}(t)) - \sum_\alpha {}^\alpha\dot{\boldsymbol{P}}(\mathcal{G}(t))], \tag{222}$$

$$\mathbf{P} = \sum_\alpha [\mathbf{P}^\alpha + \rho^\alpha (\boldsymbol{v}^\alpha - \boldsymbol{v})(\boldsymbol{v}^\alpha - \boldsymbol{v})]. \tag{223}$$

6.3 Entropy Balance

In Section 1.4.1 time rates of non-equilibrium entropies were introduced by (96) and (100). These entropies belong to the mixture as a whole. We could write down also equations for time rates of partial entropies, but partial entropy productions are in general not definite. Therefore we will not use such equations. Starting out with heat exchanges we state heat exchanges through subsurfaces are additive:

$$\dot{Q}(\partial\mathcal{G}(t)) = \sum_i \dot{Q}(\partial\mathcal{G}_i(t)) = -\oint_{\partial\mathcal{G}} \boldsymbol{q} \cdot d\boldsymbol{f}. \tag{224}$$

In contrast to heat exchanges entropy flux densities are not additive. *Proposition* : If $\{\partial\mathcal{G}_i\}$ is a division of $\partial\mathcal{G}$,

$$\mathcal{G} = \bigcup_i \mathcal{G}_i, \; \mathcal{G}_i \cap \mathcal{G}_k = \emptyset, \; i \neq k, \tag{225}$$

the inequality holds

$$\sum_i \frac{\dot{Q}(\partial\mathcal{G}_i(t))}{\Theta(\partial\mathcal{G}_i(t))} \geq \frac{\dot{Q}(\partial\mathcal{G}(t))}{\Theta(\partial\mathcal{G}(t))}. \tag{226}$$

Proof: Define subsurfaces by

$$\dot{Q}(\partial\mathcal{G}_i(t))^+ \geq 0, \qquad \dot{Q}(\partial\mathcal{G}_i(t))^- < 0. \tag{227}$$

Then we get

$$\sum_i \frac{\dot{Q}(\partial\mathcal{G}_i(t))}{\Theta(\partial\mathcal{G}_i(t))} = \sum_i \frac{\dot{Q}(\partial\mathcal{G}_i(t))^+}{\Theta(\partial\mathcal{G}_i(t))^+} + \sum_i \frac{\dot{Q}(\partial\mathcal{G}_i(t))^-}{\Theta(\partial\mathcal{G}_i(t))^-}. \tag{228}$$

Using the mean value theorem (228) yields

$$\sum_i \frac{\dot{Q}(\partial\mathcal{G}_i(t))}{\Theta(\partial\mathcal{G}_i(t))} = \frac{\dot{Q}^+}{\Theta^+} + \frac{\dot{Q}^-}{\Theta^-}. \tag{229}$$

Consequently the surface $\partial\mathcal{G}$ can be divided into two surbsurfaces $\partial\mathcal{G}^+$ and $\partial\mathcal{G}^-$. The heat exchange through $\partial\mathcal{G}^+$ is $\dot{Q}^+ \geq 0$, that through $\partial\mathcal{G}^-$ is $\dot{Q}^- < 0$. The mean value of the contact temperatures are Θ^+ on $\partial\mathcal{G}^+$ and Θ^- on $\partial\mathcal{G}^-$. Because the considered system is placed in an equilibrium environment of thermostatic temperature T we have on the subsurfaces

$$\partial\mathcal{G}_i^+ : \qquad \dot{Q}_i^+\left(\frac{1}{\Theta_i^+} - \frac{1}{T}\right) \geq 0, \tag{230}$$

$$\partial\mathcal{G}_i^- : \qquad \dot{Q}_i^-\left(\frac{1}{\Theta_i^-} - \frac{1}{T}\right) \geq 0. \tag{231}$$

Summing up gives

$$\partial\mathcal{G}^+ : \qquad \dot{Q}^+\left(\frac{1}{\Theta^+} - \frac{1}{T}\right) \geq 0, \tag{232}$$

$$\partial\mathcal{G}^- : \qquad \dot{Q}^-\left(\frac{1}{\Theta^-} - \frac{1}{T}\right) \geq 0. \tag{233}$$

Consequently the mean values Θ^+ and Θ^- are the contact temperatures of $\partial\mathcal{G}^+$ and $\partial\mathcal{G}^-$. The contact temperature Θ of $\partial\mathcal{G}$ is defined by

$$T = \Theta : \qquad \dot{Q}^+ + \dot{Q}^- = \dot{Q} = 0. \tag{234}$$

Inserting this into (232) and (233) we get

$$\Theta^+ \leq \Theta \leq \Theta^- \tag{235}$$

which is a remarkable result. Using this the following inequalities are valid

$$\frac{\dot{Q}^+}{\Theta^+} \geq \frac{\dot{Q}^+}{\Theta}, \quad \frac{\dot{Q}^-}{\Theta^-} \geq \frac{\dot{Q}^-}{\Theta}. \tag{236}$$

Adding both the inequalities we get by use of (229) the proposition (226).

□

If we interpret $\dot{Q}(\partial\mathcal{G}_i(t))/\Theta(\partial\mathcal{G}_i(t))$ as entropy flux through $\partial\mathcal{G}$, (226) shows entropy fluxes are not additive.

We now translate the dissipation inequalities (96) and (100):

$$S = \int_{(\mathcal{G}(t))} \left(\sum_\alpha \rho^\alpha \hat{s}^\alpha\right) dV, \tag{237}$$

$$\frac{\dot{Q}}{\Theta} = -\oint_{\partial\mathcal{G}} \boldsymbol{\Phi} \cdot d\boldsymbol{f}, \tag{238}$$

$$\boldsymbol{s} \cdot \dot{\boldsymbol{n}}^e = -\oint_{\partial\mathcal{G}} (\hat{s}^\alpha \dot{\boldsymbol{m}}^\alpha) \cdot d\boldsymbol{f}. \tag{239}$$

Here $\hat{s}^\alpha$ is the specific partial entropy of the α-th component,$\boldsymbol{\Phi}$ the entropy flux density. (96) yields the global entropy balance

$$\frac{d}{dt}\int_{(\mathcal{G}(t))} (\sum_\alpha \rho^\alpha \hat{s}^\alpha) dV + \oint_{\partial\mathcal{G}} (\boldsymbol{\Phi} + \hat{s}^\alpha \dot{\boldsymbol{m}}^\alpha) \cdot d\boldsymbol{f} \geq 0, \tag{240}$$

and (100) an inequality looking totally equal:

$$\frac{d}{dt}\int_{(\mathcal{G}(t))} (\sum_\alpha \rho^\alpha \hat{s}'^\alpha) dV + \oint_{\partial\mathcal{G}} (\boldsymbol{\Phi}^* + \hat{s}'^\alpha \dot{\boldsymbol{m}}^\alpha) \cdot d\boldsymbol{f} \geq 0, \tag{241}$$

Because $\hat{s}^\alpha, \boldsymbol{\Phi}, \hat{s}'^\alpha$, and $\boldsymbol{\Phi}^*$ are given by constitutive equations both the dissipation inequalities are equivalent. The different possibilities of defining entropy have no influence on the formal shape of the dissipation inequality. Therefore in continuum thermodynamics often dissipative inequalities are used without a specification of entropy.

The local formulation of (240) writes

$$\frac{\partial}{\partial t}\sum_\alpha \rho^\alpha \hat{s}^\alpha + \nabla \cdot (\boldsymbol{\Phi} + \sum_\alpha \rho^\alpha \hat{s}^\alpha \boldsymbol{v}^\alpha) \geq 0. \tag{242}$$

6.4 All Balance Equations

We now write down the balance equation without further comment [52] :

6.4.1 Mass

$$\frac{\partial}{\partial t}\rho^\alpha = -\nabla \cdot \rho^\alpha \boldsymbol{v}^\alpha + \tau^\alpha, \qquad \sum_\alpha \tau^\alpha = 0. \tag{243}$$

6.4.2 Momentum

$$\frac{\partial}{\partial t}(\rho^\alpha \boldsymbol{v}^\alpha) + \nabla \cdot (\mathbf{P}^\alpha + \rho^\alpha \boldsymbol{v}^\alpha \boldsymbol{v}^\alpha) = \rho^\alpha \boldsymbol{f}^\alpha + \boldsymbol{m}^\alpha, \tag{244}$$

$$\sum_\alpha \boldsymbol{m}^\alpha = 0. \tag{245}$$

6.4.3 Kinetic Energy

$$\frac{1}{2}\frac{\partial}{\partial t}[\rho^\alpha(\boldsymbol{v}^\alpha)^2] - \nabla\cdot[\mathbf{P}^\alpha\cdot\boldsymbol{v}^\alpha - \frac{1}{2}\rho^\alpha(\boldsymbol{v}^\alpha)^2\boldsymbol{v}^\alpha] = \tag{246}$$
$$= \mathbf{P}^\alpha : \nabla\boldsymbol{v}^\alpha + \rho^\alpha\boldsymbol{f}^\alpha\cdot\boldsymbol{v}^\alpha - \frac{1}{2}\tau^\alpha(\boldsymbol{v}^\alpha)^2 + \boldsymbol{m}^\alpha\cdot\boldsymbol{v}^\alpha.$$

6.5 Internal Energy

$$\frac{\partial}{\partial t}(\rho^\alpha\epsilon^\alpha) + \nabla\cdot(\boldsymbol{q}^\alpha + \rho^\alpha\epsilon^\alpha\boldsymbol{v}^\alpha) = -\mathbf{P}^\alpha : \nabla\boldsymbol{v}^\alpha + \omega^\alpha. \tag{247}$$

6.6 Total Energy

$$e^\alpha := \epsilon^\alpha + \frac{1}{2}(\boldsymbol{v}^\alpha)^2, \tag{248}$$

$$\frac{\partial}{\partial t}(\rho^\alpha e^\alpha) + \nabla\cdot(\boldsymbol{q}^\alpha + \mathbf{P}^\alpha\cdot\boldsymbol{v}^\alpha + \rho^\alpha e^\alpha\boldsymbol{v}^\alpha) = \rho^\alpha\boldsymbol{f}^\alpha\cdot\boldsymbol{v}^\alpha + l^\alpha, \tag{249}$$

$$l^\alpha := \omega^\alpha + \frac{1}{2}\tau^\alpha(\boldsymbol{v}^\alpha)^2 - \boldsymbol{m}^\alpha\cdot\boldsymbol{v}^\alpha, \tag{250}$$

$$\sum_\alpha l^\alpha = 0. \tag{251}$$

6.7 Relations between Partial and Mixture Quantities

$$\rho e := \sum_\alpha \rho^\alpha e^\alpha, \tag{252}$$

$$\mathbf{P} = \sum_\alpha[\mathbf{P}^\alpha + \rho^\alpha(\boldsymbol{v}^\alpha - \boldsymbol{v})(\boldsymbol{v}^\alpha - \boldsymbol{v})], \tag{253}$$

$$\rho\epsilon = \sum_\alpha \rho^\alpha[\epsilon^\alpha + \frac{1}{2}(\boldsymbol{v}^\alpha - \boldsymbol{v})(\boldsymbol{v}^\alpha - \boldsymbol{v})], \tag{254}$$

$$\boldsymbol{q} = \sum_\alpha[\boldsymbol{q}^\alpha + \rho^\alpha(\epsilon^\alpha + \frac{1}{2}(\boldsymbol{v}^\alpha - \boldsymbol{v})^2)(\boldsymbol{v}^\alpha - \boldsymbol{v}) + \mathbf{P}^\alpha\cdot(\boldsymbol{v}^\alpha - \boldsymbol{v})]. \tag{255}$$

6.8 Dissipation Inequality

$$\frac{\partial}{\partial t}\sum_\alpha \rho^\alpha\hat{s}^\alpha + \nabla\cdot(\boldsymbol{\Phi} + \sum_\alpha \rho^\alpha\hat{s}^\alpha\boldsymbol{v}^\alpha) \geq 0. \tag{256}$$

7 Usual and Extended Thermodynamics

We consider especially an 1-component system. There are five balance equations and the dissipation inequality for 23 fields:

number of fields		number of equations
4	$\dot{\rho} = -\rho\nabla \cdot \boldsymbol{v}$	1
9	$\rho\dot{\boldsymbol{v}} = -\nabla \cdot \mathbf{P} + \rho\boldsymbol{f}$	3
5	$\rho\dot{\epsilon} = -\nabla \cdot \boldsymbol{q} - \mathbf{P} : \nabla\boldsymbol{v} + r$	1
5	$\rho\dot{s} \geq -\nabla \cdot \boldsymbol{\Phi} + \psi$	(1)
23		5

(257)

We now have to determine what the fields are we are looking for. In literature we find two possibilities [3],[53], the so-called 5-field theory and the 13-field theory.

7.1 Five-Field Theory

Five-field theories are characterized by looking for five fields, namely $(\rho, \boldsymbol{v}, \epsilon)(\boldsymbol{x}, t)$ or $(\rho, \boldsymbol{v}, \Theta)(\boldsymbol{x}, t)$. They are called the *basic fields*. Because we are looking for 5 fields and 3 fields $\boldsymbol{f}(\boldsymbol{x}$,t) are given, we need additional 15 fields which have to be given by constitutive equations (5 + 3 + 15 = 23):

	number of fields
$\mathbf{P}(\boldsymbol{x}, t)$	6
$\boldsymbol{q}(\boldsymbol{x}, t)$	3
$r(\boldsymbol{x}, t)$	1
$s(\boldsymbol{x}, t)$	1
$\boldsymbol{\Phi}(\boldsymbol{x}, t)$	3
$\psi(\boldsymbol{x}, t)$	1
	15

(258)

The other possibility is a

7.2 Thirteen-Field Theory

Here we are looking for 13 basic fields [54] : $(\rho, \boldsymbol{v}, \epsilon, \mathbf{P}^0, \boldsymbol{q})(\boldsymbol{x}, t)$ $(1 + 3 + 1 + 5 + 3 = 13)$, and $\mathbf{P}^0$ is defined by

$$\mathbf{P}^0 := \mathbf{P} - (\frac{1}{3}Tr\mathbf{P})\mathbf{1}, \tag{259}$$

the traceless part of the pressure tensor. In the 5-field theory there are 5 balances, here in the 13-field theory we need beyond these 5 ones 8 more balances for the additional fields $\mathbf{P}^0$ and $\boldsymbol{q}$:

$$\dot{\mathbf{P}}^0 + \nabla \cdot \underline{\mathbf{Q}}^0 = \mathbf{F}^0, \tag{260}$$

$$\dot{\boldsymbol{q}} + \nabla \cdot \mathbf{L} = \boldsymbol{G}. \tag{261}$$

By introducing these 8 more balance equations we also introduce quantities which are determined by constitutive equations:

	number of fields	
$P(\boldsymbol{x},t) := \frac{1}{3}Tr\mathbf{P}(\boldsymbol{x},t)$	1	
$\underline{\mathbf{Q}}^0(\boldsymbol{x},t)$	15	symm. traceless
$\mathbf{F}^0(\boldsymbol{x},t)$	5	
$\mathbf{L}(\boldsymbol{x},t)$	9	
$\boldsymbol{G}(\boldsymbol{x},t)$	3	
$r(\boldsymbol{x},t)$	1	
$s(\boldsymbol{x},t)$	1	
$\boldsymbol{\Phi}(\boldsymbol{x},t)$	3	
$\psi(\boldsymbol{x},t)$	1	
	39	

(262)

Comparing the number of quantities determined by constitutive equations in 5-field theories with those of 13-field theories we see it arises from 15 to 39. So the situation is even worse in 13-field theories, but because the number of basic fields is raised from 5 to 13 the constitutive equations may be only chosen approximately. Such a choice may be

$$\mathbf{L} = -\frac{1}{\tau}\kappa\Theta, \quad \boldsymbol{G} = \frac{1}{\tau}\boldsymbol{q}, \tag{263}$$

so that (261) yields the so-called Cattaneo equation

$$\tau\dot{\boldsymbol{q}} - \nabla \cdot \kappa\Theta = \boldsymbol{q}. \tag{264}$$

This procedure is not at all a systematic treatment of the additional balance equations (260) and (261), but it may help to get better results than using the 5-field theory. A more systematic method for guessing the constitutive equations of the additional basic fields is to use transporttheoretical methods. But in transport theories the constitutive properties are chosen by a kinetic model so that in general a direct comparison with phenomenologic constitutive properties is difficult.

The basic fields are independent of each other not only in 5-field theories but also in 13-field theories. But they have nothing to do with the thermodynamical state space. As an example we mention the constitutive equation of the heat flux density

$$q = \mathcal{Q}(\nabla\Theta, ...) \tag{265}$$

in which $\nabla\Theta$ appears, but $\nabla\Theta$ is not included in the state space $(\rho, \boldsymbol{v}, \epsilon)(\boldsymbol{x}, t)$. Therefore it is obvious we need the notion of the state space also in continuum thermodynamics.

An other item not totally clear is that in the 13-field theory - which is commonly called "Extended Thermodynamics" - the state space by definition is just spanned by the basic fields, whereas in usual thermodynamics basic fields and state space are different.

Because the state space of Extended Thermodynamics does not include gradients or time derivatives the question of hyperbolicity (see below: Material axioms) can be solved quite general, if convexity of the entropy density is presupposed [55].

Only after having chosen the state space the symbols $\dot{}$ and ∇ in the balance equations get a meaning: The have to be performed by the chain rule, and we can see the balance equations determine directional derivatives in state space. Now the questions arise: Are all directional derivatives allowed, or are they restricted by the dissipation inequality? How have we to exploit the dissipation inequality? How to establish constitutive equations which exceed a simple ansatz being compatible with the dissipation? These questions lead up to formulating material axioms.

8 Material Axioms: Objectivity and Frame Indifference

The question "how to get constitutive equations?" can be answered in two different ways: We can make ansatzes compatible with the second law. Examples for such ansatzes are:

Gibbs fundamental equation
thermal equation of state
rate equations from entropy production
linear phenomenological coefficients
dissipative potentials
yield functions
multi-yield surfaces
non-identified or special defined internal variables.

The other way is to formulate material axioms. Instead of a special ansatz material axioms give standardized procedures for constructing (and classifying) classes of materials and constitutive equations belonging to them.

We now formulate

Material Axioms: Constitutive equations have to satisfy:

i) the *second law*, and if necessary its additionals,
ii) transformation properties by *changing the observer*,
iii) the *material-symmetry*.
iv) Additionally state spaces have to be chosen so that *hyperbolicity* of propagation equations is guaranteed.

In detail the second law is represented by the *dissipation inequality*, changing the observer demands covariance in relativistic theories or objectivity in non-relativistic theories. Material symmetry is described by isotropy groups. Hyperbolicity in Extended Thermodynamics is enforced by constructing a symmetric hyperbolic system of partial differential equations from the balances and the constitutive equations [56].

8.1 Changing Frames

Balance equations need a frame to which the quantities appearing in them belong. So the material velocity $\boldsymbol{v}$ is defined as the velocity of a material point $(\boldsymbol{x}, t)$ with respect to a special frame. All balances written down in sect. 1.6 are formulated in an arbitrary frame. Thus the form of the balances is frame-independent: Each observer uses the balance equations in the same shape, but the quantities appearing in them have different values due to the used frame. Therefore the balance equations are frame-invariant (or covariant), whereas the values of the quantities appearing in them are frame-dependent. So in general the force density does not only include the imposed forces but also the reaction forces which are caused by moving the considered frame with respect to the local inertial frame which we will discuss below in more detail.

The statement of frame-invariance is quite general and is also valid in Relativity Theory which is out of scope of this paper. Here in non-relativistic theory the change of frame is achieved by a local Euclidean transformation satisfying the frame-invariance of the balance equations. First of all we choose an arbitrary but fixed frame denoted by an asterisk * which we call the *standard frame of reference*. In this frame the general partial form of a scalar-valued balance equation writes

$$\frac{\partial}{\partial t}(\rho^* \Psi^*) + \nabla^* \cdot (\rho^* \boldsymbol{v}^* \Psi^* + \boldsymbol{\Phi}^*) + \Sigma^* = 0, \tag{266}$$

and according to the frame-invariance of the balance equations we get

$$\frac{\partial}{\partial t}(\rho \Psi) + \nabla \cdot (\rho \boldsymbol{v} \Psi + \boldsymbol{\Phi}) + \Sigma = 0, \tag{267}$$

where the quantities without asterisk belong to the frame we change to. Introducing the orthogonal time-dependent transformation $\mathbf{Q}(\mathrm{t})$

$$\tilde{\mathbf{Q}} \cdot \mathbf{Q} = \mathbf{Q} \cdot \tilde{\mathbf{Q}} = \mathbf{1}, \tag{268}$$

we get

$$\nabla^* \cdot \boldsymbol{v}^* = \nabla^* \cdot \mathbf{Q} \cdot \tilde{\mathbf{Q}} \cdot \boldsymbol{v}^*, \quad \boldsymbol{v}^* \cdot \nabla^* = \boldsymbol{v}^* \cdot \mathbf{Q} \cdot \tilde{\mathbf{Q}} \cdot \nabla^*, \tag{269}$$

$$\nabla^* \cdot \boldsymbol{\Phi}^* = \nabla^* \cdot \mathbf{Q} \cdot \tilde{\mathbf{Q}} \cdot \boldsymbol{\Phi}^*. \tag{270}$$

As well known the material velocity $\boldsymbol{v}^*$ transformes according to

$$\boldsymbol{v}^* = \mathbf{Q} \cdot \boldsymbol{V} = \mathbf{Q} \cdot (\boldsymbol{v} + \boldsymbol{v}^{rel}) = \mathbf{Q} \cdot (\boldsymbol{v} + \dot{\boldsymbol{c}} + \boldsymbol{\Omega} \cdot (\boldsymbol{x} - \boldsymbol{c})). \tag{271}$$

Here $\boldsymbol{v}^{rel}$ is the relative velocitiy between both the frames which decomposes into the translational part $\dot{\boldsymbol{c}}$ and the rotational part $\boldsymbol{\Omega} \cdot (\boldsymbol{x} - \boldsymbol{c})$ by use of the skew-symmetric spin tensor:

$$\boldsymbol{\Omega} := \tilde{\mathbf{Q}} \cdot \dot{\mathbf{Q}} = -\tilde{\boldsymbol{\Omega}}. \tag{272}$$

The relative velocity depends only of quantities characterizing the relative motion of both the frames considered, whereas the material velocity $\boldsymbol{v}$ is independent of the relative motion of both the frames. As the material velocity in general all quantities split into their relative part and into the part belonging to the frame to which we transform:

$$\Psi^* = \Psi + \Psi^{rel}, \quad \boldsymbol{\Phi}^* = \mathbf{Q} \cdot (\boldsymbol{\Phi} + \boldsymbol{\Phi}^{rel}). \tag{273}$$

The transformation properties of the different quantities have to be known and depend on the model we use for the considered quantity as we will discuss below. So the relative part of the mass density is zero

$$\rho^* = \rho \tag{274}$$

Also the relative part in the standard frame of reference of all quantities is identical zero by definition

$$\Psi^{*rel} \equiv 0, \quad \boldsymbol{\Phi}^{*rel} \equiv \mathbf{o}. \tag{275}$$

Defining

$$\nabla := \tilde{\mathbf{Q}} \cdot \nabla^* = \nabla^* \cdot \mathbf{Q} \tag{276}$$

we get from (269) and (270)

$$\nabla^* \cdot \boldsymbol{v}^* = \nabla \cdot \boldsymbol{V}, \qquad \boldsymbol{v}^* \cdot \nabla^* = \boldsymbol{V} \cdot \nabla, \tag{277}$$

$$\nabla^* \cdot \boldsymbol{\Phi}^* = \nabla \cdot (\boldsymbol{\Phi} + \boldsymbol{\Phi}^{rel}). \tag{278}$$

From (266) by use of (269) to (274) and (276) follows

$$\frac{\partial}{\partial t}[\rho(\Psi + \Psi^{rel})] + \nabla \cdot [\rho \boldsymbol{V}(\Psi + \Psi^{rel}) + \boldsymbol{\Phi} + \boldsymbol{\Phi}^{rel}] + \Sigma + \Sigma^{rel} = 0. \tag{279}$$

Comparison with (267) results in

$$\frac{\partial}{\partial t}(\rho \Psi^{rel}) + \nabla \cdot (\rho \boldsymbol{v}^{rel} \Psi) + \nabla \cdot (\rho \boldsymbol{V} \Psi^{rel}) + \nabla \cdot \boldsymbol{\Phi}^{rel} + \Sigma^{rel} = 0. \tag{280}$$

Because of

$$\nabla \cdot \boldsymbol{v}^{rel} = \nabla \cdot [\dot{\boldsymbol{c}} + \boldsymbol{\Omega} \cdot (\boldsymbol{x} - \boldsymbol{c})] = \boldsymbol{\Omega} : \mathbf{1} = 0 \tag{281}$$

we have

$$\nabla \cdot \boldsymbol{V} = \nabla \cdot (\boldsymbol{v} + \boldsymbol{v}^{rel}) = \nabla \cdot \boldsymbol{v} = \nabla^* \cdot \boldsymbol{v}^*. \tag{282}$$

Thus

$$\frac{\partial}{\partial t}(\rho \Psi^{rel}) + \nabla \cdot (\rho \boldsymbol{V} \Psi^{rel} + \boldsymbol{\Phi}^{rel}) + \Sigma^{rel} = -\boldsymbol{v}^{rel} \cdot \nabla \rho \Psi \tag{283}$$

follows which is a result of frame-invariance of the balance equations. (283) determines Σ^{rel}.

From (267) we get

$$\frac{\partial}{\partial t}(\rho \Psi) + \nabla \cdot (\rho \boldsymbol{V} \Psi - \rho \boldsymbol{v}^{rel} \Psi + \boldsymbol{\Phi}) + \Sigma = 0 \tag{284}$$

which results in

$$(\frac{\partial}{\partial t} + \boldsymbol{V} \cdot \nabla)(\rho \Psi) + \rho \Psi \nabla \cdot \boldsymbol{v} + \nabla \cdot (\boldsymbol{\Phi} - \rho \boldsymbol{v}^{rel} \Psi) + \Sigma = 0. \tag{285}$$

If we now define the frame-invariant *total time derivative*

$$\mathrm{D}_t^{(0)} := \frac{\partial}{\partial t} + \boldsymbol{V} \cdot \nabla = \frac{\partial}{\partial t} + \boldsymbol{v}^* \cdot \nabla^* = \mathrm{D}_t^{(0)*}, \tag{286}$$

we get using (280)

$$\mathrm{D}_t^{(0)}(\rho \Psi) + \rho \Psi \nabla \cdot \boldsymbol{v} + \nabla \cdot \boldsymbol{\Phi} + \Sigma + \mathrm{D}_t^{(0)}(\rho \Psi^{rel}) + \rho \Psi^{rel} \nabla \cdot \boldsymbol{v} + \nabla \cdot \boldsymbol{\Phi}^{rel} + \Sigma^{rel} = 0 \tag{287}$$

from which follows immediately

$$\mathrm{D}_t^{(0)*}(\rho^* \Psi^*) + \rho^* \Psi^* \nabla^* \cdot \boldsymbol{v}^* + \nabla^* \cdot \boldsymbol{\Phi}^* + \Sigma^* = 0. \tag{288}$$

As (266) and (267) also (287) and (288) are frame-invariant, when a new supply term is introduced

$$\Sigma^D := \Sigma + \Sigma^{rel} + \mathrm{D}_t^{(0)}(\rho \Psi^{rel}) + \rho \Psi^{rel} \nabla \cdot \boldsymbol{v} + \nabla \cdot \boldsymbol{\Phi}^{rel}. \tag{289}$$

which according to (275) has the property

$$\Sigma^{D*} = \Sigma^*. \tag{290}$$

Finally we get the scalar-valued frame-invariant balance equation (267) formulated by use of the partial time derivative

$$\frac{\partial}{\partial t}(\rho \Psi) + \nabla \cdot (\rho \boldsymbol{v} \Psi + \boldsymbol{\Phi}) + \Sigma = 0, \tag{291}$$

or formulated by the frame-invariant total time derivative (286)

$$\mathrm{D}_t^{(0)}(\rho\Psi) + \rho\Psi\nabla\cdot\boldsymbol{v} + \nabla\cdot\boldsymbol{\Phi} + \Sigma^D = 0. \tag{292}$$

We now write down the same relations for a vector-valued balance equation:

$$\frac{\partial}{\partial t}(\rho^*\boldsymbol{\Psi}^*) + \nabla^*\cdot(\rho^*\boldsymbol{v}^*\boldsymbol{\Psi}^* + \boldsymbol{\Phi}^*) + \boldsymbol{\Sigma}^* = \mathbf{o}. \tag{293}$$

The transformation properties are

$$\rho^* = \rho, \quad \boldsymbol{\Psi}^* = \mathbf{Q}\cdot(\boldsymbol{\Psi} + \boldsymbol{\Psi}^{rel}), \quad \boldsymbol{\Sigma}^* = \mathbf{Q}\cdot(\boldsymbol{\Sigma} + \boldsymbol{\Sigma}^{rel}), \tag{294}$$

$$\boldsymbol{\Phi}^* = \mathbf{Q}\cdot(\boldsymbol{\Phi} + \boldsymbol{\Phi}^{rel})\cdot\tilde{\mathbf{Q}}. \tag{295}$$

Inserting this into (293) we get

$$\frac{\partial}{\partial t}(\rho\mathbf{Q}\cdot\boldsymbol{\Psi}) + \nabla\cdot(\rho\boldsymbol{V}\mathbf{Q}\cdot\boldsymbol{\Psi}) + \nabla\cdot\boldsymbol{\Phi}\cdot\tilde{\mathbf{Q}} + \mathbf{Q}\cdot\boldsymbol{\Sigma} + \tag{296}$$
$$\frac{\partial}{\partial t}(\rho\mathbf{Q}\cdot\boldsymbol{\Psi}^{rel}) + \nabla\cdot(\rho\boldsymbol{V}\mathbf{Q}\cdot\boldsymbol{\Psi}^{rel}) + \nabla\cdot\boldsymbol{\Phi}^{rel}\cdot\tilde{\mathbf{Q}} + \mathbf{Q}\cdot\boldsymbol{\Sigma}^{rel} = \mathbf{o}.$$

Because the following relations are valid

$$\frac{\partial}{\partial t}(\rho\mathbf{Q}\cdot\boldsymbol{\Psi}) = \mathbf{Q}\cdot\frac{\partial}{\partial t}(\rho\boldsymbol{\Psi}) + \mathbf{Q}\cdot\boldsymbol{\Omega}\cdot\rho\boldsymbol{\Psi}, \tag{297}$$

$$\nabla\cdot(\rho\boldsymbol{V}\mathbf{Q}\cdot\boldsymbol{\Psi}) = \mathbf{Q}\cdot(\boldsymbol{V}\cdot\nabla)\rho\boldsymbol{\Psi} + \mathbf{Q}\cdot\rho\boldsymbol{\Psi}\nabla\cdot\boldsymbol{v}, \tag{298}$$

we are able to introduce a frame-invariant total time derivative for vector-valued quantities

$$\mathrm{D}_t^{(1)}\cdot \quad := (\frac{\partial}{\partial t} + \boldsymbol{V}\cdot\nabla)\mathbf{1}\cdot \quad + \boldsymbol{\Omega}\cdot. \tag{299}$$

We have

$$\mathbf{Q}\cdot\mathrm{D}_t^{(1)} = \mathbf{Q}\cdot\frac{\partial}{\partial t}\mathbf{1} + (\boldsymbol{V}\cdot\nabla)\mathbf{Q} + \mathbf{Q}\cdot\boldsymbol{\Omega} = \frac{\partial}{\partial t}\mathbf{Q} + (\boldsymbol{V}\cdot\nabla)\mathbf{Q}, \tag{300}$$

consequently

$$\mathbf{Q}\cdot\mathrm{D}_t^{(1)}\cdot\tilde{\mathbf{Q}} = \frac{\partial}{\partial t}\mathbf{1} \; + (\boldsymbol{v}^*\cdot\nabla^*)\mathbf{1} \; =: \mathrm{D}_t^{(1)*} \tag{301}$$

follows. Using the frame-invariant total time derivative (299) we get an other form of the vector-valued balance equation (293) in the standard frame

$$\mathbf{Q}\cdot\{\mathrm{D}_t^{(1)}\cdot(\rho\boldsymbol{\Psi}) + \rho\boldsymbol{\Psi}\nabla\cdot\boldsymbol{v} + \nabla\cdot\boldsymbol{\Phi} + \boldsymbol{\Sigma}\} + \tag{302}$$
$$+\mathbf{Q}\cdot\{\mathrm{D}_t^{(1)}\cdot(\rho\boldsymbol{\Psi}^{rel}) + \rho\boldsymbol{\Psi}^{rel}\nabla\cdot\boldsymbol{v} + \nabla\cdot\boldsymbol{\Phi}^{rel} + \boldsymbol{\Sigma}^{rel}\} = \mathbf{o}$$

from which follows the vector-valued balance in an abitrary frame

$$\mathrm{D}_t^{(1)}\cdot(\rho\boldsymbol{\Psi}) + \rho\boldsymbol{\Psi}\nabla\cdot\boldsymbol{v} + \nabla\cdot\boldsymbol{\Phi} + \boldsymbol{\Sigma}^D = \mathbf{o}, \tag{303}$$

$$\boldsymbol{\Sigma}^D = \boldsymbol{\Sigma} + \boldsymbol{\Sigma}^{rel} + \mathrm{D}_t^{(1)}\cdot(\rho\boldsymbol{\Psi}^{rel}) + \rho\boldsymbol{\Psi}^{rel}\nabla\cdot\boldsymbol{v} + \nabla\cdot\boldsymbol{\Phi}^{rel}. \tag{304}$$

which also has the shape of (292).

8.2 Objectivity

When changing the frame we have to know how the different quantities transform. Quantities are called *objective*, if their relative parts vanish in arbitrary frames, i.e. they transform as tensors of different orders. Thus we get for objective quantities the following balances according to (292) and (303)

$$\mathrm{D}_t^{(0)}(\rho\Psi) + \rho\Psi\nabla\cdot\boldsymbol{v} + \nabla\cdot\boldsymbol{\Phi} + \Sigma = 0. \tag{305}$$

$$\mathrm{D}_t^{(1)}\cdot(\rho\boldsymbol{\Psi}) + \rho\boldsymbol{\Psi}\nabla\cdot\boldsymbol{v} + \nabla\cdot\mathbf{\Phi} + \boldsymbol{\Sigma} = \mathbf{o}, \tag{306}$$

in which the supply quantities Σ and $\boldsymbol{\Sigma}$ are the same as in the balances formulated by the partial time derivatives.

Whether a quantity is an objective one or not depends on the microscopic or mesoscopic model we have made for this quantity. A well-known example is the heat flux densitiy $\boldsymbol{q}$ in the balance equation of the internal energy density. If no model is used, $\boldsymbol{q}$ is to be objective

$$\boldsymbol{q}^* = \mathbf{Q}\cdot\boldsymbol{q}, \quad \boldsymbol{q}^{rel} \equiv \mathbf{o}. \tag{307}$$

But if we use the model that the heat flux density is caused by the energy transport of electrons having inertia [57], or if we calculate the heat flux density in mesoscopic liquid crystal theory [58], $\boldsymbol{q}$ is not an objective quantity

$$\boldsymbol{q}^{rel} \neq \mathbf{o}. \tag{308}$$

Of course $\boldsymbol{q}^{rel}$ may be numerically small, but in general the heat flux density will not transform as a tensor of first order when changing the frame.

Non-objective quantities are not extraordinary: Material velocity and acceleration are non-objective and therefore also the force density $\boldsymbol{f}$ in the balance of momentum

$$\frac{\partial}{\partial t}(\rho\boldsymbol{v}) + \nabla\cdot(\rho\boldsymbol{v}\boldsymbol{v} - \mathbf{T}) + \boldsymbol{f} = \mathbf{o} \tag{309}$$

is not objective.

Historically a special standard frame of reference was of great importance: the inertial frame. For defining this frame one needs a rule by which we can decompose $\boldsymbol{f}$ into the so-called *imposed* and *accelerated part*

$$\boldsymbol{f} = \boldsymbol{f}^{imp} + \boldsymbol{f}^{acc}. \tag{310}$$

This decomposition cannot be achieved by measuring devices (therefore you need a rule or a model by which the decomposition can be done). The transformation properties of $\boldsymbol{f}^{imp}$ and $\boldsymbol{f}^{acc}$ are different: $\boldsymbol{f}^{imp}$ is objective

$$\boldsymbol{f}^{*imp} = \mathbf{Q}\cdot\boldsymbol{f}^{imp} \tag{311}$$

from which follows that

$$\boldsymbol{f}^{*acc} = \mathbf{Q} \cdot (\boldsymbol{f}^{acc} + \boldsymbol{f}^{rel}) \tag{312}$$

is not objective. We now can formulate the

<u>*Def.*</u>: A frame * is denoted as an *inertial frame*, if

$$\boldsymbol{f}^{*acc}_{in} = \mathbf{o}. \tag{313}$$

If an inertial frame does exist, (312) results in

$$\boldsymbol{f}^{rel} = -\boldsymbol{f}^{acc} \tag{314}$$

which is valid in arbitrary frames. Therefore (293) is often interpreted as balance equation in an inertial frame, because in this case $\boldsymbol{f}^{*}_{in}$ is equal to the imposed part of the force density $\boldsymbol{f}^{*imp}_{in}$.

As we saw, the observer-invariance of the balance equations causes that quantities in them become $\boldsymbol{\Omega}$-dependent. Here $\boldsymbol{\Omega}$ is the spin tensor of the relative motion between the standard and the arbitrary frame. The rotation of the material with respect to the standard frame of reference is described by a different quantity denoted as *material spin tensor*

$$\omega^* := \frac{1}{2}[\nabla^* \boldsymbol{v}^* - (\nabla^* \boldsymbol{v}^*)^{\top}]. \tag{315}$$

It is easy to prove its transformation properties when changing the frame:

<u>*Proposition*:</u>

$$\omega^* = \mathbf{Q} \cdot (\omega + \boldsymbol{\Omega}) \cdot \tilde{\mathbf{Q}}. \tag{316}$$

The *co-rotational rest frame* of the material is defined by

$$\mathbf{Q} = \mathbf{Q}^0, \quad \boldsymbol{v}^0 = \mathbf{o}, \quad \omega^0 = \mathbf{0}, \tag{317}$$

from which we get

$$\omega^* = \mathbf{Q}^0 \cdot \boldsymbol{\Omega}^0 \cdot \tilde{\mathbf{Q}}^0, \tag{318}$$

and $\boldsymbol{\Omega}^0$ describes the rotation of the co-rotational rest frame with respect to the standard frame of reference. Now the question arises, if and how rotation influences material properties.

8.3 Material Frame Indifference

There is no doubt about the statement:

I: *Constitutive properties do not depend on the relative motion of frames.*

Two observers in different frames looking at the same material state the same constitutive properties. This trite statement should not be taken for the so-called "principle of material frame indifference" which states

II: *Constitutive properties do not depend on the motion of the material.*

Experience shows that this statement in general is wrong: Materials perceive their motion with respect to the standard frame of reference. We will discuss this statement in more detail.

The constitutive equations in the standard frame of reference and in an arbitrary frame are in a large state space

$$M^* = \mathcal{M}^*(z^*;\omega^*), \qquad M = \mathcal{M}(z;\omega). \tag{319}$$

Changing the frames is described by a mapping B

$$BM^* = M, \qquad Bz^* = z, \qquad B = B(\mathbf{Q},\mathbf{\Omega}). \tag{320}$$

Thus we get by use of (316)

$$M = B\mathcal{M}^*(B^{-1}z;\omega^*) = \mathcal{M}(z;\tilde{\mathbf{Q}}\cdot\omega^*\cdot\mathbf{Q} - \mathbf{\Omega}). \tag{321}$$

First of all we consider

$$B(\mathbf{1},\mathbf{0}) = 1 \tag{322}$$

which results in

$$\mathcal{M}^*(z;\omega^*) = \mathcal{M}(z;\omega^*), \tag{323}$$

so that no asterisk is needed to denote the material map in the standard frame of reference.

We now have to interpret the statements I and II. We will do that in such a way that ω^* influences constitutive properties and not $\mathbf{Q}$ or $\mathbf{\Omega}$. Consequently (321) results in

$$\mathcal{M}^*(z;\omega^*) = B^{-1}\mathcal{M}(Bz^*;B\omega^*), \tag{324}$$

a relation by which we should replace the bogus principle of material frame indifference. We see that M is an isotropic function in the argument z which is essential for obtaining reduced forms of the material map, as we will see in the next part of this book.

A special class of materials is that of *Hall-free materials* which is defined by

$$\mathcal{M}^{Hf}(z;\omega^*) = \mathcal{M}^{Hf}(z;\mathbf{0}). \tag{325}$$

These are special materials for which statement II is true.

9 Exploiting Dissipation Inequality

As an example we choose the 5-field theory, but all items treated in this section are generally valid. The balance equations (257) are

$$\dot{\rho} + \rho \mathbf{1} : \nabla \boldsymbol{v} = 0, \tag{326}$$

$$\rho \dot{\epsilon} + \mathbf{1} : \nabla \boldsymbol{q} - \mathbf{P} : \nabla \boldsymbol{v} - r = 0, \tag{327}$$

$$\rho \dot{\boldsymbol{v}} + \nabla \cdot \mathbf{P} - \rho \boldsymbol{f} = \mathbf{o}, \tag{328}$$

We choose a large state space

$$Z := (\rho, \frac{1}{\Theta}, \nabla \rho, (\frac{1}{\Theta})^{\cdot}, \nabla(\frac{1}{\Theta}), \nabla \boldsymbol{v}) \tag{329}$$

which belongs to an acceleration independent fluid because the acceleration is not included into Z. The non-equilibrium variables are

$$\boldsymbol{g} := (\nabla \rho, (\frac{1}{\Theta})^{\cdot}, \nabla(\frac{1}{\Theta})), \tag{330}$$

and therefore

$$Z := (\rho, \frac{1}{\Theta}, \nabla \boldsymbol{v}, \boldsymbol{g}). \tag{331}$$

The directional derivatives are

$$\boldsymbol{y} := (\dot{\boldsymbol{g}}, \nabla \boldsymbol{g}, \dot{\boldsymbol{v}}, (\nabla \boldsymbol{v})^{\cdot}, \nabla\nabla \boldsymbol{v}). \tag{332}$$

Because the balance equations and also the dissipation inequality

$$\rho \dot{s} + \nabla \cdot \boldsymbol{\Phi} - r/\Theta =: \sigma \geq 0, \tag{333}$$

$$\boldsymbol{\Phi} := \boldsymbol{k} + \boldsymbol{q}/\Theta \tag{334}$$

are linear in the derivatives $^{\cdot}$ and ∇, we get after having applied the chain rule linear equations in the directional derivatives $\boldsymbol{y}$:

$$\mathbf{A} \cdot \boldsymbol{y} = \boldsymbol{C} \quad \text{and} \quad \boldsymbol{B} \cdot \boldsymbol{y} \geq D. \tag{335}$$

Because of the chosen large state space $\mathbf{A}, \boldsymbol{C}, \boldsymbol{B}$, and D are *state functions* and $\boldsymbol{y}$ is beyond the state space variables. $\mathbf{A}, \boldsymbol{C}, \boldsymbol{B}$, and D depend on the constitutive equations which are not fixed up to now. $\mathbf{A}, \boldsymbol{C}, \boldsymbol{B}$, and D contain derivatives of the constitutive equations to the state variables.

Definition: All constitutive equations being compatible with the chosen large state space and satisfying the balance equations and the dissipation inequality determine the *class of material*.

In principle there are two possibilities to find this class of material:

i) For fixed $\mathbf{A}, \boldsymbol{C}, \boldsymbol{B}$, and D the dissipation inequality excludes certain $\boldsymbol{y}$, i.e. certain process directions in state space are not allowed.

ii) The $\mathbf{A}, \boldsymbol{C}, \boldsymbol{B}$, and D have to be determined so that all process directions are possible, i.e. whatever $\boldsymbol{y}$ may be, the dissipation inequality is always satisfied.

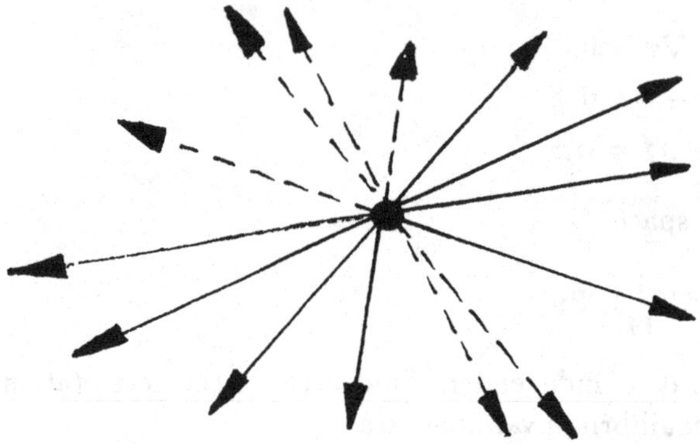

Figure 1.19 : At a fixed state Z there are directions represented by the directional derivatives which are allowed ($\longrightarrow$) and forbidden (- - -). Allowed means in agreement with the dissipation inequality, forbidden in contrast to it.

9.1 Non-Reversible-Direction Axiom

We consider an arbitrary material at a fixed state $Z(\boldsymbol{x}, t)$ [fixed position, fixed time, local]

$$\mathbf{A}(Z), \boldsymbol{C}(Z), \boldsymbol{B}(Z), D(Z). \tag{336}$$

Assumption: At Z there are allowed and forbidden directions in state space (Fig.1.19) $\boldsymbol{y}^1$ is an *allowed* directional derivative

$$\mathbf{A} \cdot \boldsymbol{y}^1 = \boldsymbol{C}, \qquad \boldsymbol{B} \cdot \boldsymbol{y}^1 \geq D, \tag{337}$$

and $\boldsymbol{y}^2$ a forbidden one

$$\mathbf{A} \cdot \boldsymbol{y}^2 = \boldsymbol{C}, \qquad \boldsymbol{B} \cdot \boldsymbol{y}^2 < D. \tag{338}$$

Proposition: A consequence of (337) and (338) is the statement: There exist a reversible direction $\alpha \boldsymbol{y}^1 + (1-\alpha)\boldsymbol{y}^2$

$$0 < \alpha := \frac{D - \boldsymbol{B} \cdot \boldsymbol{y}^2}{\boldsymbol{B} \cdot (\boldsymbol{y}^1 - \boldsymbol{y}^2)} < 1. \tag{339}$$

Proof: Multiplying (337) with $\alpha > 0$ and (338) with $\beta > 0$ we get by addition:

$$\mathbf{A} \cdot (\alpha \boldsymbol{y}^1 + \beta \boldsymbol{y}^2) = (\alpha + \beta)\boldsymbol{C}, \quad \alpha + \beta \doteq 1, \tag{340}$$

$$\boldsymbol{B} \cdot [\alpha \boldsymbol{y}^1 + (1-\alpha)\boldsymbol{y}^2] \doteq D. \tag{341}$$

By the demand $\alpha + \beta \doteq 1$ the directional derivative $\alpha \boldsymbol{y}^1 + (1-\alpha)\boldsymbol{y}^2$ in (340) satisfies the balance equations. α is determined by (341) from which immediately (339) follows. The inequalities in (339) are induced by (337) and (338). (341) shows that $\alpha \boldsymbol{y}^1 + (1-\alpha)\boldsymbol{y}^2$ is a reversible direction.

□

By this proposition we proved the existence of almost one reversible process direction, if at an arbitrary non-equilibrium state Z both kinds of allowed and forbidden directional derivatives exist. By arguments of continuity we conclude the existence not only of a reversible direction but also of a piece of a reversible trajectory. Because we are in non-equilibirium we have to exclude such reversible trajectories:

Axiom[59]: Except in equilibrium subspace *reversible process directions* in state space do not exist.

A consequence of this axiom is the fact that all process directions are either allowed or forbidden, but no non-equilibrium state exists with both kinds of process directions. By the non-reversible-direction axiom and the proposition (339) we have proved the

Proposition: If Z is no trap, the inclusion

$$\wedge \boldsymbol{y}: \quad \mathbf{A}(Z) \cdot \boldsymbol{y} = \boldsymbol{C}(Z) \longrightarrow \boldsymbol{B}(Z) \cdot \boldsymbol{y} \geq D(Z) \tag{342}$$

is valid.

By this proposition restrictions of the $\mathbf{A}, \boldsymbol{C}, \boldsymbol{B}$, and D result. These restrictions characterize the class of material we are looking for.

9.2 Coleman-Noll Technique

In this technique the inclusion (342) is enforced by:

$$\boldsymbol{B}(Z) = 0 \quad \wedge \quad D(Z) \leq 0 \tag{343}$$

$$\mathbf{A}(Z) \quad \text{and} \quad \boldsymbol{C}(Z) \quad \text{are not restricted.} \tag{344}$$

But this choice is only one possibility among others. The class of materials found by this "technique" is generally too small.

9.3 Liu Technique

We start out with

Liu's Proposition[60],[49],[61]: By use of the inclusion (342) the following statement is valid: Constitutive equations satisfy in large state spaces (21) the relations

$$\boldsymbol{B}(Z) = \boldsymbol{\lambda}(Z) \cdot \mathbf{A}(Z), \tag{345}$$

$$\boldsymbol{\lambda}(Z) \cdot \boldsymbol{C}(Z) \geq D(Z). \tag{346}$$

Here the state function $\boldsymbol{\lambda}$ is only unique, if $\mathbf{A}$ has its maximal rank. The entropy production density

$$\sigma := \boldsymbol{\lambda} \cdot \boldsymbol{C} - D \geq 0 \tag{347}$$

is independent of directional derivatives.

□

A corollary results immediately from Liu's proposition:

Corollary: For arbitrary directional deriviatives $\boldsymbol{w}$ which need not satisfy the balance equations we have:

$$\boldsymbol{B} \cdot \boldsymbol{w} - \boldsymbol{\lambda} \cdot (\mathbf{A} \cdot \boldsymbol{w} - \boldsymbol{C}) \geq D. \tag{348}$$

10 Concluding Remark

In nine sections we dicussed concepts of nonclassical thermodynamics and their connexion with other thermodynamical theories summed up in the survey. In the further four parts of this book concerned in thermodynamics of solids, also in electromagnetic fields, in Extended Thermodynamics and in material stability of solids, one can find nearly all these concepts more or less distinctly. Of course one needs something more than these basic concepts because e.g. the intermediate configuration in thermoplasticity or Lyapunov functionals in stability theory are beyond the scope of the first part. However it should be possible to recognize the discussed basic concepts in the following parts and to classify the theoretical investigations in them by applying these concepts.

11 References

1. Schottky, W.: Thermodynamik, 1. Teil Paragr.1, Springer-Verlag, Berlin (1929)

2. Axelrad, D.R.: Foundations of the Probabilistic Mechanics of Discrete Media, Pergamon Press, Oxford 1984

3. Muschik, W.: Thermodynamical Constitutive Laws - Outlines, in: Axelrad, D.R. and Muschik, W. (Eds.): Constitutive Laws and Microstructures, Springer, Berlin 1988

4. Fick, E. and Sauermann, G.: Quantenstatistik Dynamischer Systeme, Bd. I, Harri Deutsch, Thun 1983

5. Muschik, W. and Brunk, G.: Int. J. Engng. Sci 15 (1977) 377-389

6. Eu, B.C.: J. Chem. Phys. 87 (1987 1220

7. Meixner, J.: Rheol. Acta 7 (1968) 8

8. Truesdell, C. and Noll, W.: Handbook of Physics, Vol. III/3, Sect. 79, Springer, Berlin 1965

9. Muschik, W.: J. Phys. Chem. Solids 49 (1988) 709-720

10. Meixner, J.: Z. Phys. 219 (1969) 79

11. Glansdorf, P. and Nicolis, G.: Proc. Nat. Acad. Sci. USA 71 (1974) 197

12. Eringen, A.C.: Crystal Lattice Defects 7 (1977) 109

13. Liu, I.S. and Müller, I.: Arch. Rat. Mech. Anal. 83 (1983) 285

14. Noll, W.: Arch. Rat. Mech. Anal. 48 (1972) 1

15. Coleman, B.D. and Owen, D.R.: Arch. Rat. Mech Anal. 54 (1974) 1

16. Day, W.A.: The Thermodynamics of Simple Material with Fading Memory, Springer Tracts in Natural Philosophy, Vol. 22, Sect. 3.4., Springer, Berlin 1972

17. Muschik, W.: J. Non-Equilib. Thermodyn. 4 (1979) 277

18. Keller, J.U.: Physica 53 (1971) 602

19. Kestin, J. (Ed.): The Second Law of Thermodynamics, Hutchinson and Ross, Stroudsburg 1976

20. Kestin, J.: Pure appl. Chem. 22 (1970) 511

21. Muschik, W.: J. Non-Equilib. Thermodyn. 14 (1989) 173-198

22. Muschik, W.: Am. J. Phys. 58 (1990) 241

23. Clausius R.: Ann. Phys. Chem. 3 (1854)

24. Carathéodory, C.: Math. Ann. 67 (1909) 355

25. Born, M.: Phys. Z. 22 (1921) 218

26. Dunning-Davies, J.: Nuov. Cim. 64B (1969) 82

27. Landsberg, P.T.: Nature, Lond. 201 (1964) 485

28. Landsberg, P.T.: Physica Stat. Sol. 1 (1961) 120

29. Ramsey, N.F.: Phys. Rev. 103 (1956) 20

30. Dunning-Davies, J.: J. Phys. A9 (1976) 605

31. Muschik, W.: Continuum Models of Discrete Systems 4, Brulin, O. and Hsieh, R.K.T. (Eds.), p. 511, North-Holland, Amsterdam 1981

32. Muschik, W.: in: Spencer, A.J.M. (Ed.): Continuum Models of Discrete Systems, p. 39, Balkema, Rotterdam 1987

33. Falk, G. and Jung, H.: Axiomatik der Thermodynamik, Handbook of Physics, Vol. III/2, p. 119, Springer, Berlin 1959

34. Silhavy, M.: Arch. Rat. Mech. Anal. 81 (1983) 221

35. Man, C.-S.: Accad. nazn. Lincei VIII, 69 (1980) 399

36. Serrin, J.: New Perspectives in Thermodynamics, p. 3, Springer, Berlin 1986

37. Muschik, W.: Aspects of Non-Equilibrium Thermodynamics, sect. 1.1.1, World Scientific, Singapore 1990

38. Muschik, W.: J. Appl. Sci. 4 (1986) 189

39. Muschik, W.: ZAMM 63 (1983) T189

40. see 37. sect. 2.3

41. Muschik, W.: Arch. Rat. Mech. Anal. 66 (1977) 379

42. Muschik, W.: J. Non-Equilib. Thermodyn. 4 (1979) 377

43. Muschik, W. and Fang, J.: Acta Phys. Hung. 66 (1989) 39

44. Muschik, W. J. Non-Equilib. Thermodyn. 8 (1983) 219

45. Valanis, K.C.: Irreversible Thermodynamics of Continuous Media (Internal Variable Theory) CISM Courses and Lectures No. 77, Springer, 1972

46. Kluitenberg, G.A.: Plasticity and Nonequilibrium Thermodynamics, in: The Constitutive Law in Thermoplasticity, CISM Courses and Lectures No. 282, Th. Lehmann (Ed.), Springer, 1984, p. 157

47. Kestin, J.: A Course in Thermodynamics, Vol. 1, sect. 8.4.9, Hemisphere Publ. Coop., Washington 1979

48. Muschik, W.: Z. Phys. Chem. NF 68 (1969) 175

49. Muschik, W.: iin: Recent Developments in Nonequilibrium Thermodynamics by Casas-Vázquez, J., Jou, D. and Lebon, G. (Eds.): Lecture Notes in Physics 199, p. 387, Springer, 1984

50. Ponter, A.R.S., Bataille, J. and Kestin, J.: J. Méchanique 18 (1979) 511

51. Grolig, G.: Plastische Deformationen als Ausdruck mikroskopischer Instabilität, Thesis, Dep. of Mechanics, University Darmstadt, 1985

52. Müller, W.H. and Muschik, W.: J. Non-Equilib. Thermodyn. 8 (1983) 29, 47

53. Müller, I.: in Lecture Notes in Physics 199 (see 49) p. 32

54. Liu, I.S. and Müller, I.: Arch. Rat. Mech. Anal 83 (1983) 285

55. Ruggeri, T.: Continuum Mech. Thermodyn. 2 (1990) 163

56. Ruggeri, T.: Rend. Sem. Mat. Univ. Pol. Torino, Fasciolo Speciale 1988

57. Müller, I.: Acta Mechanica 24 (1976) 117

58. Blenk, S., Ehrentraut, H. and Muschik, W.:Int. J. Engng. Sci. 30 (1992) 1127

59. Muschik, W.: in: Disequilibrium and Self-Organisation, Kilmister, C.W. (Ed.), Reidel, 1986, p. 65

60. Liu, I.S.:Arch. Rat. Mech. Anal. 46 (1972) 131

61. Muschik, W. and Ellinghaus, R.: ZAMM 68 (1988) T232

THERMODYNAMICS OF SOLIDS

P. Haupt
University of Kassel, Kassel, Germany

Material properties are represented by constitutive equations, which complete the set of balance equations for momentum and energy. According to experimental observations, material behavior may be idealized to be rate independent or rate dependent. Moreover, rate dependent or rate independent hysteresis effects may occur: This implies four different categories of constitutive models, namely the theories of elasticity, viscoelasticity, plasticity and viscoplasticity. In case of elasticity the stress depends only on the present strain. For inelastic materials the stress is related to the present strain and its past history. This relationship can be represented explicitly as a functional of the strain process. An implicit representation of a functional relation is obtained from the formulation of a system of ordinary differential equations for a set of additional internal variables. These evolution equations specify the rate of change of internal variables depending on their present values and the strain (or stress) input. Different constitutive models imply different mathematical characteristics of the evolution equations. These concern the existence of equilibrium solutions and their stability properties. Some principal ideas for a modelling of material properties are outlined for the purely mechanical case and then generalized to thermodynamics. In the thermodynamic context the existence of rate independent hysteresis effects, encountered in plasticity and viscoplasticity, leads to a basic problem: It is impossible to approximate reversible processes through slow deformation and temperature histories. In the theory of viscoplasticity the evolution equations have equilibrium solutions, which depend on the process history and represent equilibrium hysteresis behavior. It is shown that the model of thermoviscoplasticity includes rate independent thermoplasticity as an asymptotic limit for infinitely slow strain and temperature processes. Thus viscoplasticity incorporates a rate dependent deviation from a sequence of equilibrium states, which becomes arbitrarily small for vanishing rates of the input processes. Taking this result as a physical motivation, a theory of finite thermoplasticity is outlined, based on the idea of an intermediate configuration and the concept of dual variables and derivatives.

[1] Valuable comments of Marc Kamlah and Stefan Hartmann are gratefully acknowledged.

1. Introduction

The object of mechanics as a part of physics is the scientific understanding of the motion of material bodies. Thermodynamics, another field of physics, investigates the transformation and exchange of energy. This includes the concept of heat and temperature and the basic observation that all natural processes are irreversible. In continuum mechanics it is convenient to consider mechanics and thermodynamics as one unified theory, based on the principal assumption that the spatial distribution of matter is continuous. Thus a material body is identified with a three-dimensional differentiable manifold; the elements of this manifold, the material points, are in one-to-one correspondence with position vectors as representations of their actual places in the Euclidean space of physical observation. Furthermore, the material body is mapped into a reference space, which must be introduced to associate mathematical "names" to the material elements. As a general consequence of those basic assumptions continuum thermomechanics is a field theory in the sense that all physical quantities depend on space and time.

All statements of continuum mechanics and thermodynamics separate into two different categories, namely into universal and individual statements: The principles of kinematics and the balance relations for mass, momentum, energy and entropy are **universal**: They are accepted as general natural laws to be true for all systems, irrespective of their material properties. The **individual** behavior of a particular material body is represented by constitutive equations, being no natural laws but rather mathematical models to represent the real behavior of matter in a certain class of processes and with a certain approximation.

A systematic construction of mathematical models to represent material properties is the objective of the **theory of material behavior**. Major results of this theory are the formulation of basic principles, which lead to a systematic elaboration of consistent models describing elasticity, plasticity, viscoelasticity and viscoplasticity.

All material properties depend on temperature. This is incorporated on the basis of general principles of thermodynamics. In each particular case a special constitutive model of thermomechanics is combined with the general balance equations of momentum and energy. This leads to initial-boundary value problems, which may be investigated analytically or solved by means of numerical methods.

In the following lectures some basic principles and results are presented, which facilitate a development of constitutive models to represent mechanical and thermodynamical material properties. This modelling should be realistic as well as consistent and compatible with basic physical and mathematical ideas and principles.

2. Balance relations

2.1 Basic notations

The position vector of a material element in its current and reference configuration is denoted by $\mathbf{x}$ and $\mathbf{X}$, respectively. According to the choice of a frame of reference, $\mathbf{x}$ is related to a point in the Euclidean space of physical observation and denotes the present position of the material element under consideration. The "name" $\mathbf{X}$ of the same material element corresponds to a point in some arbitrary reference space. $\mathbf{x}$ and $\mathbf{X}$ are related by the function

$$\mathbf{x} = \boldsymbol{\chi}_R(\mathbf{X},t) \, , \tag{2.1}$$

which is called the motion or deformation of the material body. The index R symbolizes the dependence of this function on the choice of the reference configuration. In general, the space of reference is completely independent from the space of physical observation, i.e. the reference configuration is not occupied by the body at some instant t_O. However, it is possible to identify the two spaces in special cases. If this is done, the displacement field can be defined:

$$\mathbf{u}(\mathbf{X},t) = \boldsymbol{\chi}_R(\mathbf{X},t) - \mathbf{X} \tag{2.2}$$

These considerations are summarized schematically in figure 2.1.

For each fixed t the mapping $\boldsymbol{\chi}_R$ is invertible. Accordingly, any time-dependent physical quantity can be represented either by $\mathbf{x}$ or $\mathbf{X}$ as independent variable. This corresponds to the choice of the spatial (Eulerian) or material (Lagrangian) representation, respectively. The different representations lead to different fields for one and the same physical quantity. Basically, these two fields are completely equivalent; nevertheless, particular quantities may naturally appear in the context of a spatial or material representation. For the basic quantities of continuum mechanics this is summarized below.

Material representation

In the material (Lagrangian) representation all physical quantities are referred to material elements: The value $\Psi(\mathbf{X},t)$ of a particular field Ψ represents the present state of the material point with the name $\mathbf{X}$. $\mathrm{Grad}\,\Psi(\mathbf{X},t)$ denotes the gradient operator with respect to $\mathbf{X}$.

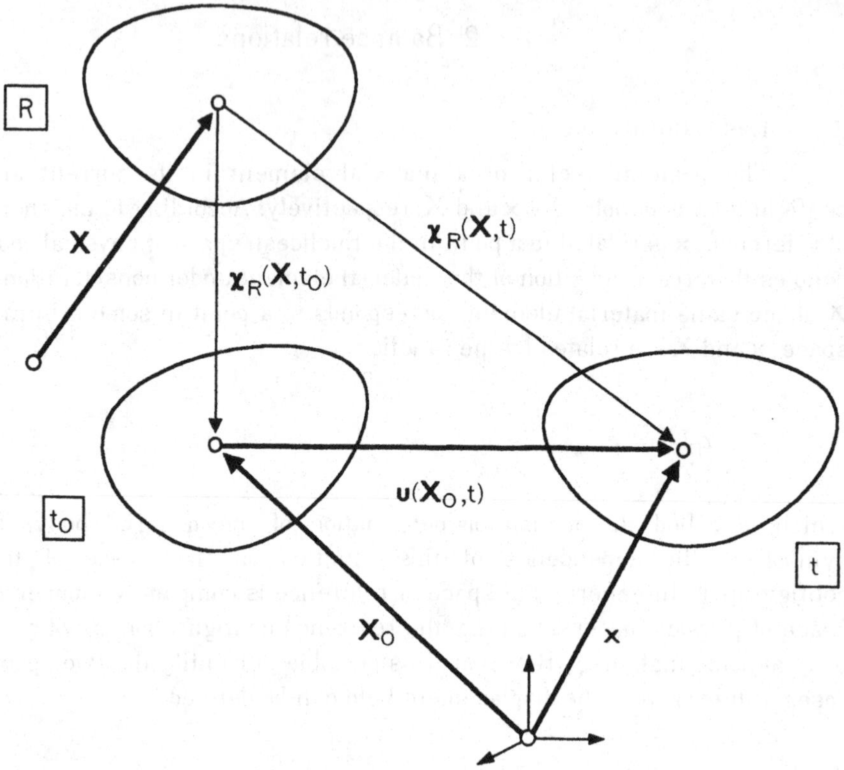

Figure 2.1: Configurations of a material body

$\mathbf{x} = \boldsymbol{\chi}_R(\mathbf{X},t)$	motion of a material body
$\mathbf{v}(\mathbf{X},t) = \frac{\partial}{\partial t}\boldsymbol{\chi}_R(\mathbf{X},t)$	velocity field
$\mathbf{F} = \text{Grad}\,\boldsymbol{\chi}_R(\mathbf{X},t)$	deformation gradient
$\mathbf{E} = \frac{1}{2}\left(\mathbf{F}^T\mathbf{F} - \mathbf{1}\right)$	Green strain tensor
$\rho_R(\mathbf{X},t)$	mass density in the reference configuration
$\mathbf{T}_R(\mathbf{X},t) = (\det\mathbf{F})\mathbf{T}\,\mathbf{F}^{T-1}$	Ist Piola-Kirchhoff stress ($\mathbf{T}$ = Cauchy stress)

$\tilde{\mathbf{T}}(\mathbf{X},t) = \mathbf{F}^{-1}\,\mathbf{T}_R$ 2nd Piola-Kirchhoff stress

$\mathbf{k}(\mathbf{X},t)$ specific volume force

$e(\mathbf{X},t)$ specific internal energy

$\mathbf{q}_R(\mathbf{X},t) = (\det\mathbf{F})\,\mathbf{F}^{-1}\mathbf{q}$ Piola-Kirchhoff heat flux vector

$r(\mathbf{X},t)$ specific heat suppy by radiation

$s(\mathbf{X},t)$ specific entropy

$\Theta(\mathbf{X},t)$ absolute temperature

$\mathbf{g}_R(\mathbf{X},t) = \mathrm{Grad}\,\Theta(\mathbf{X},t)$ temperature gradient

$\gamma(\mathbf{X},t)$ specific entropy production

Spatial representation

In the spatial (Eulerian) representation the "name" $\mathbf{X}$ of a material element is replaced by the position vector $\mathbf{x}$ of its place in the current configuration. Thus the value $\Psi(\mathbf{x},t)$ of a particular field Ψ corresponds to the material point $\mathbf{X}$ which at time t is found at position $\mathbf{x}$. $\mathrm{grad}\,\Psi(\mathbf{x},t)$ denotes the gradient operator with respect to $\mathbf{x}$.

$\rho(\mathbf{x},t)$ mass density in the current configuration

$\mathbf{v}(\mathbf{x},t)$ velocity field

$\mathbf{L}(\mathbf{x},t) =$ grad $\mathbf{v}(\mathbf{x},t)$ velocity gradient

$\mathbf{D}(\mathbf{x},t) = \frac{1}{2}(\mathbf{L} + \mathbf{L}^T)$ strain rate tensor

$\mathbf{W}(\mathbf{x},t) = \frac{1}{2}(\mathbf{L} - \mathbf{L}^T)$ vorticity tensor

$\mathbf{T}(\mathbf{x},t)$	Cauchy stress tensor
$\mathbf{k}(\mathbf{x},t)$	specific volume force
$e(\mathbf{x},t)$	specific internal energy
$\mathbf{q}(\mathbf{x},t)$	Cauchy heat flux vector
$r(\mathbf{x},t)$	specific heat supply by radiation
$s(\mathbf{x},t)$	specific entropy
$\Theta(\mathbf{x},t)$	absolute temperature
$\mathbf{g}(\mathbf{x},t) = \text{grad}\ \Theta(\mathbf{x},t)$	temperature gradient
$\gamma(\mathbf{x},t)$	specific entropy production

Throughout this text, the inner product of two vectors $\mathbf{a}$ and $\mathbf{b}$ is denoted by $\mathbf{a} \cdot \mathbf{b} = a_k b^k$ and the inner product of two tensors $\mathbf{A}$ and $\mathbf{B}$ by $\mathbf{A} \cdot \mathbf{B} = \text{tr}\left(\mathbf{A}\,\mathbf{B}^T\right) = A_{ij}B^{ij}$ (summation convention).

A superimposed dot denotes the material time derivative of a physical quantity $\mathbf{a}$: In the spatial representation it is calculated according to

$$\dot{\mathbf{a}}(\mathbf{x},t) = \frac{d}{dt}\,\mathbf{a}(\mathbf{x},t) = \frac{\partial \mathbf{a}}{\partial t} + (\text{grad}\ \mathbf{a})\,\mathbf{v}\ , \tag{2.3}$$

and in the material representation according to

$$\dot{\mathbf{a}}(\mathbf{X},t) = \frac{\partial}{\partial t}\,\mathbf{a}(\mathbf{X},t)\ . \tag{2.4}$$

2.2 Local formulation of balance relations

The following list summarizes the universal balance relations, which are accepted to be true for all systems, irrespective of their particular material properties.

Spatial representation

Mass

$$\dot{\rho} + \rho \operatorname{div} \mathbf{v} = 0 \tag{2.5}$$

Linear momentum

$$\rho \dot{\mathbf{v}} = \operatorname{div} \mathbf{T} + \rho \mathbf{k} \tag{2.6}$$

Moment of momentum

$$\mathbf{T} = \mathbf{T}^{\mathrm{T}} \tag{2.7}$$

Energy

$$\rho\left(e + \tfrac{1}{2}\mathbf{v}^2\right)^{\cdot} = \operatorname{div}\left(- \mathbf{q} + \mathbf{T}\mathbf{v}\right) + \rho\left(r + \mathbf{k} \cdot \mathbf{v}\right) \tag{2.8}$$

$$\Leftrightarrow \quad \dot{e} = - \frac{1}{\rho} \operatorname{div} \mathbf{q} + r + \frac{1}{\rho} \mathbf{T} \cdot \mathbf{D} \tag{2.9}$$

Entropy

$$\dot{s} = - \frac{1}{\rho} \operatorname{div}\left(\frac{\mathbf{q}}{\Theta}\right) + \frac{r}{\Theta} + \gamma = \tag{2.10}$$

$$= \frac{1}{\Theta}\left[- \frac{1}{\rho} \operatorname{div} \mathbf{q} + r\right] + \frac{1}{\rho\,\Theta^2}\, \mathbf{q} \cdot \operatorname{grad} \Theta + \gamma$$

$$\Leftrightarrow \quad \Theta \dot{s} = \dot{e} - \frac{1}{\rho} \mathbf{T} \cdot \mathbf{D} + \frac{1}{\rho\Theta}\, \mathbf{q} \cdot \operatorname{grad} \Theta + \Theta\gamma \tag{2.11}$$

Dissipation principle (entropy inequality)

$$\gamma \geq 0 \tag{2.12}$$

In these relations the differential operators grad and div refer to the spatial representation of all quantities, i.e. to the position vector $\mathbf{x}$ or to spatial coordinates.

Material representation

Mass

$$\frac{\partial}{\partial t}\rho_R(\mathbf{X},t) = 0\,, \quad \Leftrightarrow \quad \rho_R = \rho_R(\mathbf{X}) \tag{2.13}$$

Linear momentum

$$\rho_R \frac{\partial^2}{\partial t^2}\boldsymbol{\chi}_R(\mathbf{X},t) = \text{Div}\,\mathbf{T}_R + \rho_R\mathbf{k} \tag{2.14}$$

Moment of momentum

$$\mathbf{T}_R\,\mathbf{F}^T = \mathbf{F}\,\mathbf{T}_R^{\,T} \tag{2.15}$$

Energy

$$\rho_R\left(e + \frac{1}{2}\mathbf{v}^2\right)^{\cdot} = \text{Div}\left(-\,\mathbf{q}_R + \mathbf{T}_R\mathbf{v}\right) + \rho_R\left(r + \mathbf{k}\cdot\mathbf{v}\right) \tag{2.16}$$

$$\Leftrightarrow \quad \dot{e} = -\frac{1}{\rho_R}\text{Div}\,\mathbf{q}_R + r + \frac{1}{\rho_R}\tilde{\mathbf{T}}\cdot\dot{\mathbf{E}} \tag{2.17}$$

Entropy

$$\Theta\dot{s} = \dot{e} - \frac{1}{\rho_R}\tilde{\mathbf{T}}\cdot\dot{\mathbf{E}} + \frac{1}{\rho_R\Theta}\mathbf{q}_R\cdot\text{Grad}\,\Theta + \Theta\gamma \tag{2.18}$$

Dissipation principle (entropy inequality)

$$\gamma \geq 0 \tag{2.19}$$

In these relations the differential operators Grad and Div refer to the material representation of all quantities, i.e. to the vector $\mathbf{X}$ of a material point in the reference configuration or to material coordinates.

The above balance relations can be interpreted in the following way: Each particular material subsystem is regarded to be closed in the sense that no mass is exchanged with its environment. The material time rate of change of the mass density, velocity, energy and entropy is affected by the action of the external world, i. e. force, torque, heat supply and entropy transport.

Open systems Another equivalent version of the balance relations regards the material element as an open system, where mass, momentum, energy and entropy are exchanged with the external world. This version of the balance relations refers to the spatial representation and reads as follows:

Mass

$$\frac{\partial \rho}{\partial t} + \mathrm{div}(\rho \mathbf{v}) = 0 \tag{2.20}$$

Momentum

$$\frac{\partial}{\partial t}(\rho \mathbf{v}) + \mathrm{div}\left[\rho \mathbf{v} \times \mathbf{v} - \mathbf{T}^T\right] - \rho \mathbf{k} = \mathbf{0} \tag{2.21}$$

Moment of momentum

$$\mathbf{T} - \mathbf{T}^T = \mathbf{0} \tag{2.22}$$

Energy:

$$\frac{\partial}{\partial t}\left(\rho e + \frac{\rho}{2} \mathbf{v}^2\right) + \mathrm{div}\left[\left(\rho e + \frac{\rho}{2} \mathbf{v}^2\right)\mathbf{v} - \mathbf{T}\mathbf{v} + \mathbf{q}\right] - \rho\left(\mathbf{k} \cdot \mathbf{v} + r\right) = 0 \tag{2.23}$$

Entropy:

$$\frac{\partial}{\partial t}(\rho s) + \operatorname{div}\left(\rho s \mathbf{v} + \frac{\mathbf{q}}{\Theta}\right) - \rho\frac{r}{\Theta} = \rho\gamma \tag{2.24}$$

Entropy Inequality:

$$\gamma \geq 0 \tag{2.25}$$

Here, the local derivatives of the densities are balanced by divergence terms, which, in addition to the action of the external world, correspond to an exchange of the quantities, or to their transport across the boundary of a spatial volume. As one example, integration of the balance of mass (2.20) over a fixed spatial volume V yields

$$\iiint_V \frac{\partial\rho}{\partial t} dV + \iint_A \rho(\mathbf{v}\cdot\mathbf{n})dA = 0 ,$$

where the divergence theorem has been applied. In this formulation the principle of mass conservation states that the rate of change within the fixed spatial volume V is balanced by the total flux of the mass across the surface A(V). The same interpretations hold for the other balance equations. In these formulations the balance relations can be understood as conservation laws, which apply to open systems. An exception is the balance of entropy, which is not conserved: According to the dissipation principle there is a production γ of entropy, which must be non-negative.

3. Mechanical behavior of materials

Before investigating the thermodynamic behavior of solids, a discussion of purely mechanical material properties might be useful. In fact, the mechanical behavior of materials is the thermodynamical behavior under isothermal or adiabatic conditions. Therefore some principal characteristics of thermodynamics can be inferred from the purely mechanical situation. Moreover, it turns out that many physical and mathematical arguments, commonly applied in the theory of material behavior, are not only useful in continuum mechanics but also in the more general context of thermodynamics. Therefore some basic points of view concerning the modelling of mechanical properties of solids will be outlined in this section.

3.1 Experimental observations

In each particular case an experimental identification of material properties requires two kinds of idealization: First, we must assume that the material properties of the test specimen are uniform, i.e. homogeneously distributed in space. Then we have to take care that a homogeneous distribution of stress and strain exists within a part of the specimen. If these two assumptions are justified, applied and observed forces and displacements are equivalent to stresses and strains. Accordingly, the experimental investigation of material properties requires the existence of homogeneous states of stress and strain within homogeneous test specimen. Corresponding examples are uniaxial tension tests on slender bars or combined tests of tension and torsion on thin-walled cylindrical tubes.

Testing machines with computer-aided servohydraulic control devices provide general possibilities for a technical generation of input processes, where the material response, i.e. the corresponding output processes, can be simultaneously observed and evaluated. In this sense a test specimen can be considered as a black box or an operator relating input and output processes. This is illustrated in figure 3.1.

It is convenient to present the related input and output processes by means of stress-strain diagrams. This is sketched qualitatively in figure 3.2. The first part corresponds to to a strain controlled relaxation test: A strain process is applied with a constant strain rate and then held fixed. The material response is the stress as a certain function of time, followed by a relaxation process. The stress relaxation comes to an end, if an equilibrium point has been reached.

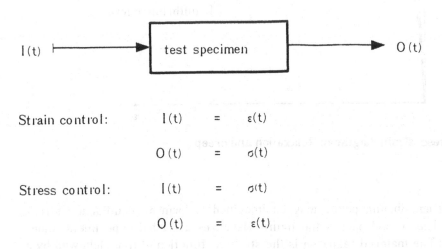

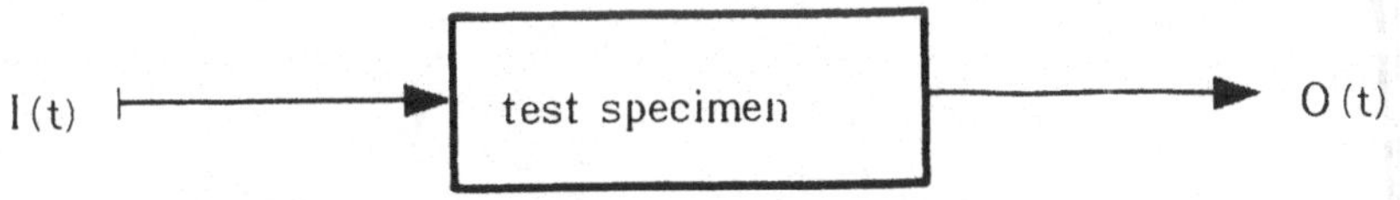

Strain control: $I(t) = \varepsilon(t)$

$O(t) = \sigma(t)$

Stress control: $I(t) = \sigma(t)$

$O(t) = \varepsilon(t)$

Figure 3.1: Test specimen as an operator (black box)

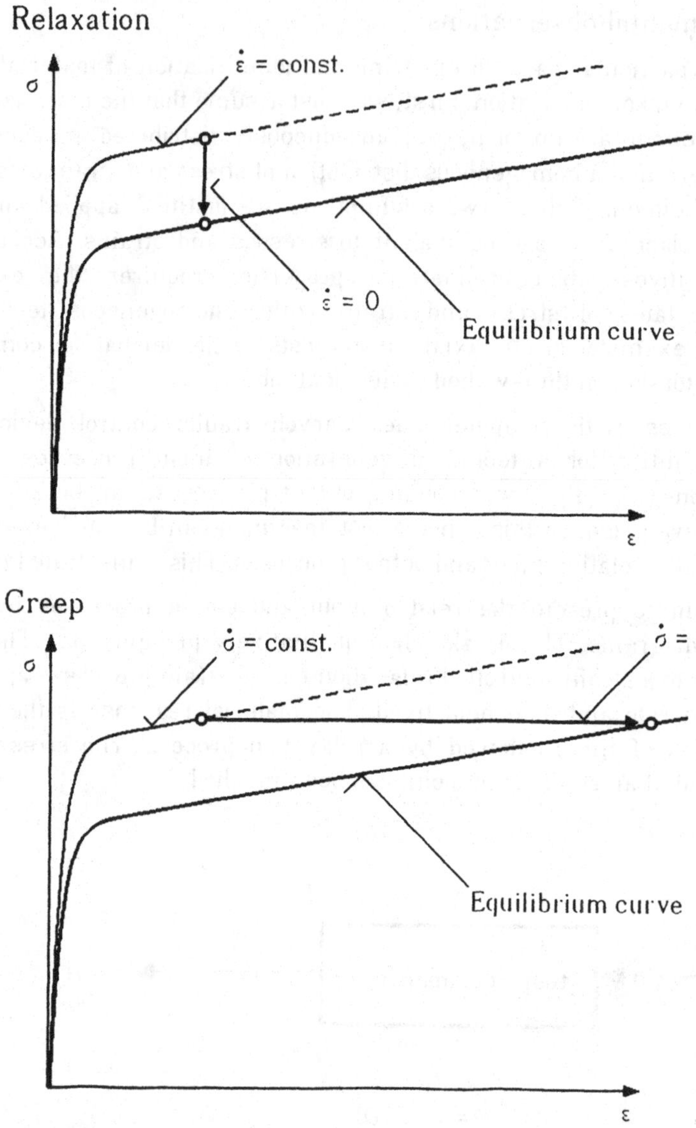

Figure 3.2: Stress-strain diagrams: Relaxation and creep

The set of all equilibrium points may be imagined to form an equilibrium stress strain curve. The second part of the figure visualizes a related experiment under stress control: The material response is the strain as function of time, followed by a creep process. Creeping comes to an end, if a corresponding point on the equilibrium stress strain curve exists. If for a given stress level there is no point of the equilibrium curve, the test specimen may be imagined to creep (possibly with a constant strain rate) to infinity or until failure takes place. This behavior is usually called secondary creep. Each experimental test delivers one pair of related input and output processes I(t) and O(t) (Figure 3.1) as a partial information about the black box, which characterizes the test specimen.

Experimental investigations of stress-strain curves should not be influenced by tacit assumptions in the sense of a special constitutive model. The objectivity of the registration of experimental data should not be affected by hasty interpretations in terms of yield stresses, elastic strains, plastic strains, creep strains or other terms, which are related to particular concepts of mathematical modelling. A general classification of the whole variety of material response is achieved in the following way:

First of all, the observed stress strain curves may be rate independent or rate dependent. In case of rate independence we have two different possibilities: There may exist a hysteresis or not. If the stress strain curves are rate dependent, we may slow down the rate of input histories to end up with quasistatic processes in the asymptotic limit. Again, we arrive at two possibilities according to the existence or nonexistence of quasistatic hysteresis effects. As a result we have four different classes of material behavior, summarized in figure 3.3: Rate dependence and rate independence with or without hysteresis and with or without quasistatic hysteresis, respectively. These four classes may help to order the variety of observations in the context of an experimental investigation and classification of the material response.

3.2 Mathematical modelling

The classification of experimentally observable material behaviors into four categories can be immediately applied to the mathematical modelling of material properties with the result that four different models of material behavior arise. These four theories are indicated in figure 3.4.

The four different mathematical models can be characterized as follows: The theory of **Elasticity** corresponds to the rate independent material behavior without hysteresis effects. The subject of the theory of **Plasticity** is the rate independent material behavior where hysteresis properties are incorporated. The theory of **Viscoelasticity** models the rate dependent material behavior without quasistatic hysteresis. Rate dependence in combination with static hysteresis is represented within the theory of **Viscoplasticity**. Figure 3.4 associates rheological models to each of these particular theories. These models symbolize some essential properties of the different constitutive models; they can easily be converted into functional relations.

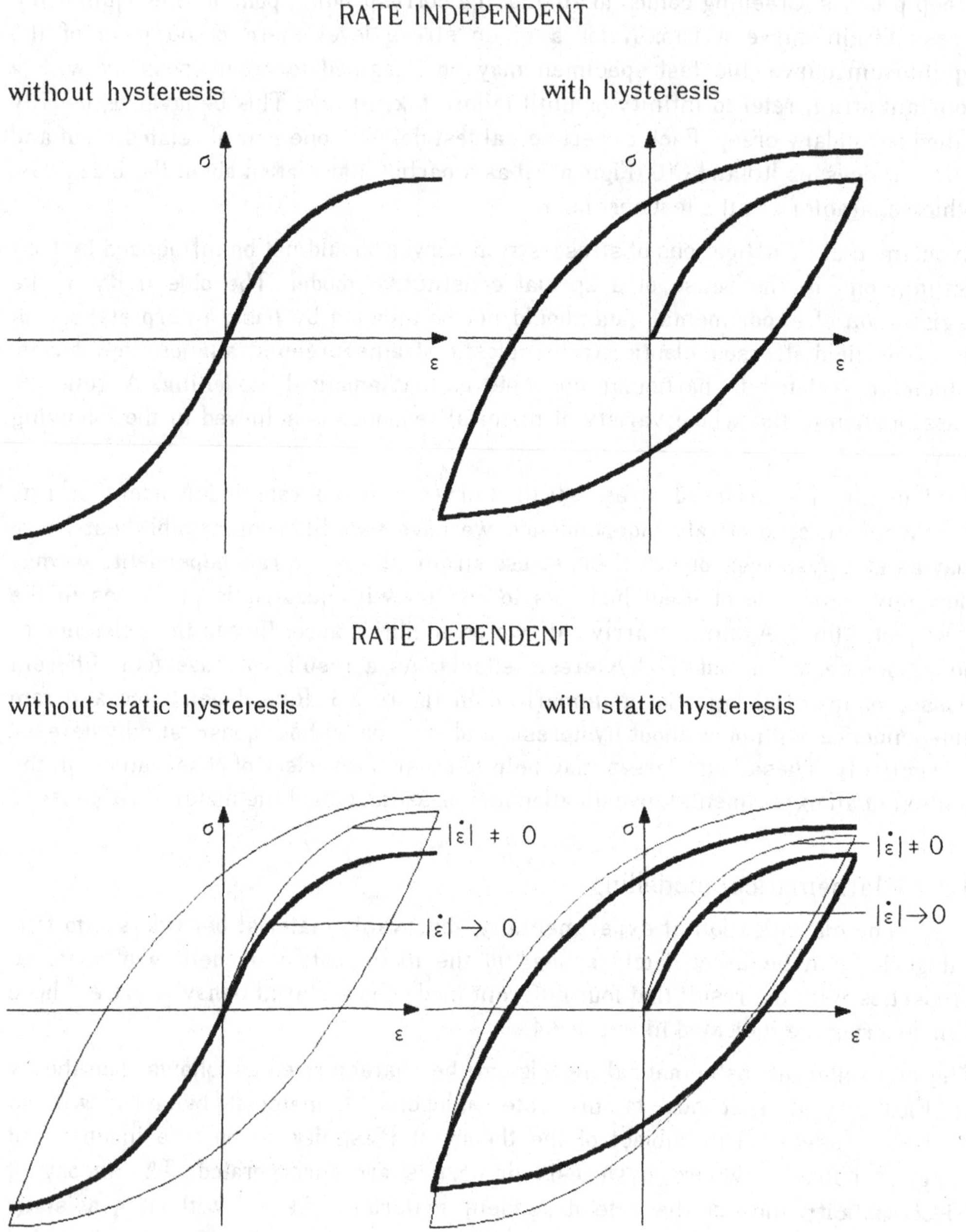

Figure 3.3: Four different classes of material response

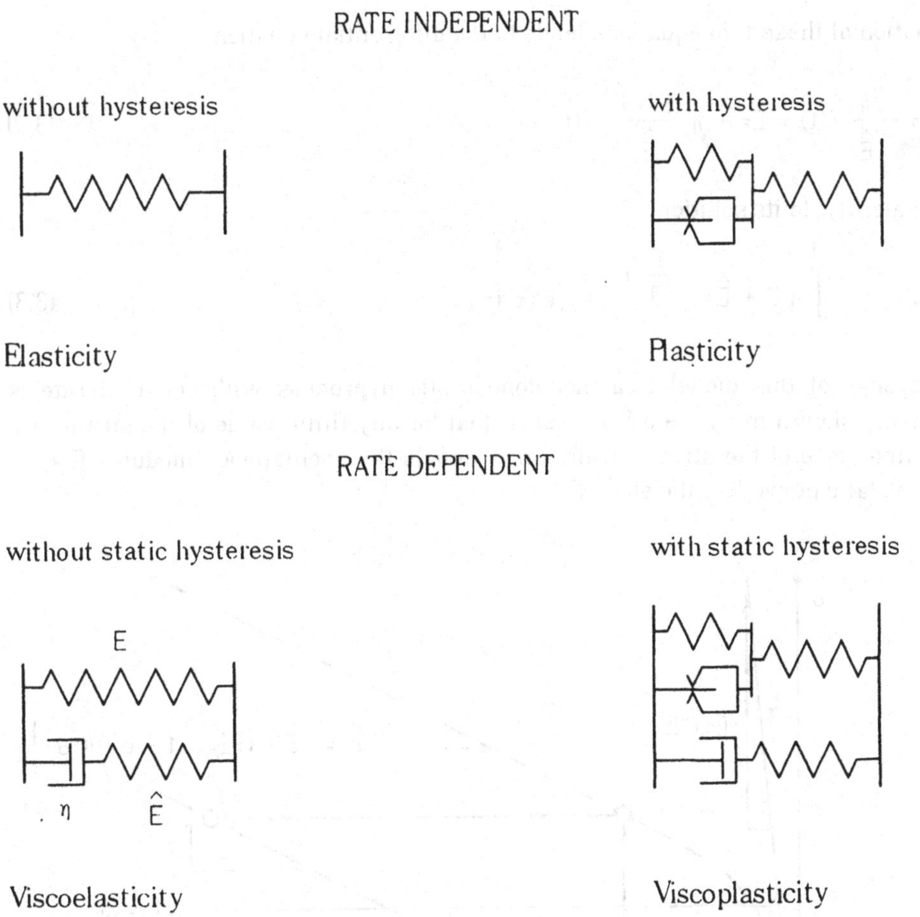

Figure 3.4: Four different models of material behavior

The most simple example to see this is the well known three parameter model of linear viscoelasticity (see figure 3.4); it consists of two parts, namely the stress strain relation

$$\sigma = E\varepsilon + \hat{E}(\varepsilon - q) \tag{3.1a}$$

and the evolution equation for one additional variable q:

$$\dot{q}(t) = \frac{\hat{E}}{\eta}(\varepsilon - q) \tag{3.1b}$$

Combination of these two equations leads to the differential equation

$$\sigma + \frac{\eta}{\hat{E}}\,\dot{\sigma}(t) = E\varepsilon + \eta\,\frac{E + \hat{E}}{\hat{E}}\,\dot{\varepsilon}(t) \tag{3.2}$$

or, equivalently, to its solution

$$\sigma(t) = \int_{-\infty}^{t} \left\{ E + \hat{E}\, e^{-\frac{\hat{E}}{\eta}(t - \tau)} \right\} \varepsilon'(\tau)\,d\tau\,. \tag{3.3}$$

The response of this model to a monotonous strain process with constant rate is qualitatively shown in figure 3.5. It is seen that for any finite value of the strain rate the starting slope of the stress-strain curve equals the spontaneous modulus $E + \hat{E}$, while the static curve has the slope E.

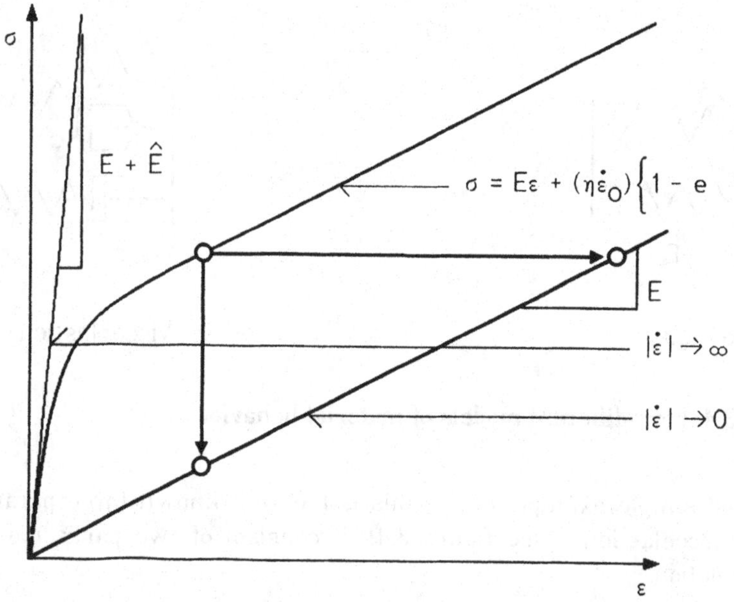

Figure 3.5: Response to a strain-controlled process with constant rate according to equations (3.1), (3.2)

According to equation (3.3) the present stress is a functional of the past strain history. This functional is the integral of the ordinary differential equation (3.1b), inserted into the stress-strain relation (3.1a). Regarding the equivalence of equations (3.1) and (3.3), we note that for the formulation of constitutive relations two possibilities exist: The stress can be represented explicitly as a functional of the strain history or this functional dependence may be implicitly defined through ordinary differential equations.

In this sense the quite special example motivates the following **generalization**:

$$\sigma = \varphi(\varepsilon, q_1, \dots, q_N) \tag{3.4a}$$

$$\dot{q}_k(t) = f_k(\varepsilon, q_1, \dots, q_N), \quad k = 1, \dots, N \tag{3.4b}$$

Of course, it is impossible to write down the general integral of the system (3.4b) in an explicit form. However, if the functions f_k have sufficient regularity properties, unique solutions $q_k(t)$ exist, if a strain process is given and initial conditions are prescribed. Thus a system of differential equations leads to an implicit representation of a functional stress strain relation. In this context two possibilities arise: The stress strain relation can be a differential equation of higher order for σ and ε only or it can be a system of first order differential equations. In the latter case additional variables are introduced, denoted by $q_1, \dots, q_N$. Traditionally, these quantities may be called **internal variables** or state variables. Equations (3.4b) are called **evolution equations.** They specify the rate of change of the internal variables in dependence on their present values and the strain input. In the context of different models of inelastic material behavior - plasticity, viscoelasticity, viscoplasticity - the evolution equations have different mathematical characteristics. In particular, these concern the existence of equilibrium solutions and their stability properties (see below).

3.3 Theory of material behavior

3.3.1 General principles

The theory of material behavior develops systematic methods to construct constitutive models, which represent mechanical properties of material systems. To this end three different kinds of statements are specified, namely:

- Constitutive Equations
- Material symmetry properties (Isotropy)
- Kinematic constraints

In its most general sense a **constitutive equation** relates the individual response of the material body to input processes, which are applied to material elements. The appropriate choice of the input and output variables depends on the specific subject. In **continuum mechanics** a constitutive equation corresponds to a relation between strain and stress tensors. The incorporation of the dependence of material properties on temperature requires a general thermodynamic theory as a principal background for the formulation of constitutive assumptions. In this context further input and output variables occur: Temperature and temperature gradient are further input quantities, internal energy, entropy and heat flux are introduced to characterize the nonmechanical response. In case of **elastic** bodies all constitutive equations are formulated by means of constitutive functions. For a representation of **inelastic** material behavior (viscoelasticity, plasticity, viscoplasticity) functional relations are needed to represent the material response in dependence on the history of the input process. Functional relations can be set up explicitly by specification of functionals or they can be defined in an implicit form specifying differential equations for additional quantities, which are called internal variables.

The concept of **material symmetry** or **isotropy** expresses the fact that constitutive relations may depend more or less on the direction of material line elements. In order to represent material symmetry, invariance properties of a constitutive equation with respect to certain changes of the reference configuration are postulated. The investigation of material symmetry leads to a distinction between solids and fluids and to a classification of the general behavior of solids into isotropy and different classes of anisotropy (see [1]).

A condition of **kinematic constraint** is a geometric restriction on the motions, which are possible for a material body. A well-known example is the constraint of incompressibility. If incompressibility is assumed, all changes of volume are excluded a priori, i.e. only isochoric motions are allowed. In the general theory of thermokinematic constraints it is assumed that only a part of the response of the material is determined by the input process history and that the undetermined part does no work or that it does not affect the energy balance or the entropy production.

The combination of a particular constitutive equation with convenient assumptions about symmetry properties and kinematic constraints leads to a set of equations, which complete the local form of the balance relations to initial-boundary value problems.

In view of a systematic construction of constitutive models some basic principles and methods can be formulated. In this context, it remains an open question to specify a list of basic axioms, which are necessary and sufficient to guarantee physical and mathematical consistency [1]. The following items may be regarded as basic principles or only as quite obvious points of view, which should be observed:

- determinism
- material objectivity
- local action
- irreversibility
- equipresence
- fading memory
- physical evidence
- mathematical consistency (e.g. invariance)
- existence and uniqueness of solutions
- possibility of experimental identification
- reasonable properties in view of numerical implementations

Not all of these items have the same importance, but some of them must be recognized to be imperative, in particular the first ones: The principle of determinism postulates that the stress is uniquely determined from the past strain history. Material objectivity as the assumption that constitutive equations should be independent of the frame of reference is most convenient in the context of reduced formulations of constitutive equations. The same is true for the principle of local action, leading to the concept of a simple material, which covers nearly all special theories of continuum mechanics. Even within a purely mechanical theory it is necessary for a constitutive model to be in accordance with the principle of irreversibility. Equipresence and fading memory are particular assumptions, not necessarily required. The last 5 items refer to implications of the constitutive modelling in the context of the resulting initial-boundary value problems, their mathematical solution and technical application.

3.3.2 Reduced forms

As a general consequence of the assumptions of determinism, material objectivity and local action the present state of stress in a simple material can be written as a functional of the past strain history:

$$\tilde{\mathbf{T}}(t) = \underset{\tau \le t}{\mathfrak{G}} \left[\mathbf{E}(\tau) \right] \tag{3.5}$$

A representation of the theory of material behavior, comprising details and basic aspects, can be found in the monograph of TRUESDELL and NOLL [1]. It is a basic result of the theory that in the most general constitutive equation (3.5) the stress is the second Piola Kirchhoff tensor,

$$\tilde{\mathbf{T}} = (\det\mathbf{F})\mathbf{F}^{-1}\,\mathbf{T}\,\mathbf{F}^{T-1}\,, \tag{3.6}$$

whereas the state of strain is represented by the Green strain tensor:

$$\mathbf{E} = \tfrac{1}{2}(\mathbf{F}^T\mathbf{F} - \mathbf{1}) \tag{3.7}$$

The functional $\mathfrak{G}$ associates a symmetric tensor $\tilde{\mathbf{T}}(t)$ to each tensorvalued function $\mathbf{E}(\tau)$, defined on the real interval $(-\infty,t]$. The above stress strain relation is called a **reduced form**: It satifies completely the assumption of objectivity in the sense that the choice of thic functional is not restricted by that assumption. Other equivalent reduced forms can be established, for example the relation

$$\tilde{\mathbf{T}}(t) = \mathbf{f}(\mathbf{E}) + \underset{s\geq 0}{\mathfrak{F}}\left[\mathbf{E}_d^t(s);\,\mathbf{E}(t)\right], \tag{3.8}$$

where

$$\mathbf{E}_d^t(s) = \mathbf{E}(t-s) - \mathbf{E}(t) \tag{3.9}$$

is the relative difference history of the Green strain, and the functional $\mathfrak{F}$ is normalized to vanish for constant (static) strain histories: For all values of the present strain $\mathbf{E}$ we have

$$\underset{s\geq 0}{\mathfrak{F}}\left[\mathbf{0}(s);\,\mathbf{E}\right] = \mathbf{0}\,. \tag{3.10}$$

As it was already mentioned in the special context of linear viscoelasticity, three different methods of representation can be distinguished, namely:

- Explicit representation by means of integrals or other explicit definitions of functional relations, e.g. (3.5) or (3.8)
- Implicit representation by ordinary differential equations of higher order for $\tilde{\mathbf{T}}(t)$ and $\mathbf{E}(t)$

- Implicit representation by systems of ordinary differential equations of first order, utilizing additional parameters (internal variables, compare eqs. (3.4)):

$$\tilde{\mathbf{T}} = \mathbf{f}(\mathbf{E}, q_1, \ldots, q_N) \tag{3.11a}$$

$$\dot{q}_k(t) = f_k\big(\mathbf{E}(t), q_1(t), \ldots, q_N(t)\big)\,, \quad k = 1, \ldots, N \tag{3.11b}$$

3.3.3 Elasticity

The model of elasticity is characterized by rate independent behavior without hysteresis effects. More specifically, in case of an elastic material the stress depends only on the present state of strain:

$$\tilde{\mathbf{T}} = \mathbf{f}(\mathbf{E}) \tag{3.12}$$

Physical reasons suggest the assumption of a strain energy function $\psi(\mathbf{E})$, defining the model of hyperelasticity:

$$\tilde{\mathbf{T}} = \rho_R \frac{d}{d\mathbf{E}} \psi(\mathbf{E}) \tag{3.13}$$

We note that the symmetry properties of the material coincide exactly with the isotropy properties of the constitutive function $\mathbf{f}(\cdot)$, i.e. the symmetry group of the material with respect to a reference configuration R is identical with the symmetry group of $\mathbf{f}(\cdot)$:

$$\mathbf{Q} \in g_R \quad \Leftrightarrow \quad \mathbf{Q}\,\mathbf{f}(\mathbf{E})\,\mathbf{Q}^T = \mathbf{f}(\mathbf{Q}\,\mathbf{E}\,\mathbf{Q}^T) \tag{3.14}$$

This result is also true for the general case of an inelastic material.

For isotropic elastic bodies there exists a reference configuration R such that the symmetry group is identical with the group of all orthogonal tensors $\mathbf{Q}$:

$$g_R = \mathrm{Orth} \tag{3.15}$$

In this case $\mathbf{f}(\mathbf{E})$ reduces to an isotropic function,

$$\tilde{\mathbf{T}} = \mathbf{f}(\mathbf{E}) = \varphi_0 \mathbf{1} + \varphi_1 \mathbf{E} + \varphi_2 \mathbf{E}^2 , \tag{3.16}$$

where the scalar-valued coefficients depend on the principal invariants of **E**:

$$\varphi_k = \varphi_k(I_{\mathbf{E}}, II_{\mathbf{E}}, III_{\mathbf{E}}) \qquad k = 1,2,3 \tag{3.17}$$

$$I_{\mathbf{E}} = \mathrm{tr}\,\mathbf{E}\,, \quad II_{\mathbf{E}} = \tfrac{1}{2}\left[(\mathrm{tr}\mathbf{E})^2 - \mathrm{tr}(\mathbf{E}^2)\right], \quad III_{\mathbf{E}} = \det \mathbf{E}$$

3.3.4 Viscoelasticity

According to its general definition the theory of viscoelasticity models the rate dependent material behavior without static hysteresis effects. Starting from the reduced form (3.8),

$$\tilde{\mathbf{T}} = \mathbf{f}(\mathbf{E}) + \underset{s \geq 0}{\mathfrak{F}}\left[\mathbf{E}_d^t(s); \mathbf{E}\right] \qquad \left(\mathbf{E}_d^t(s) = \mathbf{E}(t-s) - \mathbf{E}(t) \right), \tag{3.18}$$

it is required that the functional $\mathfrak{F}$ is continuous or differentiable in the context of the fading memory norm

$$\|\mathbf{E}_d^t(s)\|_h = \left\{ \int_0^\infty \|\mathbf{E}_d^t(s)\|^2 h^2(s)\,ds \right\}^{1/2} , \tag{3.19}$$

where $h(s)$ is an influence function of a certain order r ($s^r h(s) \longrightarrow 0$ for $s \longrightarrow \infty$, see [2]). The fading memory assumption implies the following consequences:

- The functional $\mathfrak{F}$ can be approximated by a linear functional, which admits an integral representation:

$$\tilde{\mathbf{T}} = \mathbf{f}(\mathbf{E}) + \int_0^\infty \mathbf{K}(\mathbf{E},s)\left[\mathbf{E}_d^t(s)\right]ds + \mathbf{o}\left(\|\mathbf{E}_d^t(s)\|_h\right) \tag{3.20}$$

  The kernel **K** is a fourth order tensor, depending on the strain **E** and the integration variable s. (Finite linear viscoelasticity of COLEMAN and NOLL [3])

- For retarded strain processes, defined by a retardation factor α ($0 < \alpha < 1$), the strain history can be approximated by its Taylor series, where the asymptotic error term is understood in the sense of the norm (3.19) (Retardation theorem of COLEMAN and NOLL [2]):

$$\mathbf{E}_d^t(\alpha s) = \sum_{k=1}^{n} \frac{(-\alpha s)^k}{k!} \mathbf{E}^{(k)}(0) + \mathbf{o}(\alpha^n) \tag{3.21}$$

- For sufficiently slow processes the response of the material is asymptotically elastic.

In case of an implicit representation of viscoelasticity by means of first order differential equations, i.e. eqs (3.11),

$$\tilde{\mathbf{T}} = \mathbf{f}(\mathbf{E}, q_1, \ldots, q_N)$$

$$\dot{q}_k(t) = f_k(\mathbf{E}, q_1, \ldots, q_N) \ , \ k = 1, \ldots, N \ ,$$

a corresponding fading memory assumption requires for the evolution equations the following properties:

- For each constant state of strain equilibrium solutions exist, i.e. solutions of the equations

$$f_k(\mathbf{E}, \bar{q}_1, \ldots, \bar{q}_N) = 0 \ , \ k = 1, \ldots, N \ , \tag{3.22}$$

  depending uniquely on the strain $\mathbf{E}$:

$$\bar{q}_k = g_k(\mathbf{E}) \tag{3.23}$$

- The equilibrium solutions $\bar{q}_k = g_k(\mathbf{E})$ are asymptotically stable in the large (COLEMAN and GURTIN [5]): If a static continuation of an arbitrary input process is given and initial conditions for the internal variables are prescribed, the solutions $q_k(t)$ tend asymptotically to their equilibrium values $\bar{q}_k = g_k(\mathbf{E})$, as time t goes to infinity.

These assumptions have the following consequences:

- The functional relation for the stress has the relaxation property

- Finite linear viscoelasticity is an asymptotic approximation for processes, which are not far from equilibrium.

- Equilibrium (quasistatic) processes are approximated asymptotically by sufficiently slow processes

- No static hysteresis occurs: Slow processes imply (asymptotically) elastic response

These consequences are also discussed by LUBLINER [18]. Clearly, the *relaxation property* is obvious from the assumed stability of the equilibrium states. The three other items can be verified on the basis of the Taylor expansion at the equilibrium points $\bar{q}_k$, leading to a linear approximation in the internal variables,

$$\mathbf{f}(\mathbf{E}, q_1, \dots, q_N) = \mathbf{f}(\mathbf{E}, \bar{q}_1, \dots, \bar{q}_N) + \sum_{j=1}^{N} \frac{\partial \mathbf{f}}{\partial q_j}(\mathbf{E}, \bar{q}_1, \dots, \bar{q}_N)\left[q_j - \bar{q}_j\right] + \mathbf{O}(x^2), \tag{3.24}$$

$$f_k(\mathbf{E}, q_1, \dots, q_N) = \sum_{j=1}^{N} \frac{\partial f_k}{\partial q_j}(\mathbf{E}, \bar{q}_1, \dots, \bar{q}_N)\left[q_j - \bar{q}_j\right] + O(x^2), \tag{3.25}$$

$$\left(f_k(\mathbf{E}, \bar{q}_1, \dots, \bar{q}_N) = 0\right),$$

with $x^2 = \sum_{j=1}^{N} |q_j - \bar{q}_j|^2 = \sum_{j=1}^{N} |q_j - g_j(\mathbf{E})|^2$. Inserting the functions $\bar{q}_k = g_k(\mathbf{E})$, we find

$$\tilde{\mathbf{T}} = \mathbf{f}(\mathbf{E}, g_1(\mathbf{E}), \dots, g_N(\mathbf{E})) + \sum_{j=1}^{N} \frac{\partial \mathbf{f}}{\partial q_j}(\mathbf{E}, g_1(\mathbf{E}), \dots, g_N(\mathbf{E}))\left[q_j - g_j(\mathbf{E})\right] + \mathbf{O}(x^2),$$

$$\tilde{\mathbf{T}} = \bar{\mathbf{f}}(\mathbf{E}) + \sum_{j=1}^{N} \bar{\mathbf{f}}_j(\mathbf{E})\left[q_j - g_j(\mathbf{E})\right] + \mathbf{O}(x^2), \tag{3.26a}$$

and

$$\dot{q}_k(t) = \sum_{j=1}^{N} \frac{\partial f_k}{\partial q_j}(\mathbf{E}, g_1(\mathbf{E}), \ldots, g_N(\mathbf{E}))\left[q_j - g_j(\mathbf{E})\right] + O(x^2)$$

$$\dot{q}_k(t) = -\sum_{j=1}^{N} Q_{kj}(\mathbf{E})\left[q_j - g_j(\mathbf{E})\right] + O(x^2), \tag{3.26b}$$

where for all $\mathbf{E}$ the matrix

$$Q_{kj}(\mathbf{E}) = -\frac{\partial f_k}{\partial q_j}(\mathbf{E}, g_1(\mathbf{E}), \ldots, g_N(\mathbf{E}))$$

is positive definite according to the stability property of the equilibrium state, which is the defining assumption of viscoelasticity (c.f. [18]).

Finite Linear Viscoelasticity

can be inferred from the approximated evolution equations (3.26b), if the strain-dependent coefficients $Q_{kj}(\mathbf{E})$ are replaced by the constant matrix

$$\underset{\approx}{Q} = (Q_{kj}(\mathbf{0})).$$

Then the (nonlinear) functions

$$G_k(\mathbf{E}) = \sum_{j=1}^{N} Q_{kj}(\mathbf{0}) g_j(\mathbf{E}), \quad k = 1, \ldots, N, \tag{3.26c}$$

can be defined, and we obtain for the internal variables a system of linear differential equations with constant coefficients:

$$\dot{q}_k(t) + \sum_{j=1}^{N} Q_{kj}(\mathbf{0}) q_j(t) = G_k(\mathbf{E}(t)) \quad k = 1, \ldots, N \tag{3.26d}$$

This system can be solved in closed form: If the matrix $\underset{\approx}{Q} = (Q_{kj}(\mathbf{0}))$ is positive definit and symmetric (c.f. eq. (4.46)), the solution of (3.26d) can be written as

$$q_j(t) = \int_{-\infty}^{t} \left\{ \sum_{l=1}^{N} \left[e^{-\underset{\approx}{Q}(t-\tau)} \right]_{jl} G_l(\mathbf{E}(\tau)) \right\} d\tau .$$

Integration by parts leads to

$$q_j(t) = \sum_{l=1}^{N} \left[\underset{\approx}{Q}^{-1} e^{-\underset{\approx}{Q}(t-\tau)} \right]_{jl} G_l(\mathbf{E}(\tau)) \Big|_{-\infty}^{t} -$$

$$- \int_{-\infty}^{t} \sum_{l=1}^{N} \left[\underset{\approx}{Q}^{-1} e^{-\underset{\approx}{Q}(t-\tau)} \right]_{jl} \left(\frac{dG_l}{d\mathbf{E}} \cdot \mathbf{E}'(\tau) \right) d\tau =$$

$$= \sum_{l=1}^{N} \left[\underset{\approx}{Q}^{-1} \right]_{jl} G_l(\mathbf{E}(t)) - \int_{-\infty}^{t} \cdots \, d\tau = g_j(\mathbf{E}(t)) - \int_{-\infty}^{t} \cdots \, d\tau .$$

In the last step the definition (3.26c) of G_k was applied. The result

$$q_j(t) = g_j(\mathbf{E}(t)) - \int_{-\infty}^{t} \sum_{l=1}^{N} \left[\underset{\approx}{Q}^{-1} e^{-\underset{\approx}{Q}(t-\tau)} \right]_{jl} \left(\frac{dG_l}{d\mathbf{E}} \cdot \mathbf{E}'(\tau) \right) d\tau \tag{3.27}$$

is now inserted into eq. (3.26a), leading to

$$\tilde{\mathbf{T}} = \bar{\mathbf{f}}(\mathbf{E}) - \sum_{j=1}^{N} \bar{\mathbf{f}}_j(\mathbf{E}) \left\{ \int_{-\infty}^{t} \sum_{l=1}^{N} \left[\underset{\approx}{Q}^{-1} e^{-\underset{\approx}{Q}(t-\tau)} \right]_{jl} \left(\frac{dG_l}{d\mathbf{E}} \cdot \mathbf{E}'(\tau) \right) d\tau \right\} . \tag{3.28a}$$

This equation approximates the general model (3.11) for those processes where the internal variables $q_k(t)$ remain within a close neighborhood of their equilibrium values $g_k(\mathbf{E}(t))$. The first term is the equilibrium stress; the second term is a linear functional of the past strain history. Clearly, eq. (3.28a) has the form

$$\tilde{\mathbf{T}} = \bar{\mathbf{f}}(\mathbf{E}) + \int_{-\infty}^{t} \mathbb{G}\big(\mathbf{E}(t), t-\tau\big)\big[\mathbf{E}'(\tau)\big] d\tau , \tag{3.28b}$$

where the integrand is a nonlinear function of the present strain $\mathbf{E}(t)$ and a linear function of the past strain $\mathbf{E}(\tau)$. The kernel $\mathbb{G}\big(\mathbf{E}(t), s\big)$ is a fourth order tensor- valued function, decreasing to zero as s goes to infinity: The time dependence of $\mathbb{G}$ corresponds to a sum of decreasing exponentials.

Slow motions, retardation theorem

In order to analyse slow processes within the theory of internal variables, a retarded time scale τ is defined through a retardation factor α between 0 and 1:

$$\tau = \alpha t\,, \quad 0 < \alpha < 1 \tag{3.29}$$

With $\dot{q}_k(t) = \alpha q_k'(\tau)$ $\left((\,)' = \frac{d}{d\tau}(\,)\right)$ the evolution equations and their Taylor approximations are transformed to the independent variable τ :

$$\alpha q_k'(\tau) = f_k\big(\mathbf{E}(\tau), q_1(\tau), \ldots, q_N(\tau)\big) \tag{3.30a}$$

$$\alpha q_k'(\tau) = \sum_{j=1}^{N} Q_{kj}\big(\mathbf{E}(\tau)\big)\big[q_j(\tau) - g_j\big(\mathbf{E}(\tau)\big)\big] + O\big(x^2\big) \tag{3.30b}$$

The matrix $Q_{kj} = -\dfrac{\partial f_k}{\partial q_j}$ is invertible because it is positive definite (see [18]); this yields

$$q_k(\tau) = g_k\big(\mathbf{E}(\tau)\big) + O(\alpha) + O\big(x^2\big)\,.$$

From equation (3.30a) we conclude $f_k\big(\mathbf{E}(\tau), q_1(\tau), \ldots, q_N(\tau)\big) = O(\alpha)$, which, together with $f_k\big(\mathbf{E}(\tau), q_1(\tau), \ldots, q_N(\tau)\big) = O(x)$ (eq.(3.25)), leads to $O(x) = O(\alpha)$, i.e.

$$q_k(\tau) = g_k\big(\mathbf{E}(\tau)\big) + O(\alpha)\,. \tag{3.31}$$

In accordance to the assumed stability property of the equilibrium solution this equation states that for sufficiently slow processes the internal variables q_k remain in an arbitrary close neighborhood of their equilibrium values.

Now we differentiate this equation with respect to τ, multiply with α and apply the chain rule to get

$$\dot{q}_k(t) = \frac{dg_k(\mathbf{E})}{d\mathbf{E}} \cdot \dot{\mathbf{E}}(t) + \alpha O(\alpha)\,. \tag{3.32}$$

In view of $O(x) = O(\alpha)$ or $O(x^2) = O(\alpha^2)$ eq. (3.26b) becomes

$$\dot{q}_k(t) = \sum_{j=1}^{N} Q_{kj}(\mathbf{E})\big[q_j - g_j(\mathbf{E})\big] + O(\alpha^2)\,.$$

Combination of the last two equations yields

$$\sum_{j=1}^{N} Q_{kj}(\mathbf{E})\left[q_j - g_j(\mathbf{E})\right] = \frac{dg_k(\mathbf{E})}{d\mathbf{E}} \cdot \dot{\mathbf{E}}(t) + o(\alpha)$$

$(\alpha O(\alpha) = o(\alpha))$. Solution with respect to the $q_k(t)$ gives the result

$$q_k(t) - g_k(\mathbf{E}(t)) = \sum_{j=1}^{N}\left[\underset{\approx}{Q}^{-1}(\mathbf{E})\right]_{kj} \frac{dg_j(\mathbf{E})}{d\mathbf{E}} \cdot \dot{\mathbf{E}}(t) + o(\alpha) . \tag{3.33}$$

This can be inserted into the stress-strain relation

$$\tilde{\mathbf{T}} = \bar{\mathbf{f}}(\mathbf{E}) + \sum_{j=1}^{N} \bar{\mathbf{f}}_j(\mathbf{E})\left[q_j - g_j(\mathbf{E})\right] .$$

Finally, we have

$$\tilde{\mathbf{T}} = \bar{\mathbf{f}}(\mathbf{E}) + \sum_{k=1}^{N} \bar{\mathbf{f}}_k(\mathbf{E}) \left\{ \sum_{j=1}^{N} \left[\underset{\approx}{Q}^{-1}(\mathbf{E})\right]_{kj} \left[\frac{dg_j(\mathbf{E})}{d\mathbf{E}} \cdot \dot{\mathbf{E}}(t)\right]\right\} + \mathbf{o}(\alpha) . \tag{3.34}$$

In this approximation the stress tensor depends on the strain $\mathbf{E}$ and linearly on the strain rate tensor $\dot{\mathbf{E}}$. The result corresponds to the first term of the retardation theorem (3.21) of COLEMAN and NOLL [2], applied to finite linear viscoelasticity (3.20).

3.3.5 Plasticity

The theory of plasticity is concerned with the constitutive modelling of rate independent material behavior, where hysteresis effects are incorporated. In the sequel a special class of constitutive models, widely used in applied continuum mechanics, will be considered. The definition of the model is restricted to small deformations; large deformations are discussed in chapter 5.

In the context of geometric linearization the second Piola-Kirchhoff stress $\tilde{\mathbf{T}}$ is replaced by the Cauchy stress $\mathbf{T}$ and the Green strain tensor is replaced by the linearized strain tensor

$$\mathbf{E} = \tfrac{1}{2}(\mathbf{H} + \mathbf{H}^T) , \tag{3.35}$$

where $\mathbf{H} = \mathrm{Grad}\,\mathbf{u}(\mathbf{X},t)$ is the displacement gradient. The classical model of elastoplasticity consists of different constituents: The strain is decomposed into elastic and plastic parts, the elastic strain is related to the stress through an elasticity relation and the plastic strain is governed by the flow rule on the basis of yield and loading conditions. The yield function depends on additional hardening models:

Decomposition

$$\mathbf{E} = \mathbf{E}_e + \mathbf{E}_p \tag{3.36a}$$

Elasticity relation

$$\mathbf{T}^D = 2\mu\, \mathbf{E}_e^D$$

$$\frac{1}{3}\,\mathrm{tr}\mathbf{T} = K\, \mathrm{tr}(\mathbf{E}_e) \qquad K = \frac{2\mu(1+\nu)}{3(1-2\nu)} \tag{3.36b}$$

(μ = Shear modulus, ν = Poisson ratio, K = bulk modulus)

Yield condition

$$f(\mathbf{T},\mathbf{X}) = \frac{1}{2}\left(\mathbf{T}^D - \mathbf{X}^D\right)\cdot\left(\mathbf{T}^D - \mathbf{X}^D\right) - \frac{1}{3}k^2 = 0 \tag{3.36c}$$

Flow rule

$$\dot{\mathbf{E}}_p(t) = \begin{cases} \lambda\left(\mathbf{T}^D - \mathbf{X}^D\right) & \text{for } f = 0 \text{ and } \dfrac{\partial f}{\partial \mathbf{T}}\cdot\dot{\mathbf{T}} > 0 \text{ (loading)} \\ \mathbf{0} & \text{for all other cases} \end{cases} \tag{3.36d}$$

Hardening model (example)

$$\dot{\mathbf{X}}(t) = c\,\dot{\mathbf{E}}_p(t) - b\,\dot{s}(t)\mathbf{X} \tag{3.36e}$$

Plastic arclength (rate of the accumulated plastic strain)

$$\dot{s}(t) = \sqrt{\frac{2}{3}\dot{\mathbf{E}}_p(t)\cdot\dot{\mathbf{E}}_p(t)} = \frac{2}{3}\lambda k \tag{3.36f}$$

The proportionality factor λ in the associated flow rule is not a material constant: It must be determined such that the flow rule is identically satisfied when plastic deformations occur.
Evaluation of the consistency condition $\frac{d}{dt} f(\mathbf{T}(t), \mathbf{X}(t)) = 0$ leads to

$$\lambda = \frac{\mu(\mathbf{T}^D - \mathbf{X}^D) \cdot \dot{\mathbf{E}}}{\frac{1}{3} k\{k(2\mu + c) - b\mathbf{X} \cdot (\mathbf{T}^D - \mathbf{X}^D)\}} = \frac{\mu(\mathbf{T}^D - \mathbf{X}^D) \cdot \dot{\mathbf{E}}}{N(\mathbf{T}, \mathbf{X})} . \tag{3.36g}$$

The combination of these different ingredients of the model leads to a homogeneous relation between strain rates and stress rates:

$$\dot{\mathbf{T}}^D = \begin{cases} 2\mu\left[\dot{\mathbf{E}}^D - \mu \dfrac{(\mathbf{T}^D - \mathbf{X}^D) \cdot \dot{\mathbf{E}}}{N(\mathbf{T}, \mathbf{X})} (\mathbf{T}^D - \mathbf{X}^D)\right] & \text{(loading)} \\ 2\mu \dot{\mathbf{E}}^D & \text{(all other cases)} \end{cases} \tag{3.37a}$$

$$\frac{1}{3} \operatorname{tr}\dot{\mathbf{T}} = K \operatorname{tr}\dot{\mathbf{E}} \tag{3.37b}$$

The model of rate independent plasticity can be rewritten on the basis of a new independent variable z, defined through

$$z(t) = \frac{1}{\mu} \int_0^t (\mathbf{T}^D - \mathbf{X}^D) \cdot \dot{\mathbf{E}} \, d\tau \iff \dot{z}(t) = \frac{1}{\mu} (\mathbf{T}^D - \mathbf{X}^D) \cdot \dot{\mathbf{E}} .$$

According to the flow rule (3.37a) and eq. (3.36g) $\dot{z}(t)$ is proportional to the rate of the accumulated plastic strain (plastic arclength),

$$\dot{z}(t) = \frac{3}{2} \frac{N(\mathbf{T}, \mathbf{X})}{\mu^2 k} \dot{s}(t) ,$$

where the proportionality factor depends on the present state. The new independent variable z may be interpreted as a transformed arclength. With the definitions of the derivatives with respect to z ,

$$\frac{1}{\dot{z}(t)} \dot{\mathbf{E}}_p(t) = \mathbf{E}_p'(z) , \quad \frac{1}{\dot{z}(t)} \dot{\mathbf{X}}(t) = \mathbf{X}'(z) , \tag{3.38}$$

an arclength representation of the constitutive model is derived:

$$\mathbf{T}^D = 2\mu(\mathbf{E}^D - \mathbf{E}_p) \tag{3.39a}$$

$$\mathbf{E}_p'(z) = \frac{\mu^2}{N(\mathbf{T},\mathbf{X})}(\mathbf{T}^D - \mathbf{X}^D)$$

$$\mathbf{E}_p'(z) = \frac{\mu^2}{\overline{N}(\mathbf{E}^D, \mathbf{E}_p, \mathbf{X})}\left[2\mu(\mathbf{E}^D - \mathbf{E}_p) - \mathbf{X}^D\right] \tag{3.39b}$$

$$\mathbf{X}'(z) = c\,\frac{\mu^2}{N(\mathbf{T},\mathbf{X})}\left[c(\mathbf{T}^D - \mathbf{X}^D) - \tfrac{2}{3}bk\mathbf{X}\right]$$

$$\mathbf{X}'(z) = \frac{\mu^2}{\overline{N}(\mathbf{E}^D, \mathbf{E}_p, \mathbf{X})}\left[c(2\mu(\mathbf{E}^D - \mathbf{E}_p) - \mathbf{X}^D) - \tfrac{2}{3}bk\mathbf{X}\right] \tag{3.39c}$$

These equations are a special case of the general formulation

$$\mathbf{T} = \mathbf{f}(\mathbf{E}, q_1, \dots q_N)\ , \tag{3.40a}$$

$$q_k'(z) = f_k(\mathbf{E}(z), q_1(z), \dots q_N(z)) \qquad k = 1, \dots, N\ . \tag{3.40b}$$

In eqs. (3.39) the internal variables $q_k(z)$ are identified with the plastic strain $\mathbf{E}_p$ and the hardening variable $\mathbf{X}$. With equations (3.39) and (3.40) an implicit definition of a constitutive functional is realized, which seems to be similar to the original definition (3.11); however, in this case the evolution equations are not related to the time t but to a transformed arclength, which is related to the plastic arclength. This corresponds to the well known fact that in rate independent plasticity the stress-strain relation is a rate independent functional of the strain history. It is well known that the arclength parametrization is necessary and sufficient for the rate independence of functionals. The rate independence implies that neither relaxation nor creep effects can be represented within the theory of plasticity; therefore each present state is also an equilibrium solution of the constitutive equations. However, in contrast to viscoelasticity, these equilibrium states depend both on the current strain and on the past history of the strain process.

3.3.6 Viscoplasticity

The representation of rate dependent material behavior, including also quasistatic hysteresis effects, corresponds to the theory of viscoplasticity. The model of viscoplasticity is the most general constitutive model of mechanics, because all real materials show rate dependence as well as static hysteresis behavior.

In this section two alternative concepts will be outlined, namely the model of viscoplasticity in its usual sense and the model of viscoelastoplasticity.

Viscoplasticity

A particular class of constitutive relations of viscoplasticity, widely accepted in the literature, is formulated for small strains through the following definitions:

$$\mathbf{E} = \mathbf{E}_e + \mathbf{E}_i \tag{3.41a}$$

$$\mathbf{T}^D = 2\mu\, \mathbf{E}_e^D$$

$$\frac{1}{3}\operatorname{tr}\mathbf{T} = K \operatorname{tr}(\mathbf{E}_e) \qquad K = \frac{2\mu(1+\nu)}{3(1-2\nu)} \tag{3.41b}$$

$$f(\mathbf{T},\mathbf{X}) = \frac{1}{2}(\mathbf{T}^D - \mathbf{X}^D)\cdot(\mathbf{T}^D - \mathbf{X}^D) - \frac{1}{3}k^2 \tag{3.41c}$$

$$\dot{\mathbf{E}}_i(t) = \frac{1}{\eta}\langle f(\mathbf{T},\mathbf{X})\rangle(\mathbf{T}^D - \mathbf{X}^D) \tag{3.41d}$$

$$\dot{\mathbf{X}} = c\,\dot{\mathbf{E}}_i(t) - b\,|\dot{\mathbf{E}}_i(t)|\,\mathbf{X} \tag{3.41e}$$

$$|\dot{\mathbf{E}}_i(t)| = \sqrt{\frac{2}{3}\operatorname{tr}(\dot{\mathbf{E}}_i^2)} \tag{3.41f}$$

$$\langle f(\mathbf{T},\mathbf{X})\rangle = \begin{cases} f(\mathbf{T},\mathbf{X}) & \text{for } f(\mathbf{T},\mathbf{X}) > 0 \\ 0 & \text{for } f(\mathbf{T},\mathbf{X}) \leq 0 \end{cases} \tag{3.41g}$$

The essential difference to the preceding model of plasticity is the flow rule (3.41d): Inelastic strains are produced for strictly positive values of the yield function $f(\mathbf{T}, \mathbf{X})$, whereas in plasticity those values are excluded by definition. According to eq. (3.41d) the inelastic strain rate depends on the distance $f(\mathbf{T}, \mathbf{X}) > 0$ from the static yield surface $f(\mathbf{T}, \mathbf{X}) = 0$; the static yield surface represents a domain in the stress space, where purely elastic response occurs. Summarizing the formal structure of the above constitutive model, we note that this model is a special case of the general implicit definition of a functional utilizing a system of differential equations (3.11b) with respect to time as independent variable:

$$\mathbf{T} = \mathbf{f}(\mathbf{E}, q_1, \ldots, q_N)$$

$$\dot{q}_k(t) = f_k(\mathbf{E}, q_1, \ldots, q_N)\,, \quad k = 1, \ldots, N$$

Viscoelastoplasticity

In the following alternative formulation a viscoelastic overstress is superimposed on an equilibrium stress, which is a rate independent functional of the total strain history. Originally, this model was proposed by KORZEN [14, 17]. It was further developed as well as experimentally identified by LION [15]. The rate dependent functional for the overstress is implicitly defined by the differential equation

$$\dot{\mathbf{T}} = \mathfrak{G}_0[\dot{\mathbf{E}}] - \frac{1}{z_0 M(\|\mathbf{T}^D - \mathbf{T}^D_\infty\|)}(\mathbf{T}^D - \mathbf{T}^D_\infty)\,, \tag{3.42a}$$

where the nonlinearity is incorporated by means of the scaling function

$$\sigma \longmapsto M(\sigma)\,, \quad \sigma = \|\mathbf{T}^D - \mathbf{T}^D_\infty\| \tag{3.42b}$$

with the properties $M(0) = 1$ and $M'(\sigma) \geq 0$.

$\mathfrak{G}_0$ is the spontaneous elasticity tensor, consisting in the isotropic case of two elasticity constants:

$$\mathfrak{G}_0[\dot{\mathbf{E}}] = G_{10}\dot{\mathbf{E}}^D + G_{20}(\mathrm{tr}\dot{\mathbf{E}})\mathbf{1} \tag{3.42c}$$

The equilibrium stress $\mathbf{T}_\infty$ is a rate independent functional of the strain history,

$$\mathbf{T}_\infty = \underset{\tau \leq t}{\mathfrak{X}}[\mathbf{E}(\tau)]\,, \tag{3.42d}$$

defined implicitly within the usual framework of rate independent plasticity: With the yield function

$$f(\mathbf{T}_\infty, \mathbf{X}) = \frac{1}{2}(\mathbf{T}_\infty^D - \mathbf{X}^D)\cdot(\mathbf{T}_\infty^D - \mathbf{X}^D) - \frac{1}{3}k^2 \tag{3.42e}$$

we have for $f(\mathbf{T}_\infty, \mathbf{X}) = 0$ and $(\mathbf{T}_\infty^D - \mathbf{X}^D)\cdot\dot{\mathbf{T}}_\infty^D > 0$ (loading) the following constitutive equations:

$$\dot{\mathbf{T}}_\infty = \mathbb{G}_\infty[\dot{\mathbf{E}}] - \frac{G_{1\infty}^2(\mathbf{T}_\infty^D - \mathbf{X}^D)\cdot\dot{\mathbf{E}}}{N(\mathbf{T}_\infty, \mathbf{X})}(\mathbf{T}_\infty^D - \mathbf{X}^D) \tag{3.42f}$$

$$\dot{\mathbf{X}}_1 = c_1\left(\dot{\mathbf{E}}^D - \frac{1}{G_{1\infty}}\dot{\mathbf{T}}_\infty^D\right) - b_1\left\|\dot{\mathbf{E}}^D - \frac{1}{G_{1\infty}}\dot{\mathbf{T}}_\infty^D\right\|\mathbf{X}_1 \tag{3.42g}$$

$$\dot{\mathbf{X}}_2 = c_2\left(\dot{\mathbf{E}}^D - \frac{1}{G_{1\infty}}\dot{\mathbf{T}}_\infty^D\right) - b_2\left\|\dot{\mathbf{E}}^D - \frac{1}{G_{1\infty}}\dot{\mathbf{T}}_\infty^D\right\|\mathbf{X}_2 \tag{3.42h}$$

$$\mathbf{X} = \mathbf{X}_1 + \mathbf{X}_2 \tag{3.42i}$$

$$N(\mathbf{T}_\infty, \mathbf{X}) = \frac{2}{3}k^2\sum_{i=1}^{2}\left\{c_i - \sqrt{\frac{2}{3}}\frac{b_i}{k}(\mathbf{T}_\infty^D - \mathbf{X}^D)\cdot\mathbf{X}_i\right\} \tag{3.42j}$$

$\mathbb{G}_\infty$ is the equilibrium elasticity tensor with components $G_{1\infty}$ and $G_{2\infty}$.
For $f(\mathbf{T}_\infty, \mathbf{X}) < 0$ or $(\mathbf{T}_\infty^D - \mathbf{X}^D)\cdot\dot{\mathbf{T}}_\infty^D \leq 0$ (unloading or neutral loading) the equilibrium elasticity relation holds:

$$\dot{\mathbf{T}}_\infty = \mathbb{G}_\infty[\dot{\mathbf{E}}] = G_{1\infty}\dot{\mathbf{E}}^D + G_{2\infty}(\mathrm{tr}\dot{\mathbf{E}})\mathbf{1}, \quad \dot{\mathbf{X}}_i = \mathbf{0}, \quad i = 1,2 \tag{3.42k}$$

We note that this constitutive model has neither the structure of eqs. (3.11) nor the structure of eqs. (3.40): Some evolution equations refer to the time, whereas the other are related to the plastic arclength as independent variable.

3.4 Asymptotic behavior for slow processes

In this section the asymptotic behavior of rate dependent constitutive models for slow deformation processes will be analysed in more detail. The objective of this analysis is to understand a principal thermodynamic point of view, which can be inferred already from the isothermal or purely mechanical case: If the independent kinematic variables of a constitutive model (i.e. the strains in the isothermal case) are kept constant after an arbitrary process, the dependent variables will develop in time and tend asymptotically to equilibrium solutions, i.e. an equilibrium state is approached during a relaxation process. Many constitutive models have the following property: If a kinematic process is sufficiently slow, all dependent variables (i.e. the solution of the differential equations) remain within a close neighborhood of their equilibrium values. In fact, this is the consequence of a stability property. In this context two possibilities exist:

- The equilibrium values are unique functions of the present state of strain. In this case infinitely slow (quasistatic) processes are called **reversible processes**, whereas processes with a finite rate are called **irreversible**. This notion of irreversibility is related to the non-existence of static hysteresis effects. By definition, all models within the theory of viscoelasticity have this property.
- The present deformation does not determine uniquely the equilibrium values of the dependent variables; these depend on the past history of the strain process. In this case a quasistatic hysteresis exists, and it is impossible to approximate reversible processes by slow deformations. This basic property is encountered in all models of plasticity and viscoplasticity.

3.4.1 Viscoelasticity

As already mentioned, the theory of viscoelasticity does not allow for static hysteresis effects: Sufficiently slow (quasistatic) processes are asymptotically elastic or - in the language of thermodynamics - reversible. This was outlined in section 3.3.4; following the ideas of COLEMAN [4] and COLEMAN and GURTIN [5], it is straight forward to generalize the mechanical theory of viscoelasticity to thermodynamics (see chapter 4).

3.4.2 Viscoelastoplasticity

The model of viscoelastoplasticity, discussed in section 3.3.6, incorporates static hysteresis properties. This is evident from the construction of the model: For slow deformation processes the overstress, defined by eqs. (3.42a-c), vanishes and the total stress is approximately equal to the equilibrium stress, defined as a rate independent functional of the strain history. The rate independent functional for the equilibrium stress, implicitly defined through equations (3.42e) to (3.42k), was constructed within the theory of rate independent plasticity. From this construction and from section 3.3.5 it is clear that all differential equations of rate independent plasticity are related to an arclength instead of the time as independent variable: They have the structure of equations (3.40) instead of (3.11). This is a fundamental difference: In contrast to viscoelasticity, reversible processes cannot be realized with slow deformations.

3.4.3 Viscoplasticity

In this section the usual model (3.41) of viscoplasticity will be discussed in the context of slow deformation processes. The following investigation is based on a slight generalization of equations (3.41):

$$\mathbf{E} = \mathbf{E}_e + \mathbf{E}_i \tag{3.43a}$$

$$\mathbf{T} = \mathbb{C}[\mathbf{E}_e] = \mathbb{C}[\mathbf{E} - \mathbf{E}_i] \tag{3.43b}$$

$$\dot{\mathbf{E}}_i(t) = \frac{1}{\eta} \langle f(\mathbf{T},\mathbf{X}) \rangle \frac{\partial f}{\partial \mathbf{T}} \tag{3.43c}$$

$$|\dot{\mathbf{E}}_i(t)| = \sqrt{\tfrac{2}{3}\operatorname{tr}(\dot{\mathbf{E}}_i^2)} \tag{3.43d}$$

$$\dot{\mathbf{X}}(t) = c\,\dot{\mathbf{E}}_i(t) - b|\dot{\mathbf{E}}_i(t)|\mathbf{X} \tag{3.43e}$$

$$\langle f(\mathbf{T},\mathbf{X}) \rangle = \begin{cases} f(\mathbf{T},\mathbf{X}) & \text{for } f(\mathbf{T},\mathbf{X}) > 0 \\ 0 & \text{for } f(\mathbf{T},\mathbf{X}) \leq 0 \end{cases} \tag{3.43f}$$

Obviously, this model is a special case of the general implicit definition (3.11) of a functional utilizing a system of differential equations:

$$\mathbf{T} = \mathbf{f}(\mathbf{E}(t), q_1(t), \ldots, q_N(t))$$

$$\dot{q}_k(t) = f_k(\mathbf{E}(t), q_1(t), \ldots, q_N(t)), \quad k = 1, \ldots, N$$

In the sequel the stability properties of the model (3.43) will be analysed. This analysis follows the ideas of KRATOCHVIL and DILLON [6], investigated and generalized by HAUPT, KAMLAH and TSAKMAKIS [12].

A basic identity for the yield function

As a first step, we differentiate the yield function $f(t) = f(\mathbf{T}(t), \mathbf{X}(t))$ and insert the elasticity relation and the evolution equation for the back stress $\mathbf{X}$:

$$\dot{f}(t) = \frac{d}{dt} f(\mathbf{T}(t), \mathbf{X}(t)) = \frac{\partial f}{\partial \mathbf{T}} \cdot \dot{\mathbf{T}}(t) + \frac{\partial f}{\partial \mathbf{X}} \cdot \dot{\mathbf{X}}(t) =$$

$$= \frac{\partial f}{\partial \mathbf{T}} \cdot \mathbb{C}\left[\dot{\mathbf{E}} - \dot{\mathbf{E}}_i(t)\right] + \frac{\partial f}{\partial \mathbf{X}} \cdot \left\{c\dot{\mathbf{E}}_i - b|\dot{\mathbf{E}}_i|\mathbf{X}\right\}$$

If the inelastic strain rate is expressed through the flow rule (3.43c), we arrive at a differential identity, which must hold for each solution of the constitutive equations, namely

$$\dot{f}(t) = \mathbb{C}\left[\frac{\partial f}{\partial \mathbf{T}}\right] \cdot \dot{\mathbf{E}}(t) - \frac{1}{\eta}\langle f(t)\rangle \left\{\frac{\partial f}{\partial \mathbf{T}} \cdot \mathbb{C}\left[\frac{\partial f}{\partial \mathbf{T}}\right] - c\,\frac{\partial f}{\partial \mathbf{X}} \cdot \frac{\partial f}{\partial \mathbf{T}} + b\left(\frac{\partial f}{\partial \mathbf{X}} \cdot \mathbf{X}\right)\left|\frac{\partial f}{\partial \mathbf{T}}\right|\right\},$$

or

$$\dot{f}(t) + \frac{1}{\eta}\, K(t)\langle f(t)\rangle = \mathbb{C}\left[\frac{\partial f}{\partial \mathbf{T}}\right] \cdot \dot{\mathbf{E}}(t) . \tag{3.44}$$

This identitiy has been discussed by KRATOCHVIL and DILLON in their paper from 1969 [6]. Following these authors we assume that the function

$$K(t) = \frac{\partial f}{\partial \mathbf{T}} \cdot \mathbb{C}\left[\frac{\partial f}{\partial \mathbf{T}}\right] - c\,\frac{\partial f}{\partial \mathbf{T}} \cdot \frac{\partial f}{\partial \mathbf{X}} + b\left(\mathbf{X} \cdot \frac{\partial f}{\partial \mathbf{X}}\right)\left|\frac{\partial f}{\partial \mathbf{T}}\right| , \tag{3.45}$$

defined for each solution of the constitutive equations (3.43), is strictly positive. In its physical consequence this requirement corresponds to a

Stability condition: $K(t) \geq K_o > 0$ (3.46)

An important implication of this condition is the following **relaxation property**:
If after an arbitrary process the strain is held constant, the yield function f(t) tends to zero:

$$\dot{\mathbf{E}}(t) = \mathbf{0} \;\Longrightarrow\; f(t) \longrightarrow 0$$

Accordingly, the inelastic strain rate tends to zero as well as the rate of the back stress:

$$\dot{\mathbf{E}}(t) = \mathbf{0} \;\Longrightarrow\; \dot{\mathbf{E}}_i(t) \longrightarrow \mathbf{0} \quad \text{and} \quad \dot{\mathbf{X}}(t) \longrightarrow \mathbf{0}$$

Therefore the solution of the constitutive model tends to an equilibrium solution, however, this equilibrium solution is not a function of the present strain: It depends on the history of the foregoing process.

An integral formulation of the differential identity (3.44), based on $f(t_0) = 0$ and $f(t) > 0$ (inelastic process), is given by

$$f(t) = \int_{t_0}^{t} \mathbf{L}(t,\tau) \cdot \mathbf{E}'(\tau)d\tau\,, \tag{3.47a}$$

where the kernel $\mathbf{L}$ is a nonlinear functional of the stress and back stress, defined by

$$\mathbf{L}(t,\tau) = \left\{ e^{\frac{1}{\eta}\int_{\tau}^{t} K(\sigma)d\sigma} \right\} \mathbf{C}\left[\frac{\partial f}{\partial \mathbf{T}}\right](\tau)\,. \tag{3.47b}$$

The time-dependence of this functional is of exponential type. The yield function may be equivalently expressed in terms of the difference history (3.9), $\mathbf{E}^t_d(s) = \mathbf{E}(t-s) - \mathbf{E}(t)$. Then, integration by parts transforms eq. (3.47a) into

$$f(t) = \mathbf{L}(t,t_o)\cdot\left[\mathbf{E}(t) - \mathbf{E}(t_o)\right] + \int_{0}^{t-t_o} \frac{\partial}{\partial s}\mathbf{L}(t,t-s) \cdot \mathbf{E}^t_d(s)ds\,. \tag{3.48}$$

This relation will now be investigated for slow deformation processes.

Slow deformation processes

On the basis of the retardation factor α $(0 < \alpha < 1)$ the **retarded strain history** is defined as follows (compare [2]):

$$\mathbf{E}^t_{d(\alpha)}(s) = \mathbf{E}^t_d(\alpha s) = \mathbf{E}(t-\alpha s) - \mathbf{E}(t) \tag{3.49}$$

In fact, this definition implies a retardation of the present strain rate: Similar to

$$\dot{\mathbf{E}}(t) = -\frac{d}{ds}\mathbf{E}^t_d(s)\Big|_{s=0} \tag{3.50}$$

we have

$$-\frac{d}{ds}\mathbf{E}^t_{d(\alpha)}(s)\Big|_{s=0} = \dot{\mathbf{E}}_{(\alpha)}(t) = \alpha\dot{\mathbf{E}}(t)\,. \tag{3.51}$$

The duration of the retarded process is enlarged according to

$$t_{o(\alpha)} = t - \frac{t - t_o}{\alpha} \,. \tag{3.52}$$

The yield function $f_{(\alpha)}(t)$ corresponding to the retarded process is calculated from equation (3.48), rewritten for slow deformation processes:

$$f_{(\alpha)}(t) = \mathbf{L}_{(\alpha)}(t,t_{o(\alpha)}) \cdot \left[\mathbf{E}(t) - \mathbf{E}(t_{o(\alpha)})\right] + \int_0^{t-t_{o(\alpha)}} \frac{\partial}{\partial s} \mathbf{L}_{(\alpha)}(t,t-s) \cdot \mathbf{E}^t_d(\alpha s)\,ds \tag{3.53}$$

The objective of the following considerations is now to determine the limit

$$\lim_{\alpha \to 0} \frac{1}{\alpha} f_{(\alpha)}(t) \,.$$

The motivation for this purpose originates from the stability condition (3.46) and the related relaxation property: One might suspect that $f_{(\alpha)}(t)$ becomes small for slow strain processes, such that an undetermined expression arises, which might tend to a unique limit as the retardation factor α goes to zero.

In order to prepare the computation of the limit, some intermediate calculations are necessary: The first term on the right hand side of (3.53), divided by α, is governed by

$$\frac{1}{\alpha} \mathbf{L}_{(\alpha)}(t,t_{o(\alpha)}) = \frac{1}{\alpha} \left\{ e^{\frac{1}{\eta}\int_{t_{o(\alpha)}}^{t} K(\sigma)\,d\sigma} \right\} \mathbf{C}\left[\frac{\partial f_{(\alpha)}}{\partial \mathbf{T}}\right].$$

An implication of the stability assumption (3.46) is the inequality

$$\frac{1}{\alpha} e^{\frac{1}{\eta}\int_{t_{o(\alpha)}}^{t} K(\sigma)\,d\sigma} \leq \frac{1}{\alpha} e^{\frac{1}{\eta} K_o \frac{t - t_o}{\alpha}} \,;$$

this tends to zero with vanishing α. Thus we have the relation

$$\lim_{\alpha \to 0} \frac{1}{\alpha} \mathbf{L}_{(\alpha)}(t, t_{o(\alpha)}) \cdot \left[\mathbf{E}(t) - \mathbf{E}(t_{o(\alpha)})\right] = 0 \,,$$

and we are left with the problem to calculate the limit

$$\lim_{\alpha\to 0} \frac{1}{\alpha} f_{(\alpha)}(t) = \lim_{\alpha\to 0} \frac{1}{\alpha} \int_0^{t-t_{o(\alpha)}} \left[\frac{\partial}{\partial s} \mathbf{L}_{(\alpha)}(t, t-s) \cdot \mathbf{E}_d^t(\alpha s)\right] ds . \tag{3.54}$$

Now, the retardation theorem of Coleman and Noll [2] can be applied to replace the retarded strain history $\mathbf{E}^t_{d(\alpha)}(s) = \mathbf{E}^t_d(\alpha s) = \mathbf{E}(t-\alpha s) - \mathbf{E}(t)$ by its Taylor expansion of first order: According to (3.21) we have for $n = 1$ the asymptotic relation

$$\mathbf{E}^t_d(\alpha s) = - \alpha s\, \dot{\mathbf{E}}(t) + \mathbf{o}(\alpha) , \tag{3.55}$$

where the asymptotic vanishing of the error term $\mathbf{o}(\alpha)$ is understood in the sense of the fading memory norm (3.19).

The retardation theorem, which was proved in the original paper [2] in the context of viscoelasticity, is also applicable in the present case, since the kernel $\frac{\partial}{\partial s}\mathbf{L}$ is of exponential type and therefore decreases with respect to an influence function of arbitrary order. Then the application of the retardation theorem leads to

$$\lim_{\alpha\to 0} \frac{1}{\alpha} f_{(\alpha)}(t) = - \lim_{\alpha\to 0} \left\{ \int_0^{t-t_{o(\alpha)}} s \frac{\partial}{\partial s} \mathbf{L}_{(\alpha)}(t, t-s) ds \right\} \cdot \dot{\mathbf{E}}(t) . \tag{3.56}$$

The assumption that the integral be finite, i.e.

$$\left\{ \int_0^{t-t_{o(\alpha)}} s \frac{\partial}{\partial s} \mathbf{L}_{(\alpha)}(t, t-s) ds \right\} \cdot \dot{\mathbf{E}}(t) < \infty ,$$

implies as a first consequence the asymptotic behavior

$$f_{(\alpha)}(t) = O(\alpha) . \tag{3.57}$$

This intermediate result corroborates the conjecture that for slow deformations the yield function and hence the inelastic strain rate becomes asymptotically small. From the differential identity (3.44), written down for slow processes,

$$\dot{f}_{(\alpha)}(t) + \frac{1}{\eta} K_{(\alpha)}(t) \langle f_{(\alpha)}(t) \rangle = \mathbb{C}\left[\frac{\partial f_{(\alpha)}}{\partial \mathbf{T}}\right] \cdot \dot{\mathbf{E}}_{(\alpha)}(t) ,$$

and from eq. (3.51) we conclude in addition

$$\dot{f}_{(\alpha)}(t) = O(\alpha) . \tag{3.58}$$

Applying integration by parts to the right hand side of eq. (3.56) yields

$$-\lim_{\alpha \to 0}\left\{\int_0^{t-t_{o(\alpha)}} s\,\frac{\partial}{\partial s}\mathbf{L}_{(\alpha)}(t,t-s)\,ds\right\} = \lim_{\alpha \to 0}\left\{\int_0^{t-t_{o(\alpha)}} \mathbf{L}_{(\alpha)}(t,t-s)\,ds\right\}, \quad (3.59)$$

i.e. the limit of

$$\int_0^{t-t_{o(\alpha)}} \mathbf{L}_{(\alpha)}(t,t-s)ds = \int_0^{t-t_{o(\alpha)}} e^{-\frac{1}{\eta}\int_0^s K_{(\alpha)}(t-\sigma)d\sigma}\,\mathbf{C}\left[\frac{\partial f_{(\alpha)}}{\partial \mathbf{T}}\right](t-s)ds$$

must be computed or, equivalently, the limit of

$$\int_0^{t-t_{o(\alpha)}} \mathbf{L}_{(\alpha)}(t,t-s)ds =$$

$$= \int_0^{t-t_{o(\alpha)}} e^{-\frac{1}{\eta}K_{(\alpha)}(t)s}\left\{e^{-\frac{1}{\eta}\int_0^s |K_{(\alpha)}(t-\sigma)-K_{(\alpha)}(t)|d\sigma}\,\mathbf{C}\left[\frac{\partial f_{(\alpha)}}{\partial \mathbf{T}}\right](t-s)\right\}ds\,. \quad (3.60)$$

The calculation is carried out on the basis of a further integration by parts:

$$\int_0^{t-t_{o(\alpha)}} e^{-\frac{1}{\eta}K_{(\alpha)}(t)s}\left\{e^{-\frac{1}{\eta}\int_0^s |K_{(\alpha)}(t-\sigma)-K_{(\alpha)}(t)|d\sigma}\,\mathbf{C}\left[\frac{\partial f_{(\alpha)}}{\partial \mathbf{T}}\right](t-s)\right\}ds =$$

$$= \frac{\eta}{K_{(\alpha)}(t)}\,\mathbf{C}\left[\frac{\partial f_{(\alpha)}}{\partial \mathbf{T}}\right](t) - \frac{\eta}{K_{(\alpha)}(t)}\,e^{-\frac{1}{\eta}\int_0^{t-t_{o(\alpha)}} K_{(\alpha)}(t-\sigma)d\sigma}\,\mathbf{C}\left[\frac{\partial f_{(\alpha)}}{\partial \mathbf{T}}\right](t_{o(\alpha)}) +$$

$$+ \frac{\eta}{K_{(\alpha)}(t)}\int_0^{t-t_{o(\alpha)}} e^{-\frac{1}{\eta}\int_0^s K_{(\alpha)}(t-\sigma)d\sigma}\,\frac{\partial}{\partial s}\mathbf{C}\left[\frac{\partial f_{(\alpha)}}{\partial \mathbf{T}}\right](t-s)ds +$$

$$- \frac{1}{K_{(\alpha)}(t)}\int_0^{t-t_{o(\alpha)}} e^{-\frac{1}{\eta}\int_0^s K_{(\alpha)}(t-\sigma)d\sigma}\left[K_{(\alpha)}(t-s)-K_{(\alpha)}(t)\right]\mathbf{C}\left[\frac{\partial f_{(\alpha)}}{\partial \mathbf{T}}\right](t-s)ds$$

(3.61)

The stability condition (3.46) implies

$$e^{\frac{1}{\eta}\int_0^s K_{(\alpha)}(t-\sigma)\,d\sigma} \le e^{\frac{1}{\eta}K_0 s} .$$

Therefore the second term on the right-hand side of eq. (3.61) vanishes with α. The third term is governed by the tensor-valued factor

$$\frac{\partial}{\partial s}\mathbb{C}\Big[\frac{\partial f_{(\alpha)}}{\partial \mathbf{T}}\Big](t-s) = -\,\mathbb{C}\Big[\frac{d}{dt}\frac{\partial f_{(\alpha)}}{\partial \mathbf{T}}\Big](t-s) = \mathbf{O}(\alpha) , \tag{3.62}$$

which goes to zero since the application of the chain rule generates stress rates

$$\frac{d}{dt}\frac{\partial f_{(\alpha)}}{\partial \mathbf{T}} = \frac{\partial^2 f_{(\alpha)}}{\partial \mathbf{T}\partial \mathbf{T}}\Big[\dot{\mathbf{T}}_{(\alpha)}\Big] + \frac{\partial^2 f_{(\alpha)}}{\partial \mathbf{T}\partial \mathbf{X}}\Big[\dot{\mathbf{X}}_{(\alpha)}\Big] ,$$

vanishing with α: $\dot{\mathbf{T}}_{(\alpha)} = \mathbf{O}(\alpha) = \dot{\mathbf{X}}_{(\alpha)}$

The last term can be discussed in the same manner: Observing the relation

$$K_{(\alpha)}(t-s) - K_{(\alpha)}(t) = \frac{d}{ds}K_{(\alpha)}(\xi) ,$$

which is valid for an appropriate ξ between 0 and s, the time derivative of $K_{(\alpha)}$ must be calculated from the definition (3.45), observing the chain rule. Again, stress rates vanishing with α occur, and the relation

$$K_{(\alpha)}(t-s) - K_{(\alpha)}(t) = O(\alpha) \tag{3.63}$$

is concluded. Accordingly, the last term vanishes as α goes to zero. Thus we arrive at the following result:

$$\lim_{\alpha \to 0}\frac{1}{\alpha}f_{(\alpha)}(t) = \lim_{\alpha \to 0}\frac{\eta}{K_{(\alpha)}(t)}\,\mathbb{C}\Big[\frac{\partial f_{(\alpha)}}{\partial \mathbf{T}}\Big]\cdot\dot{\mathbf{E}}(t) \tag{3.64}$$

This establishes the static hysteresis property of the model, namely the derivation of the theory of rate independent plasticity as the asymptotic limit of viscoplasticity for slow deformation processes.

Plasticity as an asymptotic limit of viscoplasticity

The flow rule

$$\dot{\mathbf{E}}_i(t) = \frac{1}{\eta}\left\langle f(t)\right\rangle \frac{\partial f}{\partial \mathbf{T}},$$

evaluated for slow processes, takes the form

$$\frac{1}{\alpha}\,\dot{\mathbf{E}}_{i(\alpha)}(t) = \left\{\frac{1}{\alpha\eta} f_{(\alpha)}(t)\right\} \frac{\partial f_{(\alpha)}}{\partial \mathbf{T}}, \qquad (3.65)$$

which can be interpreted as follows: According to the flow rule of viscoplasticity the inelastic strain rate tends to zero for slow processes, since the yield function f tends to zero. However, the inelastic strain rate divided by the retardation factor α, (i.e. the rescaled inelastic atrain rate) tends to a finite value. If we define this limit value as the rate of the plastic strain $\mathbf{E}_p$,

$$\lim_{\alpha > 0} \frac{1}{\alpha}\,\dot{\mathbf{E}}_{i(\alpha)}(t) = \dot{\mathbf{E}}_p(t), \qquad (3.66)$$

we get

$$\dot{\mathbf{E}}_p(t) = \left\{\frac{1}{K_{(0)}(t)}\,\mathbb{C}\left[\frac{\partial f_{(0)}}{\partial \mathbf{T}}\right]\cdot \dot{\mathbf{E}}(t)\right\} \frac{\partial f_{(0)}}{\partial \mathbf{T}} \qquad (3.67)$$

or, omitting the index zero, the flow rule

$$\dot{\mathbf{E}}_p(t) = \frac{\mathbb{C}\left[\frac{\partial f}{\partial \mathbf{T}}\right]\cdot \dot{\mathbf{E}}(t)}{\frac{\partial f}{\partial \mathbf{T}}\cdot \mathbb{C}\left[\frac{\partial f}{\partial \mathbf{T}}\right] - c\,\frac{\partial f}{\partial \mathbf{T}}\cdot\frac{\partial f}{\partial \mathbf{X}} + b\left(\mathbf{X}\cdot\frac{\partial f}{\partial \mathbf{X}}\right)\left|\frac{\partial f}{\partial \mathbf{T}}\right|}\,\frac{\partial f}{\partial \mathbf{T}}. \qquad (3.68)$$

This evolution equation coincides exactly with the flow rule of rate independent plasticity, calculated in the usual way on the basis of the consistency condition: Differentiation of $f(\mathbf{T},\mathbf{X}) = 0$ yields

$$0 = \frac{d}{dt} f(\mathbf{T}(t),\mathbf{X}(t)) = \frac{\partial f}{\partial \mathbf{T}}\cdot\dot{\mathbf{T}} + \frac{\partial f}{\partial \mathbf{X}}\cdot\dot{\mathbf{X}} =$$

$$= \frac{\partial f}{\partial \mathbf{T}}\cdot\mathbb{C}\left[\dot{\mathbf{E}} - \dot{\mathbf{E}}_p\right] + \frac{\partial f}{\partial \mathbf{X}}\cdot\left[c\,\dot{\mathbf{E}}_p(t) - b\left|\dot{\mathbf{E}}_p(t)\right|\mathbf{X}\right],$$

and combination with the associated flow rule of rate independent plasticity, $\dot{\mathbf{E}}_p = \lambda \frac{\partial f}{\partial \mathbf{T}}$, leads to

$$\mathbf{C}\left[\frac{\partial f}{\partial \mathbf{T}}\right] \cdot \dot{\mathbf{E}} - \lambda \left\{ \frac{\partial f}{\partial \mathbf{T}} \cdot \mathbf{C}\left[\frac{\partial f}{\partial \mathbf{T}}\right] - c \frac{\partial f}{\partial \mathbf{T}} \cdot \frac{\partial f}{\partial \mathbf{X}} + b \left(\mathbf{X} \cdot \frac{\partial f}{\partial \mathbf{X}}\right) \left|\frac{\partial f}{\partial \mathbf{T}}\right| \right\} = 0 .$$

This determines the factor λ according to eq. (3.68). It can be shown (see [12]) that the validity of the loading condition can be included into this analysis, i.e. in the limit we have

$$\lim_{\alpha > 0} \frac{\partial f_{(\alpha)}}{\partial \mathbf{T}} \cdot \frac{1}{\alpha} \dot{\mathbf{T}}_{(\alpha)}(t) \geq 0 . \tag{3.69}$$

The result can be interpreted as follows: At first sight the constitutive model of rate independent plasticity seems to be artificial because, according to the general structure (3.40) or the more special models (3.39) and (3.42e-k), the time t as independent variable is replaced by the arclength s. In view of its physical and thermodynamical meaning, this choice of variables and the related representation theory may appear to be rather formal. However, it was proved by the above argumentation that viscoplasticity implies plasticity and thus the arclength representation can be understood to be the asymptotic limit of the model of viscoplasticity for slow deformation processes. In contrast to rate independent plasticity the model of viscoplasticity has a natural representation within the theory of internal variables in the sense that the evolution equations determine the time rate of the internal variables as functions of their present values and the present state of strain.

4. Thermomechanical behavior of solids

In order to embed the modelling of mechanical properties into a more general theory of thermodynamic behavior, the lists of independent and dependent variables must be extended in an appropriate manner. More specifically, the strain tensor is supplemented by nonmechanical "kinematic" variables, namely the temperature and the temperature gradient. Conjugate to the temperature, the concept of entropy must be introduced. In addition to the stress tensor an energy density and a heat flux vector has to be modelled by constitutive assumptions. There are thermodynamic theories, where the entropy transport is represented by additional constitutive equations, but these will not be discussed here.

4.1 Elastic materials

The following very brief review of thermoelasticity starts with the introduction of the free energy

$$\psi = e - \Theta s \, . \tag{4.1}$$

If this energy density ψ, the specific entropy s, the heat flux vector $\mathbf{q}$ and the stress tensor $\mathbf{T}$ are assumed to be the dependent variables, a list of constitutive equations, generalizing the stress-strain relation $\tilde{\mathbf{T}} = \mathbf{f}(\mathbf{E})$ to thermodynamics, may be set up in the following form:

$$\psi = \hat{\psi}(\mathbf{E}, \Theta, \mathbf{g}_R) \tag{4.2}$$

$$\tilde{\mathbf{T}} = \hat{\tilde{\mathbf{T}}}(\mathbf{E}, \Theta, \mathbf{g}_R) \tag{4.3}$$

$$s = \hat{s}(\mathbf{E}, \Theta, \mathbf{g}_R) \tag{4.4}$$

$$\mathbf{q}_R = \hat{\mathbf{q}}_R(\mathbf{E}, \Theta, \mathbf{g}_R) \tag{4.5}$$

In these equations the postulate of objectivity has already been taken into consideration.

If the entropy inequality is invoked, the above constitutive equations of thermoelasticity turn out to be not independent from each other. Inserting the free energy into the entropy balance (2.18), the dissipation principle (entropy inequality) (2.19) reads

$$-\dot{\psi} + \frac{1}{\rho_R}\tilde{\mathbf{T}} \cdot \dot{\mathbf{E}} - \Theta\dot{s} - \frac{1}{\rho_R \Theta}\mathbf{q}_R \cdot \mathbf{g}_R \geq 0 \, . \tag{4.6}$$

This inequality must hold for all thermodynamic processes satisfying the balance relations and the constitutive equations. If the chain rule is applied to compute the total rate of change of the free energy,

$$\dot{\psi}(t) = \frac{\partial\hat{\psi}}{\partial\mathbf{E}} \cdot \dot{\mathbf{E}} + \frac{\partial\hat{\psi}}{\partial\Theta}\dot{\Theta} + \frac{\partial\hat{\psi}}{\partial\mathbf{g}_R} \cdot \dot{\mathbf{g}}_R ,$$

we get the inequality

$$-\frac{\partial\hat{\psi}}{\partial\mathbf{g}_R} \cdot \dot{\mathbf{g}}_R + \left[\frac{1}{\rho_R}\tilde{\mathbf{T}} - \frac{\partial\hat{\psi}}{\partial\mathbf{E}}\right] \cdot \dot{\mathbf{E}} - \left[\hat{s} + \frac{\partial\hat{\psi}}{\partial\Theta}\right]\dot{\Theta} - \frac{1}{\rho_R\Theta}\mathbf{q}_R \cdot \mathbf{g}_R \geq 0 , \tag{4.7}$$

which, according to a standard argumentation, must hold for arbitrary choices of the rates $\dot{\mathbf{g}}_R$, $\dot{\mathbf{E}}$ and $\dot{\Theta}$. Since the corresponding coefficients are independent from these quantities, they must vanish. This implies the following consequences:

- Independence of the free energy from the temperature gradient

$$\frac{\partial\hat{\psi}}{\partial\mathbf{g}_R} = \mathbf{0} \iff \psi = \hat{\psi}(\mathbf{E}, \Theta) \tag{4.8}$$

- Stress relation

$$\frac{1}{\rho_R}\tilde{\mathbf{T}} = \frac{\partial\hat{\psi}}{\partial\mathbf{E}}(\mathbf{E}, \Theta) \tag{4.9}$$

- Entropy relation

$$s = -\frac{\partial\hat{\psi}}{\partial\Theta}(\mathbf{E}, \Theta) \tag{4.10}$$

With these relations the entropy inequality reduces to the

- Heat conduction inequality

$$\mathbf{q}_R \cdot \mathbf{g}_R \leq 0 . \tag{4.11}$$

A further consequence is the

- Gibbs relation

$$\dot{\psi}(t) = \frac{1}{\rho_R}\tilde{\mathbf{T}} \cdot \dot{\mathbf{E}} - s\dot{\Theta} , \tag{4.12}$$

and the

- Nonexistence of a piezocaloric effect (no heat flux without temperature gradient):

$$\hat{\mathbf{q}}_R(\mathbf{E}, \Theta, \mathbf{0}) = \mathbf{0} \tag{4.13}$$

The latter result follows from the restrictive condition (4.11) for the constitutive equation of the heat flux vector, $\mathbf{g}_R \cdot \hat{\mathbf{q}}_R(\mathbf{E}, \Theta, \mathbf{g}_R) \leq 0$, which must hold identically in $\mathbf{g}_R$. This relation can be understood as a condition for a relative maximum for the function $f(\mathbf{g}_R) = \mathbf{g}_R \cdot \hat{\mathbf{q}}_R(\mathbf{g}_R)$ at $\mathbf{g}_R = \mathbf{0}$. Accordingly, the derivative must vanish:

$$\left[\hat{\mathbf{q}}_R(\mathbf{E}, \Theta, \mathbf{g}_R) + \left(\frac{\partial \hat{\mathbf{q}}_R}{\partial \mathbf{g}_R}\right)^T \mathbf{g}_R\right]\Bigg|_{\mathbf{g}_R = \mathbf{0}} = \mathbf{0}$$

This implies the relation (4.13). Summarizing the results, the properties of a thermoelastic material are completely specified if two material functions are given, namely the free energy depending on strain and temperature and the heat flux vector as a function of strain, temperature and temperature gradient, i.e.

$$\psi = \hat{\psi}(\mathbf{E}, \Theta) \tag{4.14a}$$

and

$$\mathbf{q}_R = \hat{\mathbf{q}}_R(\mathbf{E}, \Theta, \mathbf{g}_R) \, . \tag{4.14b}$$

The free energy is a thermodynamic potential: It is the potential of the strain tensor and the entropy. The relation (4.1) between the internal energy and the entropy is called a Legendre transformation. In contrast to the free energy the natural (canonical) representation for the internal energy is a constitutive function of the strain tensor and the entropy:

$$e = \hat{e}(\mathbf{E}, s) \tag{4.15}$$

This is derived from the evaluation of the entropy inequality written in terms of the internal energy, which turns out to be the potential of the stress and the temperature. There are two more thermodynamic potentials, namely the enthalpy

$$h = \frac{1}{\rho_R} \tilde{\mathbf{T}} \cdot \mathbf{E} - e = \hat{h}(\tilde{\mathbf{T}}, s) \tag{4.16}$$

and the free enthalpy

$$g = \frac{1}{\rho_R} \tilde{\mathbf{T}} \cdot \mathbf{E} - \psi = \hat{g}(\tilde{\mathbf{T}}, \Theta) \tag{4.17}$$

The enthalpy is the potential of the strain tensor and the temperature and the free enthalpy is the potential of the strain and the entropy. The four thermodynamic potentials and their basic properties are summarized in table 4.1.

Potential	Representation	Property	Gibbs Relation
Internal energy	$e = \hat{e}(\mathbf{E}, s)$	$\frac{1}{\rho_R}\tilde{\mathbf{T}} = \frac{\partial \hat{e}}{\partial \mathbf{E}}$	$\dot{e}(\mathbf{E}, s) = \frac{1}{\rho_R} \tilde{\mathbf{T}} \cdot \dot{\mathbf{E}} + \Theta \dot{s}$
		$\Theta = \frac{\partial \hat{e}}{\partial s}$	
Free energy	$\psi = e - \Theta s$	$\frac{1}{\rho_R}\tilde{\mathbf{T}} = \frac{\partial \hat{\psi}}{\partial \mathbf{E}}$	$\dot{\psi}(\mathbf{E}, \Theta) = \frac{1}{\rho_R} \tilde{\mathbf{T}} \cdot \dot{\mathbf{E}} - s\dot{\Theta}$
	$\psi = \hat{\psi}(\mathbf{E}, \Theta)$	$s = -\frac{\partial \hat{\psi}}{\partial \Theta}$	
Enthalpy	$h = \frac{1}{\rho_R} \tilde{\mathbf{T}} \cdot \mathbf{E} - e$	$\frac{1}{\rho_R}\mathbf{E} = \frac{\partial \hat{h}}{\partial \tilde{\mathbf{T}}}$	$\dot{h}(\tilde{\mathbf{T}}, s) = \frac{1}{\rho_R} \mathbf{E} \cdot \dot{\tilde{\mathbf{T}}} - \Theta \dot{s}$
	$h = \hat{h}(\tilde{\mathbf{T}}, s)$	$\Theta = -\frac{\partial \hat{h}}{\partial s}$	
Free enthalpy	$g = \frac{1}{\rho_R} \tilde{\mathbf{T}} \cdot \mathbf{E} - \psi$	$\frac{1}{\rho_R}\mathbf{E} = \frac{\partial \hat{g}}{\partial \tilde{\mathbf{T}}}$	$\dot{g}(\tilde{\mathbf{T}}, \Theta) = \frac{1}{\rho_R} \mathbf{E} \cdot \dot{\tilde{\mathbf{T}}} + s \dot{\Theta}$
	$g = \hat{g}(\tilde{\mathbf{T}}, \Theta)$	$s = \frac{\partial \hat{g}}{\partial \Theta}$	

Table 4.1 Thermodynamic potentials

It is seen from table 4.1 that for isothermal processes $(\dot{\Theta} = 0)$ the free energy coincides with the isothermal strain energy function, whereas the free enthalpy is equal to the complementary energy function. For isentropic processes the internal energy and the enthalpy reduce to the adiabatic strain energy and complementary energy, respectively.

The term *adiabatic* is related to the fact that, if the entropy is materially constant, no heat is exchanged between a material element and its neighborhood. In order to demonstrate this interpretation the energy balance is considered: Combination of (2.17),

$$\dot{e} = -\frac{1}{\rho_R}\mathrm{Div}\,\mathbf{q}_R + r + \frac{1}{\rho_R}\tilde{\mathbf{T}}\cdot\dot{\mathbf{E}}\ ,$$

with the Gibbs relation

$$\dot{e}(\mathbf{E}, s) = \frac{1}{\rho_R}\tilde{\mathbf{T}}\cdot\dot{\mathbf{E}} + \Theta\dot{s}$$

(table 4.1) yields the **equation of heat conduction**:

$$\Theta\dot{s} = -\frac{1}{\rho_R}\mathrm{Div}\,\mathbf{q}_R + r \tag{4.18}$$

Two versions of this equations are useful: With the entropy relation

$$s = -\frac{\partial\hat{\psi}}{\partial\Theta}(\mathbf{E}, \Theta)$$

the specific heat at constant deformation is defined as

$$c_d(\mathbf{E}, \Theta) = \Theta\frac{\partial s}{\partial\Theta}(\mathbf{E}, \Theta) = -\Theta\frac{\partial^2\hat{\psi}}{\partial\Theta^2}(\mathbf{E}, \Theta)\ , \tag{4.19}$$

to give

$$c_d\dot{\Theta} - \Theta\frac{\partial^2\hat{\psi}}{\partial\Theta\,\partial\mathbf{E}}\cdot\dot{\mathbf{E}} = -\frac{1}{\rho_R}\mathrm{Div}\,\mathbf{q}_R + r\ . \tag{4.20}$$

On the other hand the specific heat at constant stress, based on $s = \frac{\partial\hat{g}}{\partial\Theta}(\tilde{\mathbf{T}},\Theta)$, is defined as

$$c_s(\tilde{\mathbf{T}}, \Theta) = \Theta \frac{\partial s}{\partial \Theta}(\tilde{\mathbf{T}}, \Theta) = \Theta \frac{\partial^2 \hat{g}}{\partial \Theta^2}(\tilde{\mathbf{T}}, \Theta). \tag{4.21}$$

With this definition the second version for the equation of heat conduction is given by

$$c_s \dot{\Theta} + \Theta \frac{\partial^2 \hat{g}}{\partial \Theta \, \partial \tilde{\mathbf{T}}} \cdot \dot{\tilde{\mathbf{T}}} = -\frac{1}{\rho_R} \operatorname{Div} \mathbf{q}_R + r. \tag{4.22}$$

The term

$$\Theta \frac{\partial^2 \hat{g}}{\partial \Theta \, \partial \tilde{\mathbf{T}}} \cdot \dot{\tilde{\mathbf{T}}} \text{ or, alternatively, } \Theta \frac{\partial^2 \hat{\psi}}{\partial \Theta \, \partial \mathbf{E}} \cdot \dot{\mathbf{E}}$$

represents thermoelastic coupling effects. These terms are neglected in most applications.

4.2 Inelastic materials: Thermodynamics with Internal Variables

Similar to the purely mechanical theory, the thermodynamic behavior of inelastic materials can be modelled within the functional approach or utilizing systems of ordinary differential equations, which are evolution equations for an additional set of internal variables. Only the second approach will be outlined in the sequel.

4.2.1 Viscoelastic materials

The theory of internal variables as formulated by COLEMAN and GURTIN [5] is a generalization of thermoelasticity in the sense that a set of additional variables is introduced to represent the inelastic material behavior. These may be assumed to be the inelastic strain tensor $\mathbf{E}_i$ and a further set of quantities $q_1, q_2, \dots, q_N$ to be specified in the particular context.

Generalizing thermoelasticity, a thermoviscoelastic material may be defined through a set of constitutive equations for the free energy, the stress, the entropy and the heat flux vector as functions of the strain tensor, the temperature and N additional internal variables:

$$\psi = \hat{\psi}(\mathbf{E}, \Theta, \mathbf{g}_R, q_1, \dots, q_N) \tag{4.23}$$

$$\tilde{\mathbf{T}} = \hat{\tilde{\mathbf{T}}}(\mathbf{E}, \Theta, \mathbf{g}_R, q_1, \dots, q_N) \tag{4.24}$$

$$s = \hat{s}(\mathbf{E}, \Theta, \mathbf{g}_R, q_1, \dots, q_N) \tag{4.25}$$

$$\mathbf{q}_R = \hat{\mathbf{q}}_R(\mathbf{E}, \Theta, \mathbf{g}_R, q_1, \dots, q_N) \tag{4.26}$$

These constitutive equations are completed by a set of evolution equations

$$\dot{q}_k(t) = f_k(\mathbf{E}, \Theta, q_1, \dots, q_N) \quad k = 1, \dots, N , \tag{4.27}$$

where a dependence on the temperature gradient $\mathbf{g}_R$ has been omitted. For thermoviscoelastic materials the following properties are postulated (compare section 3.3.4):

- For each constant state of strain and temperature there are equilibrium solutions, i.e. solutions of the equations

$$f_k(\mathbf{E}, \Theta, \bar{q}_1, \dots, \bar{q}_N) = 0 , \quad k = 1, \dots, N , \tag{4.28}$$

  depending uniquely on the strain $\mathbf{E}$ and the temperature Θ:

$$\bar{q}_k = g_k(\mathbf{E}, \Theta) \tag{4.29}$$

- The equilibrium solutions $\bar{q}_k = g_k(\mathbf{E}, \Theta)$ are asymptotically stable in the large (COLEMAN and GURTIN [5]): If a static continuation of an arbitrary strain-temperature process $\{\mathbf{E}(\tau), \Theta(\tau)\}$ is given and any initial conditions for the internal variables are prescribed, the solutions $q_k(t)$ tend asymptotically to their equilibrium values $\bar{q}_k = g_k(\mathbf{E}, \Theta)$, as time t goes to infinity.

The total time derivative of the free energy,

$$\dot{\psi}(t) = \frac{\partial\hat{\psi}}{\partial\mathbf{E}} \cdot \dot{\mathbf{E}} + \frac{\partial\hat{\psi}}{\partial\Theta}\dot{\Theta} + \frac{\partial\hat{\psi}}{\partial\mathbf{g}_R} \cdot \dot{\mathbf{g}}_R + \sum_{j=1}^{N}\frac{\partial\hat{\psi}}{\partial q_j}\dot{q}_j(t) ,$$

must now be inserted into the entropy inequality (4.6),

$$-\dot{\psi} + \frac{1}{\rho_R}\tilde{\mathbf{T}}\cdot\dot{\mathbf{E}} - \Theta\dot{s} - \frac{1}{\rho_R\Theta}\mathbf{q}_R\cdot\mathbf{g}_R \geq 0 .$$

This yields the inequality

$$-\frac{\partial\hat{\psi}}{\partial\mathbf{g}_R}\cdot\dot{\mathbf{g}}_R + \left[\frac{1}{\rho_R}\hat{\tilde{\mathbf{T}}} - \frac{\partial\hat{\psi}}{\partial\mathbf{E}}\right]\cdot\dot{\mathbf{E}} - \left[\hat{s} + \frac{\partial\hat{\psi}}{\partial\Theta}\right]\dot{\Theta} - \sum_{j=1}^{N}\frac{\partial\hat{\psi}}{\partial q_j}\dot{q}_j(t) - \frac{1}{\rho_R\Theta}\mathbf{q}_R\cdot\mathbf{g}_R \geq 0 , \quad (4.30)$$

which must hold for arbitrary choices of the rates $\dot{\mathbf{g}}_R$, $\dot{\mathbf{E}}$ and $\dot{\Theta}$. According to the standard argumentation this leads to the following consequences:

- Independence of the free energy from the temperature gradient

$$\frac{\partial\hat{\psi}}{\partial\mathbf{g}_R} = \mathbf{0} \Leftrightarrow \psi = \hat{\psi}(\mathbf{E}, \Theta, q_1, \dots, q_N) \quad (4.31)$$

- Stress relation

$$\frac{1}{\rho_R}\tilde{\mathbf{T}} = \frac{\partial\hat{\psi}}{\partial\mathbf{E}}(\mathbf{E}, \Theta, q_1, \dots, q_N) \quad (4.32)$$

- Entropy relation

$$s = -\frac{\partial\hat{\psi}}{\partial\Theta}(\mathbf{E}, \Theta, q_1, \dots, q_N) \quad (4.33)$$

Therefore a thermoviscoelastic material is completely defined, if the following constitutive equations are given:

$$\psi = \hat{\psi}(\mathbf{E}, \Theta, q_1, \dots, q_N)$$

$$\mathbf{q}_R = \hat{\mathbf{q}}_R(\mathbf{E}, \Theta, \mathbf{g}_R, q_1, \dots, q_N)$$

$$\dot{q}_k(t) = f_k(\mathbf{E}, \Theta, q_1, \dots, q_N) , \quad k = 1, \dots, N$$

With the above relations the entropy inequality reduces to the

• General dissipation inequality

$$-\sum_{j=1}^{N} \frac{\partial \hat{\psi}}{\partial q_j} \dot{q}_j(t) - \frac{1}{\rho_R \Theta} \mathbf{q}_R \cdot \mathbf{g}_R \geq 0 . \tag{4.34}$$

Evaluating (4.34) for $\mathbf{g}_R = \mathbf{0}$ leads to the

• Internal dissipation inequality

$$\delta \geq 0 , \tag{4.35}$$

where

$$\delta = - \sum_{j=1}^{N} \frac{\partial \hat{\psi}}{\partial q_j} \dot{q}_j(t) \tag{4.36}$$

is the internal dissipation. A further consequence is the

• Generalized Gibbs relation:

$$\dot{\psi}(t) = \frac{1}{\rho_R} \tilde{\mathbf{T}} \cdot \dot{\mathbf{E}} - s\dot{\Theta} - \delta \tag{4.37}$$

From (4.37) and (4.35) we conclude the inequality

$$\dot{\psi}(t) \leq \frac{1}{\rho_R} \tilde{\mathbf{T}} \cdot \dot{\mathbf{E}} - s\dot{\Theta} . \tag{4.38}$$

For an isothermal static continuation of an arbitrary process $(\dot{\mathbf{E}} = \mathbf{0},\ \dot{\Theta} = 0)$ this implies

$$\dot{\psi}(t) \leq 0 : \tag{4.39}$$

The rate of change of the free energy during an isothermal static continuation is not positive. Since ψ cannot increase during isothermal static continuations, the free energy must tend to a minimum if the continuation is sufficiently long, such that a state of equilibrium is approached. In this sense we have the following

• Minimum property of the free energy

In thermodynamic equilibrium, approached through an isothermal static continuation the free energy takes a relative minimum:

$$\bar{\psi}(\mathbf{E}, \Theta) = \hat{\psi}\big(\mathbf{E}, \Theta, g_1(\mathbf{E}, \Theta), \dots, g_N(\mathbf{E}, \Theta)\big) = \text{Min.} \tag{4.40}$$

$$\frac{\partial \hat{\psi}}{\partial q_k}(\mathbf{E}, \Theta, q_1, \dots, q_N)\Big|_{q_l = \bar{q}_l = g_l(\mathbf{E}, \Theta)} = 0\,, \quad k = 1, \dots, N \tag{4.41}$$

A consequence of the stability property of the equilibrium state is the

• Relaxation property

During isothermal static continuations the stress and the entropy tend to equilibrium values; these satify the potential relations of thermoelasticity:

$$\frac{1}{\rho_R}\bar{\tilde{\mathbf{T}}} = \frac{\partial \hat{\psi}}{\partial \mathbf{E}}\left(\mathbf{E}, \Theta, \bar{q}_1, \dots, \bar{q}_N\right) = \frac{\partial \bar{\psi}}{\partial \mathbf{E}}\left(\mathbf{E}, \Theta\right) \tag{4.42}$$

$$\bar{s} = \hat{\psi}\left(\mathbf{E}, \Theta, \bar{q}_1, \dots, \bar{q}_N\right) = -\frac{\partial \bar{\psi}}{\partial \Theta}\left(\mathbf{E}, \Theta\right) \tag{4.43}$$

Similar to the considerations of section 3.3.4 it can be shown that thermodynamic equilibrium states including the potential relations of thermoelasticity are asymptotically approximated for slow strain and temperature processes.

Finally, obeying eq. (4.1) the generalized Gibbs relation (4.37) is inserted into the energy balance (2.17). This yields the

• Equation of heat conduction,

$$\Theta\dot{s} = -\frac{1}{\rho_R}\,\text{Div}\,\mathbf{q}_R + r + \delta\,. \tag{4.44}$$

With the specific heat at constant deformation,

$$c_d(\mathbf{E}, \Theta, q_1, ..., q_N) = - \Theta \frac{\partial^2 \hat{\psi}}{\partial \Theta^2}(\mathbf{E}, \Theta, q_1, ..., q_N),$$

and

$$\Theta \dot{s} = c_d \dot{\Theta} - \Theta \frac{\partial^2 \hat{\psi}}{\partial \Theta \, \partial \mathbf{E}} \cdot \dot{\mathbf{E}} - \Theta \sum_{j=1}^{N} \frac{\partial^2 \hat{\psi}}{\partial \Theta \, \partial q_j} \dot{q}_j(t)$$

the equation of heat conduction takes the form

$$c_d \dot{\Theta} - \Theta \frac{\partial^2 \hat{\psi}}{\partial \Theta \, \partial \mathbf{E}} \cdot \dot{\mathbf{E}} = - \frac{1}{\rho_R} \mathrm{Div}\, \mathbf{q}_R + r - \sum_{j=1}^{N} \frac{\partial}{\partial q_j} (\hat{\psi} + \Theta s) \dot{q}_j(t) . \tag{4.45}$$

This shows that in addition to the thermoelastic coupling term and the heat supply the temperature change is influenced by the internal dissipation.

According to eq. (4.35) the internal dissipation must be non-negative:

$$\delta = - \sum_{j=1}^{N} \frac{\partial \hat{\psi}}{\partial q_j} \dot{q}_j(t) \geq 0$$

This is a restrictive condition for the evolution equations (4.27):

$$\dot{q}_k(t) = f_k(\mathbf{E}, \Theta, q_1, ..., q_N) , \quad k = 1, ..., N$$

In the theory of viscoelasticity this condition can easily be satisfied: A simple choice for the evolution equations, which is sufficient for the fulfilment of the internal dissipation inequality (however not necessary), is given by

$$f_k(\mathbf{E}, \Theta, q_1, ..., q_N) = - \eta_k \frac{\partial \hat{\psi}}{\partial q_k} , \quad k = 1, ..., N , \tag{4.46}$$

where the η_k are positive-valued material constants. We note that this potential relation implies the symmetry of the matrix Q_{kj}, which led to the representation (3.28) of finite linear viscoelasticity.

The constitutive equations (4.31 - 33), (4.26, 27) and (4.46) include a variety of special cases and systematic possibilities to construct constitutive models of viscoelasticity, which are thermodynamically consistent in the sense that they satisfy the dissipation principle by construction. In particular, the generalization of finite linear viscoelasticity to thermodynamics (see section 3.3.4) is included in the above equations.

4.2.2 Viscoplastic materials

To concentrate on the basic physical ideas, only small strains will be considered in this section. In the next chapter a possibility for a formal generalization to arbitrarily large strains will be outlined.

The following considerations start from an additive decomposition of the strain rate tensor $\dot{\mathbf{E}}$ into elastic and inelastic parts:

$$\dot{\mathbf{E}} = \dot{\mathbf{E}}_e + \dot{\mathbf{E}}_i \tag{4.47}$$

Accordingly, the stress power decomposes, i.e.

$$w = \frac{1}{\rho}\mathbf{T}\cdot\dot{\mathbf{E}} = \frac{1}{\rho}\mathbf{T}\cdot\dot{\mathbf{E}}_e + \frac{1}{\rho}\mathbf{T}\cdot\dot{\mathbf{E}}_i = w_e + w_i\,. \tag{4.48}$$

Similar to thermoelasticity two other thermodynamic quantities are defined on the basis of the specific internal energy e by Legendre transformations, namely the **free energy** and the **free enthalpy**:

$$\psi = e - \Theta s \tag{4.49}$$

$$g = \frac{1}{\rho}\mathbf{T}\cdot\mathbf{E}_e - \psi \tag{4.50}$$

It is important to recognize that the free enthalpy is defined on the basis of the elastic part $\mathbf{E}_e$ of the total strain. For the free energy the following general constitutive equation is now postulated (compare (4.31)):

$$\psi = \hat{\psi}\left(\mathbf{E}_e, \Theta, q_1, \ldots, q_N\right) \tag{4.51}$$

Here, it is assumed that the free energy depends on the total strain $\mathbf{E}$, the temperature Θ and the internal variables $\mathbf{E}_i, q_1, \ldots, q_N$; however, the dependence on $\mathbf{E}$ and $\mathbf{E}_i$ is restricted to the dependence on the difference $\mathbf{E} - \mathbf{E}_i = \mathbf{E}_e$. Standard arguments suggest that the free energy is the potential of the stress tensor and the entropy (compare section 4.2.1):

$$\frac{1}{\rho}\mathbf{T} = \frac{\partial \hat{\psi}}{\partial \mathbf{E}_e} \tag{4.52}$$

$$s = -\frac{\partial \hat{\psi}}{\partial \Theta} \tag{4.53}$$

A corresponding constitutive assumption for the free enthalpy is related to the stress-space formulation. In this representation the free enthalpy is the potential of the elastic strain and the temperature (compare table 4.1.) :

$$g = \hat{g}\left(\mathbf{T}, \Theta, q_1, \ldots, q_N\right) \tag{4.54}$$

$$\frac{1}{\rho}\mathbf{E}_e = \frac{\partial \hat{g}}{\partial \mathbf{T}} \tag{4.55}$$

$$s = \frac{\partial \hat{g}}{\partial \Theta} \tag{4.56}$$

The entropy inequality reduces to the remaining **general dissipation inequality**,

$$\frac{1}{\rho}\mathbf{T}\cdot\dot{\mathbf{E}}_i + \sum_{k=1}^{N}\frac{\partial \hat{g}}{\partial q_k}\dot{q}_k - \frac{1}{\rho\Theta}\mathbf{q}\cdot\mathbf{g} \geq 0\,, \tag{4.57a}$$

which for $\mathbf{g} = \mathbf{0}$ implies the **internal dissipation inequality**

$$\frac{1}{\rho}\mathbf{T}\cdot\dot{\mathbf{E}}_i + \sum_{k=1}^{N}\frac{\partial \hat{g}}{\partial q_k}\dot{q}_k \geq 0\,. \tag{4.57b}$$

The energy balance implies (c.f. eq. (4.44)

$$\Theta\dot{s} = -\frac{1}{\rho}\operatorname{div}\mathbf{q} + r + \frac{1}{\rho}\mathbf{T}\cdot\dot{\mathbf{E}}_i + \sum_{k=1}^{N}\frac{\partial \hat{g}}{\partial q_k}\dot{q}_k\,. \tag{4.58}$$

With the specific heat at constant stress,

$$c_s\left(\mathbf{T}, \Theta, q_1, \ldots, q_N\right) = \Theta\frac{\partial^2 \hat{g}}{\partial \Theta^2}\,, \tag{4.59}$$

the following version of the **equation of heat conduction** is derived (compare eq. (4.45) and eq. (4.22)):

$$c_s\dot{\Theta} + \Theta \frac{\partial^2 \hat{g}}{\partial\Theta\,\partial\mathbf{T}} \cdot \dot{\mathbf{T}} = -\frac{1}{\rho}\,\mathrm{div}\,\mathbf{q} + r + \frac{1}{\rho}\mathbf{T}\cdot\dot{\mathbf{E}}_i + \sum_{k=1}^{N}\frac{\partial}{\partial q_k}(g - \Theta s)\dot{q}_k \tag{4.60}$$

This equation states that (apart from the thermoelastic coupling on the left-hand side) the change of temperature is caused by the local heating, the inelastic stress power and, finally, the change of the internal variables expressed in terms of the enthalpy h = g - Θs.

The above constitutive equations are completed by evolution equations for the inelastic strain rate and the other internal variables:

$$\dot{\mathbf{E}}_i(t) = \mathbf{F}(\mathbf{T}, \Theta, q_1, \ldots, q_N) \tag{4.61}$$

$$\dot{q}_k(t) = f_k(\mathbf{T}, \Theta, q_1, \ldots, q_N) \tag{4.62}$$

The whole set of evolution equations (4.61, 62) together with a constitutive equation for the heat flux vector $\mathbf{q}$ must satisfy the inequalities (4.57), which are restrictive conditions for the constitutive equations specifying $\dot{\mathbf{E}}_i$, $\dot{q}_k$ and $\mathbf{q}$.

The mathematical characteristics of these evolution equations must be less restrictive in comparison to viscoelasticity. This results from the existence of static hysteresis effects, which are essential for viscoplasticity and impossible in viscoelasticity. In chapter 3 it was shown that evolution equations of viscoplasticity are as well characterized by certain stability properties. In the next section the analysis of section 3.4.3 will be generalized to thermodynamics.

4.3 Thermoplasticity as an asymptotic limit of thermoviscoplasticity

The theory of internal variables is qualified for a consistent representation of viscoelastic as well as viscoplastic behavior. A thermoviscoplastic material is completely defined, if the following constitutive equations are specified: A free enthalpy (or free energy) function, a flow rule, evolution equations for further internal variables and, finally, a constitutive law for the heat flux vector. For the following discussion it suffices to assume a special case of viscoplasticity, namely a free enthalpy function

$$\hat{g}\,(\mathbf{T}, \Theta, q_1, \ldots, q_N) = \hat{g}\,(\mathbf{T}, \Theta, \mathbf{X})\ , \tag{4.63a}$$

such that relation (4.55) is compatible with the linear thermoelasticity relation

$$\dot{\mathbf{T}}(t) = \mathbb{C}\left[\dot{\mathbf{E}}_e(t)\right] + \mathbf{C}\dot{\Theta}(t)\,, \tag{4.63b}$$

where $\mathbb{C}$ and $\mathbf{C}$ are fourth and second order elasticity tensors, (compare (3.43b)). The evolution equations for the internal variables are assumed to be the associated flow rule,

$$\dot{\mathbf{E}}_i(t) = \frac{1}{\eta}\left\langle f(\mathbf{T}, \mathbf{X}, \Theta)\right\rangle \frac{\partial f}{\partial \mathbf{T}}\,, \tag{4.63c}$$

and the Armstrong-Frederick hardening model $\left(|\dot{\mathbf{E}}_i| = \sqrt{\frac{2}{3}\mathrm{tr}\left(\dot{\mathbf{E}}_i\right)^2}\right)$:

$$\dot{\mathbf{X}} = c\,\dot{\mathbf{E}}_i - b\,|\dot{\mathbf{E}}_i|\,\mathbf{X}\,. \tag{4.63d}$$

η, c and b are further material constants (compare (3.43c-e). The yield function f, appearing in the flow rule is a scalar-valued material function of stress and temperature. The condition $f(\mathbf{T}, \mathbf{X}, \Theta) = 0$ defines a temperature-dependent surface in the stress space, the **static yield surface**. A special case would be the v. Mises yield function $\left(\mathbf{T}^D\text{: deviator of } \mathbf{T}, k\text{: uniaxial static yield stress}\right)$

$$f(\mathbf{T}, \mathbf{X}, \Theta) = \frac{1}{2}\left(\mathbf{T}^D - \mathbf{X}^D\right)\cdot\left(\mathbf{T}^D - \mathbf{X}^D\right) - \frac{1}{3}k^2(\Theta)\,.$$

In this case $\mathbf{X}$ is the center of the static yield surface. Usually, $\mathbf{X}$ is called the **back stress** tensor. The brackets $\langle\ \rangle$ in the flow rule (4.63c) are defined according to eq. (3.41g):

$$\langle x\rangle = \begin{cases} x \text{ for } x \geq 0 \\ 0 \text{ for } x < 0 \end{cases}$$

Thus the static yield surface forms the boundary of the elastic region, where the inelastic strain $\mathbf{E}_i$ is constant in time. Inelastic strains are produced for stresses outside the static yield surface, i.e. for positive values of the yield function f. If a positive value of f is interpreted as a distance from the static yield surface, the inelastic strain rate is proportional to that distance and the parameter η corresponds to the **viscosity** of the material. Finally, the center of the static yield surface may move according to the evolution equation (4.63d) or another hardening rule (compare section 3.3.6).

Obviously, the above constitutive model of a thermoviscoplastic material is a generalization of the model (3.43) to thermodynamics. On the other hand it is a special case of the general theory of internal variables. To recognize this, we identify the back stress tensor $\mathbf{X}$ with the internal variables $q_1, \ldots, q_N$ and insert the flow rule into the hardening model.

Since the model (4.63) of thermoviscoplasticity is a straight forward generalization of the model (3.43), the analysis carried out in section 3.4.3 for the case of slow processes can be transferred to the thermodynamic situation. This will be sketched very briefly in the sequel. It may be noted that for these considerations (as well as for those of section 3.4.3) it is not important that the associated flow rule has been assumed: In the flow rule (4.63c) the gradient of the yield function may be replaced by a general tensor valued material function $\mathbf{G} = \mathbf{G}(\mathbf{T}, \mathbf{X}, \Theta)$. The special choice of an associated flow rule is merely a simple possibility to guarantee the stability property of the constitutive model.

Now, let a strain and temperature process

$$\mathbf{\Lambda}(t) = \{\mathbf{E}(t), \Theta(t)\} \tag{4.64}$$

be given and initial conditions for $\mathbf{E}_i$, $\mathbf{T}$ and $\mathbf{X}$ be prescribed such that the static yield condition is satisfied for $t = t_o$. We assume that the set of constitutive equations determines a unique solution $\mathbf{E}_i(t)$, $\mathbf{T}(t)$ and $\mathbf{X}(t)$. Accordingly, the yield function f is also determined as a function of time, i.e.

$$f(t) = f\left(\mathbf{T}(t), \mathbf{X}(t), \Theta(t)\right).$$

In complete analogy to section 3.4.3 the relation between viscoplasticity and plasticity can be studied, if we calculate the total time derivative $\dot{f}(t)$. Observing the chain rule, the decomposition of the total strain rate and the constitutive assumptions, we arrive at the identity

$$\dot{f}(t) + \frac{1}{\eta} K(t) \langle f(t) \rangle = \mathbf{\Sigma}(t) \cdot \dot{\mathbf{\Lambda}}(t) \,, \tag{4.65}$$

where the following definitions have been introduced:

$$K(t) = \frac{\partial f}{\partial \mathbf{T}} \cdot \mathbf{C}\left[\frac{\partial f}{\partial \mathbf{T}}\right] - c \, \frac{\partial f}{\partial \mathbf{T}} \cdot \frac{\partial f}{\partial \mathbf{X}} + b \left(\mathbf{X} \cdot \frac{\partial f}{\partial \mathbf{X}}\right) \left| \frac{\partial f}{\partial \mathbf{T}} \right| \tag{4.66}$$

$$\mathbf{\Sigma}(t) = \left\{\mathbb{C}\left[\frac{\partial f}{\partial \mathbf{T}}\right], \mathbf{C} \cdot \frac{\partial f}{\partial \mathbf{T}} + \frac{\partial f}{\partial \Theta}\right\} \tag{4.67}$$

$$\mathbf{\Sigma}(t) \cdot \dot{\mathbf{\Lambda}}(t) = \mathbb{C}\left[\frac{\partial f}{\partial \mathbf{T}}\right] \cdot \dot{\mathbf{E}}(t) + \left(\mathbf{C} \cdot \frac{\partial f}{\partial \mathbf{T}} + \frac{\partial f}{\partial \Theta}\right)\dot{\Theta}(t) \tag{4.68}$$

It is important to note that K and $\mathbf{\Sigma}$ depend on the current state only. For inelastic processes ($f > 0$) we have

$$\dot{f}(t) + \frac{1}{\eta} K(t) f(t) = \mathbf{\Sigma}(t) \cdot \dot{\mathbf{\Lambda}}(t) \ , t \geq t_o \ , \tag{4.69}$$

or, equivalently,

$$f(t) = \int_{t_o}^{t} e^{-\frac{1}{\eta}\int_{\tau}^{t} K(\sigma) d\sigma} \mathbf{\Sigma}(\tau) \cdot \dot{\mathbf{\Lambda}}(\tau) d\tau \ . \tag{4.70}$$

The above equations correspond exactly to eqs. (3.44) and (3.47). Therefore they can be analysed in the same way on the basis of the stability condition

$$K(t) \geq K_o > 0 \ . \tag{4.71}$$

The first result is the **relaxation property**: A static continuation of an arbitrary strain-temperature process implies $\dot{\mathbf{\Lambda}} = \{\mathbf{0}, 0\}$ and, because of the stability condition, f(t) tends to zero. Consequently, both the inelastic strain rate $\dot{\mathbf{E}}_i$ and the rate $\dot{\mathbf{X}}$ of the back stress tend to zero. Thus an equilibrium solution of the system (4.63) of differential equations is approached during a relaxation process at constant strain and temperature. As a characteristic feature of viscoplasticity this equilibrium state depends on the foregoing process history: This is due to the particular structure of the flow rule and the evolution equation for the back stress $\mathbf{X}$.

In complete analogy with the purely mechanical case it can be shown that for inelastic processes corresponding to very small strain and temperature rates the values of f(t) will be very close to zero. In the asymptotic limit the flow rule of viscoplasticity transforms into the flow rule of rate independent thermoplasticity, computed in the usual way from the consistency condition: Starting from the definition (3.66),

$$\dot{\mathbf{E}}_p(t) = \lim_{\alpha \to 0} \frac{1}{\alpha} \dot{\mathbf{E}}_{i(\alpha)}(t) = \lim_{\alpha \to 0} \left\{\frac{1}{\alpha\eta} f_{(\alpha)}(t)\right\} \frac{\partial f_{(\alpha)}}{\partial \mathbf{T}} \ ,$$

the flow rule

$$\dot{\mathbf{E}}_p = \lambda \frac{\partial f}{\partial \mathbf{T}} \tag{4.72}$$

is obtained, with

$$\lambda = \frac{\mathbf{C}\left[\frac{\partial f}{\partial \mathbf{T}}\right] \cdot \dot{\mathbf{E}}(t) + \left(\mathbf{C} \cdot \frac{\partial f}{\partial \mathbf{T}} + \frac{\partial f}{\partial \Theta}\right)\dot{\Theta}(t)}{\left\{\frac{\partial f}{\partial \mathbf{T}} \cdot \mathbf{C}\left[\frac{\partial f}{\partial \mathbf{T}}\right] - c\,\frac{\partial f}{\partial \mathbf{T}} \cdot \frac{\partial f}{\partial \mathbf{X}} + b\left(\mathbf{X} \cdot \frac{\partial f}{\partial \mathbf{X}}\right)\left|\frac{\partial f}{\partial \mathbf{T}}\right|\right\}} \tag{4.73}$$

(compare (3.68)). The result may be summarized in the following conclusion: It is physically meaningful to formulate a theory of rate independent thermoplasticity in terms of the arclength representation, i.e. through constitutive equations of the following structure:

$$g = \hat{g}(\mathbf{T}, \Theta, q_1, \dots, q_N) \tag{4.74a}$$

$$\dot{\mathbf{E}}_p = \lambda \frac{\partial f}{\partial \mathbf{T}} \tag{4.74b}$$

$$q_k'(s) = f_k\left(\mathbf{T}, \frac{d\mathbf{T}}{ds}, \Theta, q_1, \dots, q_N\right) \tag{4.74c}$$

for $f(\mathbf{T}, \Theta, q_1, \dots, q_N) = 0$ and loading

$$\dot{\mathbf{E}}_p = \mathbf{0}\,, \quad q_k'(s) = 0 \quad \text{for all other cases} \tag{4.74d}$$

Here, s is the plastic arclength (accumulated plastic strain) or eventually a transformed arclength (see [16]).

It was shown that constitutive equations of the structure (4.74) can be regarded to be the asymptotic limit of models like eqs. (4.54), (4.61) and (4.62) for slow thermodynamic processes.

5. Thermoplasticity at finite strains

In this chapter a formulation of rate independent thermoplasticity is developed, which incorporates large elastoplastic deformations. The basic kinematic formalism of this presentation was developed in [9, 10, 11]. The physical motivation may be taken from the preceding chapter.

5.1 Intermediate Configuration

The general theory of rate independent elastoplasticity and thermoplasticity is based on the multiplicative decomposition of the deformation gradient:

$$\mathbf{F} = \hat{\mathbf{F}}_c \mathbf{F}_p \tag{5.1}$$

In fact, all relevant decompositions of the deformation into elastic and plastic parts can be derived from this formula (see [11]). The decomposition of the deformation gradient can be imagined as a result of a local unloading process, which does not take place in reality and which leads to the concept of an intermediate configuration (see figure 5.1). The plastic part $\mathbf{F}_p(\mathbf{X},t)$ is not the gradient of a vector field.
Furthermore, the rotational part $\mathbf{R}_p$ of $\mathbf{F}_p$ $(\mathbf{F}_p = \mathbf{R}_p\mathbf{U}_p)$ is not unique, because

$$\hat{\mathbf{F}}_c\mathbf{F}_p = \hat{\mathbf{F}}_c\mathbf{Q}^T\mathbf{Q}\mathbf{F}_p$$

holds for all orthogonal tensors $\mathbf{Q}$ In principle, it can be assumed that the plastic Green strain is determined by the past history of the total strain and the temperature (see [9]):

$$\mathbf{E}_p = \frac{1}{2}(\mathbf{F}_p^T\mathbf{F}_p - \mathbf{1}) = \underset{\tau \leq t}{\mathfrak{P}}\left[\mathbf{E}(\tau), \Theta(\tau)\right] \tag{5.2}$$

Another interpretation of eq. (5.1) is that a material element relaxes into a non-Euclidean space, characterized by a metric with a non-vanishing tensor of curvature. The metric of this space corresponds to a field of plastic strain, which does not satisfy the compatibility conditions.

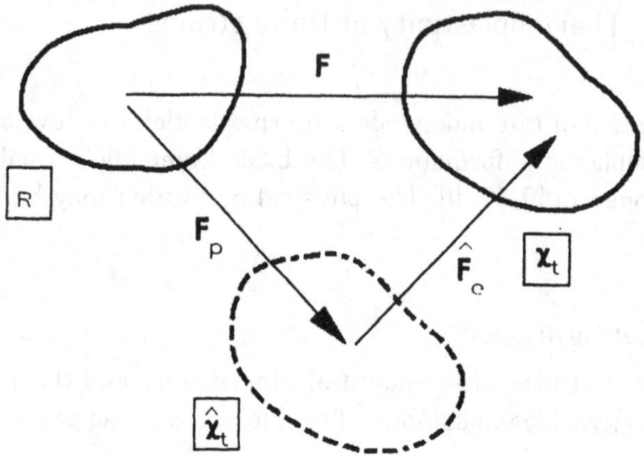

Fig. 5.1: Intermediate Configuration

In the sequel the concept of dual variables is applied as it was developed by HAUPT and TSAKMAKIS [11]. Dual variables are generally defined as affine transformations of the Green strain tensor

$$\mathbf{\Pi} = \mathbf{\Psi}^{T-1}\mathbf{E}\,\mathbf{\Psi}^{-1}, \tag{5.3}$$

$$\mathbf{E} = \frac{1}{2}\left(\mathbf{F}^T\mathbf{F} - \mathbf{1}\right), \tag{5.4}$$

and the 2nd Piola-Kirchhoff stress tensor

$$\mathbf{\Sigma} = \mathbf{\Psi}\,\tilde{\mathbf{T}}\,\mathbf{\Psi}^T, \tag{5.5}$$

$$\tilde{\mathbf{T}} = (\det\mathbf{F})\,\mathbf{F}^{-1}\mathbf{T}\,\mathbf{F}^{T-1} = \mathbf{F}^{-1}\mathbf{S}\,\mathbf{F}^{T-1}. \tag{5.6}$$

$\mathbf{S}$ is the weighted Cauchy stress tensor. The associated dual strain and stress rates are given by the same transformations:

$$\overset{\triangle}{\mathbf{\Pi}} = \mathbf{\Psi}^{T-1}\dot{\mathbf{E}}\,\mathbf{\Psi}^{-1} = \dot{\mathbf{\Pi}} + \left(\dot{\mathbf{\Psi}}\mathbf{\Psi}^{-1}\right)^T\mathbf{\Pi} + \mathbf{\Pi}\left(\dot{\mathbf{\Psi}}\mathbf{\Psi}^{-1}\right) \tag{5.7}$$

$$\overset{\nabla}{\boldsymbol{\Sigma}} = \boldsymbol{\Psi}\,\dot{\tilde{\mathbf{T}}}\,\boldsymbol{\Psi}^T = \dot{\boldsymbol{\Sigma}} - \left(\dot{\boldsymbol{\Psi}}\boldsymbol{\Psi}^{-1}\right)\boldsymbol{\Sigma} - \boldsymbol{\Sigma}\left(\dot{\boldsymbol{\Psi}}\boldsymbol{\Psi}^{-1}\right)^T \tag{5.8}$$

These tensors and time rates define the stress power,

$$w = \frac{1}{\rho_R}\tilde{\mathbf{T}} \cdot \dot{\mathbf{E}} = \frac{1}{\rho_R}\boldsymbol{\Sigma} \cdot \overset{\Delta}{\boldsymbol{\Pi}}\,, \tag{5.9}$$

the complementary stress power

$$w_c = \frac{1}{\rho_R}\dot{\tilde{\mathbf{T}}} \cdot \mathbf{E} = \frac{1}{\rho_R}\overset{\nabla}{\boldsymbol{\Sigma}} \cdot \boldsymbol{\Pi}\,, \tag{5.10}$$

and the incremental stress power

$$w_i = \frac{1}{\rho_R}\dot{\tilde{\mathbf{T}}} \cdot \dot{\mathbf{E}} = \frac{1}{\rho_R}\overset{\nabla}{\boldsymbol{\Sigma}} \cdot \overset{\Delta}{\boldsymbol{\Pi}}\,, \tag{5.11}$$

all being invariants with respect to affine transformations of the form (5.3) and (5.5), respectively. In the following, all dual tensorial quantities will be referred to the intermediate configuration. This corresponds to the choice $\boldsymbol{\Psi} = \mathbf{F}_p$.
The total strain tensor, operating on the intermediate configuration, is given by

$$\hat{\boldsymbol{\Gamma}} = \mathbf{F}_p^{T-1}\,\mathbf{E}\,\mathbf{F}_p^{-1} = \frac{1}{2}\left(\hat{\mathbf{F}}_e^T\,\hat{\mathbf{F}}_e - \mathbf{F}_p^{T-1}\,\mathbf{F}_p^{-1}\right). \tag{5.12}$$

The dual stress tensor

$$\hat{\mathbf{S}} = \mathbf{F}_p\,\tilde{\mathbf{T}}\,\mathbf{F}_p^T = \hat{\mathbf{F}}_e^{-1}\,\mathbf{S}\,\hat{\mathbf{F}}_e^{T-1} \tag{5.13}$$

is the 2nd Piola Kirchhoff stress tensor, referred to the intermediate configuration. The dual strain and stress rates are Oldroyd derivatives, calculated on the basis of the plastic "velocity gradient"

$$\hat{\mathbf{L}}_p = \dot{\mathbf{F}}_p\mathbf{F}_p^{-1}\,: \tag{5.14}$$

$$\overset{\Delta}{\hat{\boldsymbol{\Gamma}}} = \mathbf{F}_p^{T-1}\left(\mathbf{F}_p^T\,\hat{\boldsymbol{\Gamma}}\,\mathbf{F}_p\right)^{\cdot}\,\mathbf{F}_p^{-1} = \dot{\hat{\boldsymbol{\Gamma}}} + \hat{\mathbf{L}}_p^T\hat{\boldsymbol{\Gamma}} + \hat{\boldsymbol{\Gamma}}\hat{\mathbf{L}}_p \tag{5.15}$$

$$\overset{\nabla}{\hat{\mathbf{S}}} = \mathbf{F}_p\left(\mathbf{F}_p^{-1}\hat{\mathbf{S}}\mathbf{F}_p^{T-1}\right)^{\cdot}\mathbf{F}_p^T = \dot{\hat{\mathbf{S}}} - \hat{\mathbf{L}}_p\hat{\mathbf{S}} - \hat{\mathbf{S}}\hat{\mathbf{L}}_p^T \tag{5.16}$$

The definition of the total strain tensor related to the intermediate configuration implies the additive decomposition of the total strain into elastic and plastic parts; eq. (5.12) is equivalent to

$$\hat{\mathbf{\Gamma}} = \hat{\mathbf{E}}_e + \hat{\mathbf{A}}_p, \tag{5.17}$$

with the elastic Green strain

$$\hat{\mathbf{E}}_e = \frac{1}{2}\left(\hat{\mathbf{F}}_e^T\hat{\mathbf{F}}_e - \mathbf{1}\right) \tag{5.18}$$

and the plastic Almansi strain

$$\hat{\mathbf{A}}_p = \frac{1}{2}\left(\mathbf{1} - \mathbf{F}_p^{T-1}\mathbf{F}_p^{-1}\right). \tag{5.19}$$

The dual derivative of eq. (5.17) is calculated to

$$\overset{\triangle}{\hat{\mathbf{\Gamma}}} = \overset{\triangle}{\hat{\mathbf{E}}}_e + \overset{\triangle}{\hat{\mathbf{A}}}_p = \hat{\mathbf{D}}_e + \hat{\mathbf{D}}_p, \tag{5.20}$$

with the elastic strain rate

$$\hat{\mathbf{D}}_e = \overset{\triangle}{\hat{\mathbf{E}}}_e = \dot{\hat{\mathbf{E}}}_e + \hat{\mathbf{L}}_p^T\hat{\mathbf{E}}_e + \hat{\mathbf{E}}_e\hat{\mathbf{L}}_p \tag{5.21}$$

and the plastic strain rate

$$\hat{\mathbf{D}}_p = \overset{\triangle}{\hat{\mathbf{A}}}_p = \dot{\hat{\mathbf{A}}}_p + \hat{\mathbf{L}}_p^T\hat{\mathbf{A}}_p + \hat{\mathbf{A}}_p\hat{\mathbf{L}}_p = \frac{1}{2}\left(\hat{\mathbf{L}}_p + \hat{\mathbf{L}}_p^T\right). \tag{5.22}$$

5.2 Stress Power

As a consequence of the decomposition (5.20) of the strain rate, the stress power decomposes additively into elastic and plastic parts:

$$w = w_c + w_p \tag{5.23}$$

$$w_c = \frac{1}{\rho_R} \hat{\mathbf{S}} \cdot \hat{\mathbf{D}}_c \tag{5.24}$$

$$w_p = \frac{1}{\rho_R} \hat{\mathbf{S}} \cdot \hat{\mathbf{D}}_p \tag{5.25}$$

In particular, the stress power as well as its elastic part can be expressed in terms of the complementary stress power:

$$w = \frac{1}{\rho_R} \hat{\mathbf{S}} \cdot \overset{\triangle}{\hat{\Gamma}} = \frac{1}{\rho_R} (\hat{\mathbf{S}} \cdot \hat{\Gamma})^{\cdot} - \frac{1}{\rho_R} \overset{\triangledown}{\hat{\mathbf{S}}} \cdot \hat{\Gamma} \tag{5.26}$$

$$w_c = \frac{1}{\rho_R} \hat{\mathbf{S}} \cdot \hat{\mathbf{D}}_c = \frac{1}{\rho_R} (\hat{\mathbf{S}} \cdot \hat{\mathbf{E}}_c)^{\cdot} - \frac{1}{\rho_R} \overset{\triangledown}{\hat{\mathbf{S}}} \cdot \hat{\mathbf{E}}_c \tag{5.27}$$

This latter formula is important in view of Legendre transformations, which will be applied in the next section.

5.3 Thermodynamic Potentials

The decomposition of the stress power influences the energy balance and the entropy inequality: The balance of energy (2.17) reads

$$\dot{e} = w_c + w_p - \frac{1}{\rho_R} \operatorname{Div} \mathbf{q}_R + r\,, \tag{5.28}$$

where the decomposition (5.23) has been applied. The entropy inequality (c.f.(2.18)) is given by

$$-\dot{e} + \Theta \dot{s} + w_c + w_p - \frac{1}{\rho_R \Theta} \mathbf{q}_R \cdot \operatorname{Grad} \Theta \geq 0\,. \tag{5.29}$$

The considerations of chapter 4 motivate the assumption that the intermediate configuration is a stable equilibrium state. Then the definition of thermodynamic potentials is analogous to thermoelasticity. The important difference to thermoelasticity is the fact that the intermediate state (i.e. the plastic strain) must be incorporated as a parameter and that additional internal variables are included to allow for hardening effects and other inelastic material properties. In fact, all these parameters remain constant during purely elastic processes or within the elastic domain, which is limited by means of a yield surface. In this sense a constitutive equation for the internal energy can be specified as

$$e = \hat{e}\left(\hat{\mathbf{E}}_c, s, \hat{\mathbf{Y}}, \mu\right), \tag{5.30}$$

where $\hat{\mathbf{Y}}$ and μ are tensor-valued and real-valued parameters, respectively. These parameters incorporate kinematic and isotropic hardening behavior. In particular, $\hat{\mathbf{Y}}$ is the center of the yield surface in the strain space; it is a Green strain tensor with respect to the intermediate configuration. The constitutive assumption (5.30) can be inserted into the entropy inequality (5.29). To this end the total time derivative of e has to be calculated in terms of dual derivatives, and it will be proved that the time rate of e is given by

$$\dot{e} = \frac{\partial\hat{e}}{\partial\hat{\mathbf{E}}_c}\cdot\overset{\triangle}{\hat{\mathbf{E}}}_c + \frac{\partial\hat{e}}{\partial\hat{\mathbf{Y}}}\cdot\overset{\triangle}{\hat{\mathbf{Y}}} + \frac{\partial\hat{e}}{\partial s}\dot{s} + \frac{\partial\hat{e}}{\partial\mu}\dot{\mu} - 2\left[\frac{\partial\hat{e}}{\partial\hat{\mathbf{E}}_c}\hat{\mathbf{E}}_c + \frac{\partial\hat{e}}{\partial\hat{\mathbf{Y}}}\hat{\mathbf{Y}}\right]\cdot\hat{\mathbf{D}}_p, \tag{5.31}$$

where the dual derivative

$$\overset{\triangle}{\hat{\mathbf{Y}}} = \dot{\hat{\mathbf{Y}}} + \hat{\mathbf{L}}_p^T\hat{\mathbf{Y}} + \hat{\mathbf{Y}}\hat{\mathbf{L}}_p \tag{5.32}$$

for the strain-like hardening parameter $\hat{\mathbf{Y}}$ is applied, in analogy to eq. (5.15):

To verify eq. (5.31), the generalized chain rule is applied to eq. (5.30):

$$\dot{e}(t) = \frac{d}{dt}\hat{e}\left(\hat{\mathbf{E}}_c(t), s(t), \hat{\mathbf{Y}}(t), \mu(t)\right) = \frac{\partial\hat{e}}{\partial\hat{\mathbf{E}}_c}\cdot\dot{\hat{\mathbf{E}}}_c + \frac{\partial\hat{e}}{\partial\hat{\mathbf{Y}}}\cdot\dot{\hat{\mathbf{Y}}} + \frac{\partial\hat{e}}{\partial s}\dot{s} + \frac{\partial\hat{e}}{\partial\mu}\dot{\mu} \tag{5.33}$$

Together with (5.21) and (5.32) we have

$$\dot{e}(t) = \frac{\partial\hat{e}}{\partial\hat{\mathbf{E}}_c}\cdot\overset{\triangle}{\hat{\mathbf{E}}}_c + \frac{\partial\hat{e}}{\partial\hat{\mathbf{Y}}}\cdot\overset{\triangle}{\hat{\mathbf{Y}}} + \frac{\partial\hat{e}}{\partial s}\dot{s} + \frac{\partial\hat{e}}{\partial\mu}\dot{\mu} - 2\left[\frac{\partial\hat{e}}{\partial\hat{\mathbf{E}}_c}\hat{\mathbf{E}}_c + \frac{\partial\hat{e}}{\partial\hat{\mathbf{Y}}}\hat{\mathbf{Y}}\right]\cdot\hat{\mathbf{L}}_p. \tag{5.34}$$

Now, as an alternative to eq. (5.30), the energy relation can also been stated in the form

$$e(t) = \tilde{e}\left(\mathbf{E}_c(t), s(t), \mathbf{Y}(t), \mathbf{E}_p(t), \mu(t)\right), \tag{5.35}$$

where the tensors $\mathbf{E}_e$, $\mathbf{E}_p$ and $\mathbf{Y}$ are related to the reference configuration in the sense of

$$\mathbf{E}_e = \mathbf{F}_p^T \hat{\mathbf{E}}_e \mathbf{F}_p \,, \mathbf{Y} = \mathbf{F}_p^T \hat{\mathbf{Y}} \mathbf{F}_p \,, \tag{5.36}$$

and

$$\dot{\mathbf{E}}_e = \mathbf{F}_p^T \overset{\triangle}{\mathbf{E}}_e \mathbf{F}_p \,, \dot{\mathbf{Y}} = \mathbf{F}_p^T \overset{\triangle}{\mathbf{Y}} \mathbf{F}_p \,, \tag{5.37}$$

according to eqs. (5.12) and (5.15), respectively. Differentiation of (5.35) with respect to time yields

$$\dot{e}(t) = \frac{\partial \tilde{e}}{\partial \mathbf{E}_e} \cdot \dot{\mathbf{E}}_e + \frac{\partial \tilde{e}}{\partial \mathbf{Y}} \cdot \dot{\mathbf{Y}} + \frac{\partial \tilde{e}}{\partial \mathbf{E}_p} \cdot \dot{\mathbf{E}}_p + \frac{\partial \tilde{e}}{\partial s} \dot{s} + \frac{\partial \tilde{e}}{\partial \mu} \dot{\mu} =$$

$$= \left[\mathbf{F}_p \frac{\partial \tilde{e}}{\partial \mathbf{E}_e} \mathbf{F}_p^T \right] \cdot \left(\mathbf{F}_p^{T-1} \dot{\mathbf{E}}_e \mathbf{F}_p^{-1} \right) + \cdots + \left[\mathbf{F}_p \frac{\partial \tilde{e}}{\partial \mathbf{E}_p} \mathbf{F}_p^T \right] \cdot \left(\mathbf{F}_p^{T-1} \dot{\mathbf{E}}_p \mathbf{F}_p^{-1} \right) + \cdots =$$

$$= \left[\mathbf{F}_p \frac{\partial \tilde{e}}{\partial \mathbf{E}_e} \mathbf{F}_p^T \right] \cdot \overset{\triangle}{\mathbf{E}}_e + \cdots + \left[\mathbf{F}_p \frac{\partial \tilde{e}}{\partial \mathbf{E}_p} \mathbf{F}_p^T \right] \cdot \hat{\mathbf{D}}_p + \cdots .$$

Now, in view of eq. (5.36), the first factor in brackets can be written as

$$\mathbf{F}_p \frac{\partial \tilde{e}}{\partial \mathbf{E}_e} \mathbf{F}_p^T = \frac{\partial \hat{e}}{\partial \left(\mathbf{F}_p^{T-1} \mathbf{E}_e \mathbf{F}_p^{-1} \right)} = \frac{\partial \hat{e}}{\partial \hat{\mathbf{E}}_e} \,;$$

the same is true for the other terms. With this in mind, we have as an alternative to eq. (5.34),

$$\dot{e}(t) = \frac{\partial \hat{e}}{\partial \hat{\mathbf{E}}_e} \cdot \overset{\triangle}{\mathbf{E}}_e + \frac{\partial \hat{e}}{\partial \hat{\mathbf{Y}}} \cdot \overset{\triangle}{\mathbf{Y}} + \frac{\partial \hat{e}}{\partial s} \dot{s} + \frac{\partial \hat{e}}{\partial \mu} \dot{\mu} + \left[\mathbf{F}_p \frac{\partial \tilde{e}}{\partial \mathbf{E}_p} \mathbf{F}_p^T \right] \cdot \hat{\mathbf{D}}_p \,,$$

valid for all $\hat{\mathbf{D}}_p$ and $\hat{\mathbf{L}}_p$, respectively. Comparison with eq. (5.34) yields

$$- 2\left[\frac{\partial\hat{e}}{\partial\hat{\mathbf{E}}_e}\hat{\mathbf{E}}_e + \frac{\partial\hat{e}}{\partial\hat{\mathbf{Y}}}\hat{\mathbf{Y}}\right]\cdot\left(\hat{\mathbf{D}}_p + \hat{\mathbf{W}}_p\right) = \left[\mathbf{F}_p\frac{\partial\tilde{e}}{\partial\mathbf{E}_p}\mathbf{F}_p^T\right]\cdot\hat{\mathbf{D}}_p\,,$$

which must hold for all symmetric $\hat{\mathbf{D}}_p$ and all antisymmetric $\hat{\mathbf{W}}_p$. This implies

$$\left[\frac{\partial\hat{e}}{\partial\hat{\mathbf{E}}_e}\hat{\mathbf{E}}_e + \frac{\partial\hat{e}}{\partial\hat{\mathbf{Y}}}\hat{\mathbf{Y}}\right]\cdot\hat{\mathbf{W}}_p = 0\,.$$

and completes the proof of eq. (5.31).

Applying eqs. (5.29), (5.31), (5.21) and (5.24), we now have

$$\left(-\frac{\partial\hat{e}}{\partial\hat{\mathbf{E}}_e} + \frac{1}{\rho_R}\hat{\mathbf{S}}\right)\cdot\overset{\triangle}{\hat{\mathbf{E}}}_e + \left(\Theta - \frac{\partial\hat{e}}{\partial s}\right)\dot{s} - \frac{\partial\hat{e}}{\partial\hat{\mathbf{Y}}}\cdot\overset{\triangle}{\hat{\mathbf{Y}}} - \frac{\partial\hat{e}}{\partial\mu}\dot{\mu} +$$

$$+ 2\left[\frac{\partial\hat{e}}{\partial\hat{\mathbf{E}}_e}\hat{\mathbf{E}}_e + \frac{\partial\hat{e}}{\partial\hat{\mathbf{Y}}}\hat{\mathbf{Y}}\right]\cdot\hat{\mathbf{D}}_p + w_p - \frac{1}{\rho_R\Theta}\mathbf{q}_R\cdot\operatorname{Grad}\Theta \geq 0\,. \tag{5.38}$$

This inequality can be evaluated in the usual way to give the potential relations

$$\frac{1}{\rho_R}\hat{\mathbf{S}} = \frac{\partial\hat{e}}{\partial\hat{\mathbf{E}}_e}\left(\hat{\mathbf{E}}_e, s, \hat{\mathbf{Y}}, \mu\right), \tag{5.39}$$

$$\Theta = \frac{\partial}{\partial s}\hat{e}\left(\hat{\mathbf{E}}_e, s, \hat{\mathbf{Y}}, \mu\right), \tag{5.40}$$

and the remaining dissipation inequality

$$w_p - \frac{\partial\hat{e}}{\partial\hat{\mathbf{Y}}}\cdot\overset{\triangle}{\hat{\mathbf{Y}}} - \frac{\partial\hat{e}}{\partial\mu}\dot{\mu} + 2\left[\frac{\partial\hat{e}}{\partial\hat{\mathbf{E}}_e}\hat{\mathbf{E}}_e + \frac{\partial\hat{e}}{\partial\hat{\mathbf{Y}}}\hat{\mathbf{Y}}\right]\cdot\hat{\mathbf{D}}_p - \frac{1}{\rho_R\Theta}\mathbf{q}_R\cdot\operatorname{Grad}\Theta \geq 0 \tag{5.41a}$$

$\Leftrightarrow$

$$\left\{\frac{1}{\rho_R}\hat{\mathbf{S}} + 2\left[\frac{\partial\hat{e}}{\partial\hat{\mathbf{E}}_e}\hat{\mathbf{E}}_e + \frac{\partial\hat{e}}{\partial\hat{\mathbf{Y}}}\hat{\mathbf{Y}}\right]\right\}\cdot\hat{\mathbf{D}}_p - \frac{\partial\hat{e}}{\partial\hat{\mathbf{Y}}}\cdot\overset{\triangle}{\hat{\mathbf{Y}}} - \frac{\partial\hat{e}}{\partial\mu}\dot{\mu} - \frac{1}{\rho_R\Theta}\mathbf{q}_R\cdot\operatorname{Grad}\Theta \geq 0 \tag{5.41b}$$

From thermoelasticity the concept of the Legendre transformation is well known, which leads to a change of independent variables. For technical applications it may be more convenient to have the temperature as an independent variable instead of the entropy. Moreover, the elasticity relation (5.39) can be written in terms of the elastic strain as a function of the stress. This can be achieved by introducing the free enthalpy

$$g = - e + \frac{1}{\rho_R} \hat{\mathbf{S}} \cdot \hat{\mathbf{E}}_c + \Theta s \,. \tag{5.42}$$

(compare eq. (4.50) and see also LEHMANN [7,8]). The constitutive assumption

$$g = \hat{g}(\hat{\mathbf{S}}, \Theta, \hat{\mathbf{X}}, \varkappa) \tag{5.43}$$

is most natural in this case. Here, $\hat{\mathbf{X}}$ is a stress tensor for the representation of kinematic hardening, i.e. the center of the yield surface in the stress space. It is a 2nd Piola Kirchhoff stress tensor with respect to the intermediate configuration. Differentiation of eq. (5.43) with respect to time yields

$$\dot{g}(t) = \frac{\partial \hat{g}}{\partial \hat{\mathbf{S}}} \cdot \overset{\triangledown}{\hat{\mathbf{S}}} + \frac{\partial \hat{g}}{\partial \Theta} \dot{\Theta} + \frac{\partial \hat{g}}{\partial \hat{\mathbf{X}}} \cdot \overset{\triangledown}{\hat{\mathbf{X}}} + \frac{\partial \hat{g}}{\partial \varkappa} \dot{\varkappa} + 2\left[\frac{\partial \hat{g}}{\partial \hat{\mathbf{S}}} \hat{\mathbf{S}} + \frac{\partial \hat{g}}{\partial \hat{\mathbf{X}}} \hat{\mathbf{X}}\right] \cdot \hat{\mathbf{D}}_p \,,$$

and combination of this constitutive equation with the entropy inequality (5.29) leads to

$$\left(\frac{\partial \hat{g}}{\partial \hat{\mathbf{S}}} - \frac{1}{\rho_R} \hat{\mathbf{E}}_c\right) \cdot \overset{\triangledown}{\hat{\mathbf{S}}} + \left(\frac{\partial \hat{g}}{\partial \Theta} - s\right)\dot{\Theta} + \frac{\partial \hat{g}}{\partial \hat{\mathbf{X}}} \cdot \overset{\triangledown}{\hat{\mathbf{X}}} + \frac{\partial \hat{g}}{\partial \varkappa} \dot{\varkappa} +$$

$$+ 2\left[\frac{\partial \hat{g}}{\partial \hat{\mathbf{S}}} \hat{\mathbf{S}} + \frac{\partial \hat{g}}{\partial \hat{\mathbf{X}}} \hat{\mathbf{X}}\right] \cdot \hat{\mathbf{D}}_p + w_p - \frac{1}{\rho_R \Theta} \mathbf{q}_R \cdot \mathrm{Grad}\, \Theta \geq 0 \,. \tag{5.44}$$

The usual evaluation of this inequality yields the potential relations

$$\frac{1}{\rho_R} \hat{\mathbf{E}}_c = \frac{\partial}{\partial \hat{\mathbf{S}}} \hat{g}(\hat{\mathbf{S}}, \Theta, \hat{\mathbf{X}}, \varkappa) \,, \tag{5.45}$$

$$s = \frac{\partial}{\partial \Theta} \hat{g}(\hat{\mathbf{S}}, \Theta, \hat{\mathbf{X}}, \varkappa) \tag{5.46}$$

and the remaining dissipation inequality

$$w_p + \frac{\partial \hat{g}}{\partial \hat{\mathbf{X}}} \cdot \overset{\triangledown}{\hat{\mathbf{X}}} + \frac{\partial \hat{g}}{\partial \varkappa} \dot{\varkappa} + 2\left[\frac{\partial \hat{g}}{\partial \hat{\mathbf{S}}} \hat{\mathbf{S}} + \frac{\partial \hat{g}}{\partial \hat{\mathbf{X}}} \hat{\mathbf{X}}\right] \cdot \hat{\mathbf{D}}_p - \frac{1}{\rho_R \Theta} \mathbf{q}_R \cdot \mathrm{Grad}\, \Theta \geq 0 \tag{5.47a}$$

$\Leftrightarrow$

$$\left\{\frac{1}{\rho_R}\hat{\mathbf{S}} + 2\left[\frac{\partial\hat{g}}{\partial\hat{\mathbf{S}}}\hat{\mathbf{S}} + \frac{\partial\hat{g}}{\partial\hat{\mathbf{X}}}\hat{\mathbf{X}}\right]\right\}\cdot\hat{\mathbf{D}}_p + \frac{\partial\hat{g}}{\partial\hat{\mathbf{X}}}\cdot\overset{\nabla}{\hat{\mathbf{X}}} + \frac{\partial\hat{g}}{\partial\varkappa}\dot{\varkappa} - \frac{1}{\rho_R\Theta}\mathbf{q}_R\cdot\operatorname{Grad}\Theta \geq 0\,. \tag{5.47b}$$

5.4 Equation of heat conduction

The thermoelastic potentials have to be introduced into the energy balance (5.28) in order to get the equation of heat conduction. Utilizing the free enthalpy we have

$$\Theta\dot{s} = -\frac{1}{\rho_R}\operatorname{Div}\mathbf{q}_R + r + w_p + \frac{\partial\hat{g}}{\partial\hat{\mathbf{X}}}\cdot\overset{\nabla}{\hat{\mathbf{X}}} + \frac{\partial\hat{g}}{\partial\varkappa}\dot{\varkappa} + 2\left[\frac{\partial\hat{g}}{\partial\hat{\mathbf{S}}}\hat{\mathbf{S}} + \frac{\partial\hat{g}}{\partial\hat{\mathbf{X}}}\hat{\mathbf{X}}\right]\cdot\hat{\mathbf{D}}_p\,. \tag{5.48}$$

The material derivative of the entropy is calculated to

$$\dot{s} = \frac{d}{dt}\frac{\partial\hat{g}}{\partial\Theta}(\hat{\mathbf{S}},\Theta,\hat{\mathbf{X}},\varkappa) =$$

$$= \frac{\partial^2\hat{g}}{\partial\Theta^2}\dot{\Theta} + \frac{\partial^2\hat{g}}{\partial\hat{\mathbf{S}}\partial\Theta}\cdot\overset{\nabla}{\hat{\mathbf{S}}} + \frac{\partial^2\hat{g}}{\partial\hat{\mathbf{X}}\partial\Theta}\cdot\overset{\nabla}{\hat{\mathbf{X}}} + \frac{\partial^2\hat{g}}{\partial\varkappa\partial\Theta}\dot{\varkappa} + 2\left[\frac{\partial^2\hat{g}}{\partial\hat{\mathbf{S}}\partial\Theta}\hat{\mathbf{S}} + \frac{\partial^2\hat{g}}{\partial\hat{\mathbf{X}}\partial\Theta}\hat{\mathbf{X}}\right]\cdot\hat{\mathbf{D}}_p\,.$$

Introducing the specific heat at constant stress,

$$c_p = \hat{c}_p(\hat{\mathbf{S}},\Theta,\hat{\mathbf{X}},\varkappa) = \Theta\frac{\partial^2\hat{g}}{\partial\Theta^2}\,, \tag{5.49}$$

we finally arrive at the following **equation of heat conduction:**

$$c_p\dot{\Theta} + \frac{\partial^2\hat{g}}{\partial\hat{\mathbf{S}}\partial\Theta}\cdot\overset{\nabla}{\hat{\mathbf{S}}} = -\frac{1}{\rho_R}\operatorname{Div}\mathbf{q}_R + r + \frac{\partial}{\partial\hat{\mathbf{X}}}\left[\left(\hat{g} - \Theta\frac{\partial\hat{g}}{\partial\Theta}\right)\right]\cdot\overset{\nabla}{\hat{\mathbf{X}}} + \frac{\partial}{\partial\varkappa}\left(\hat{g} - \Theta\frac{\partial\hat{g}}{\partial\Theta}\right)\dot{\varkappa} +$$

$$+ \left\{\frac{1}{\rho_R}\hat{\mathbf{S}} + 2\left[\frac{\partial}{\partial\hat{\mathbf{S}}}\left(\hat{g} - \Theta\frac{\partial\hat{g}}{\partial\Theta}\right)\right]\hat{\mathbf{S}} + 2\left[\frac{\partial}{\partial\hat{\mathbf{X}}}\left(\hat{g} - \Theta\frac{\partial\hat{g}}{\partial\Theta}\right)\right]\hat{\mathbf{X}}\right\}\cdot\hat{\mathbf{D}}_p \tag{5.50}$$

This type of heat conduction equation can also be found in the papers of LEHMANN, where a different notation is used (see, e.g. [7,8]).

5.5 Constitutive Equations

Constitutive equations should be formulated in terms of dual variables and derivatives related to the intermediate configuration [11]. In this sense, a rate independent elastoplastic material can be defined by the following set of constitutive assumptions:

- **Energy relation** $e = \hat{e}(\hat{\mathbf{E}}_e, s, \hat{\mathbf{Y}}, \mu)$ or $g = \hat{g}(\hat{\mathbf{S}}, \Theta, \hat{\mathbf{X}}, \varkappa)$

- **Yield condition** $\hat{h}(\hat{\mathbf{E}}_e, s, \hat{\mathbf{Y}}, \mu) = 0$ or $\hat{f}(\hat{\mathbf{S}}, \Theta, \hat{\mathbf{X}}, \varkappa) = 0$

- **Flow rule** e.g. $\hat{\mathbf{D}}_p = \lambda \dfrac{\partial \hat{f}}{\partial \hat{\mathbf{S}}}$ for $\hat{f} = 0$ and $\dfrac{\partial \hat{f}}{\partial \hat{\mathbf{S}}} \cdot \overset{\nabla}{\hat{\mathbf{S}}} > 0$ (loading)

- **Hardening model** e.g. $\overset{\nabla}{\hat{\mathbf{X}}} = c\, \hat{\mathbf{D}}_p - b|\hat{\mathbf{D}}_p|\, \hat{\mathbf{X}}$.

All constitutive assumptions must be compatible with the remaining dissipation inequality (5.41) or (5.47). In this context, further work has to be done.

6. References

1 TRUESDELL, C.A.; NOLL, W.: The Non-Linear Field theories of Mechanics. **Encyclopedia of Physics**, Vol. III/3, Springer 1965.

2 COLEMAN, B. D.; NOLL, W.: An Approximation Theorem for Functionals with Applications in Continuum Mechanics. **Arch. Rat. Mech. Anal. 6** (1960), 355 - 370

3 COLEMAN, B. D.; NOLL, W.: Foundations of Linear Viscoelasticity. **Reviews of Modern Physics, 33**, (1961), 239 - 249.

4 COLEMAN, B. D.: Thermodynamics of Materials with Memory. **Arch. Rat. Mech. Anal. 6** (1964), 1 - 46.

5 COLEMAN, B. D.; GURTIN, M. E.: Thermodynamics with Internal State Variables. **The Journal of Chemical Physics 47** (1967), 597 - 613.

6 KRATOCHVIL, J.; DILLON, O. W.: Thermodynamics of Elastic-Plastic Materials as a Theory with Internal State Variables. **J. Appl. Phys. 40** (1969), 3207 - 3218.

7 LEHMANN, TH.: On a Generalized Constitutive Law in Thermoplasticity. **Plasticity Today** (1983), 115 - 134.

8 LEHMANN, TH.: On Thermodynamically-Consistent Constitutive Laws in Plasticity and Viscoplasticity. **Archive of Mechanics 40** (1988), 415 - 431.

9 HAUPT, P.: On the Concept of an Intermediate Configuration and its Application to a Representation of Viscoelastic-Plastic Material Behavior. **International Journal of Plasticity 1** (1985), 303 - 316.

10 HAUPT, P.; TSAKMAKIS, CH.: On the Principle of Virtual Work and Rate-Independent Plasticity, **Archive of Mechanics 40** (1988), 403 - 414.

11 HAUPT, P.; TSAKMAKIS, CH.: On the Application of Dual Variables in Continuum Mechanics, **Journal of Continuum Mechanics and Thermodynamics 1** (1989), 165 - 196.

12 HAUPT, P.; KAMLAH, M.; TSAKMAKIS, CH.: On the Thermodynamics of Rate-Independent Plasticity as an Asymptotic Limit of Viscoplasticity for Slow Processes. In BESDO, D.; STEIN, E. (Eds.): **Finite Inelastic Deformations - Theory and Applications**, IUTAM - Symposium, Hannover 1991, Springer 1992.

13 HAUPT, P.: On the Thermodynamics of Rate-Independent Elastoplastic Materials, in: BRÜLLER, O.; MANNL, V.; NAJAR, J. (Eds.): **Advances in Continuum Mechanics**, Springer 1992.

14 KORZEN, M.: Beschreibung des inelastischen Materialverhaltens im Rahmen der Kontinuumsmechanik: Vorschlag einer Materialgleichung vom viskoelastisch-plastichen Typ. **Dissertation**, Darmstadt 1988.

15 HAUPT, P.; SCHREIBER, L.; LION, A.: Experimentelle Untersuchung des geschwindigkeitsabhängigen Materialverhaltens bei nichtradialen Belastungen. **Zeitschrift für Angewandte Mathematik und Mechanik (ZAMM)** (1993). To appear

16 HAUPT, P.; KAMLAH, M.; TSAKMAKIS, CH.: Continuous Representation of Hardening Properties in Cyclic Plasticity. To appear in the **International Journal of Plasticity 8** (1992)

17 HAUPT, P.; KORZEN, M.: On the Mathematical Modelling of Material Behavior in Continuum Mechanics. In.: JINGHONG, F.; MURAKAMI, S. (Eds.): **Advances in Constitutive Laws for Engineering Materials**, Pergamon Press, 1989.

18 LUBLINER, J.: On Fading Memory in Materials of Evolutionary Type. **Acta Mech. 8** (1969), 75 - 81

EXTENDED THERMODYNAMICS

G. Lebon
Liège University, Liège, Belgium

ABSTRACT

The basic ideas underlying Extended Irreversible Thermodynamics are reviewed. This formalism is an extension of the Classical Theory of Irreversible Thermodynamics : it consists essentially of extending the space of the state variables by including the thermodynamical fluxes, like the heat flux, the inelastic stress tensor, mass flux, ... among the set of variables. As illustrations, heat conduction in a rigid body and an elastic body will be explicitly treated. The microscopic foundations of Extended Thermodynamics are discussed in the framework of the kinetic theory of gases. It is also shown that Extended Thermodynamics is particularly well suited for obtaining the constitutive equations of rheology in general and, viscoelastic materials in particular. Comparison with Classical Irreversible Thermodynamics, Rational Thermodynamics and the Internal Variable Theory will also be outlined.

1. INTRODUCTION

After the second world war, essentially two lines of thought have been developed in the field on nonequilibrium thermodynamics. The first is known as the classical thermodynamic theory of irreversible processes, in short classical irreversible thermodynamics (CIT) [1], the second one is referred to by its founders Truesdell, Coleman and Noll as rational thermodynamics (RT) [2]. The foundations of CIT were laid down by Lars Onsager in two celebrated papers published in 1931 [3,4]. But the theory owes much of its success to Ilya Prigogine and the Brussels school of thermodynamics. It is worth recalling that both Onsager and Prigogine were awarded the Nobel prize in chemistry in 1968 and 1977 respectively.

Rational and classical thermodynamics aim at he same objective : to derive constitutive equations of material systems driven out of equilibrium; by constitutive equations are meant relations expressing the thermomechanical properties of a body in terms of basic quantities. As examples, let us mention Fourier's law of heat conduction relating the heat flux to the temperature difference or Newton's law of fluid mechanics expressing the stress tensor as a linear function of the velocity gradient.

CIT borrows several results from classical thermodynamics and is concerned with the class of states and processes described by the local equilibrium hypothesis. The latter states that the local and instantaneous relations between the thermal and mechanical properties of a material system are the same as for a uniform system at equilibrium. CIT is mainly applicable to situations "not too far" from equilibrium. More explicitly stated, is is supposed that the constitutive equations are expressed by means of linear relations between cause (also named force) and effect (the flux, according to the CIT terminology). An important pillar of the theory is provided by the Onsager symmetry relations between the transport coefficients of coupling processes. Onsager's symmetry relations were obtained from the statistical theory of gases and the theory of linear response; they were also comforted by several experimental observations mainly in thermoelectricity and thermodiffusion.

The objective of RT is more ambitious than that of CIT as it is wished to describe a wider class of materials driven far form equilibrium. But this goal is achieved at the price of a greater mathematical complexity. In RT, absolute temperature and entropy are introduced as primitive concepts without a sound physical interpretation. The notion of state, expressing that any property evaluated at time t can be written in terms of the state parameters given at the same time t is given up and replaced by the notion of history or memory. Accordingly, the properties of any system are not only determined by the values of the variables at the present time but may also depend on their values in the past. As a consequence, the constitutive equations take the form of time-functionals. The latter cannot however take any arbitrary possible form : there are restrictions placed by the second law of thermodynamics like the positiveness of the heat conductivity and the viscosity and the criterion of material frame-indifference, demanding that the constitutive equations are independent of the motion of any observer. The golden age of RT can be traced back to the sixties and seventies. The theory has met an impressive success among mathematicians and theoretical mechanicians because of its generality and its mathematical rigour.

Let us briefly report about the contents of CIT and RT.

1.1 Classical Irreversible Thermodynamics

a) Local equilibrium

The fundamental hypothesis underlying CIT is that of **local equilibrium**. It postulates that the local and instantaneous relations beween the thermal and mechanical properties of a physical system are the same as for a uniform system at equilibrium. It is assumed that the system under study can be cut into a series of cells sufficiently large to allow averaging, but sufficiently small so that equilibrium is very close to being realized in each cell.

The local equilibrium hypothesis implies that :

1. The equilibrium is stable.
2. The values defined in thermostatics remain significant. In particular, the notions of temperature and entropy are rigorously and unambiguously the same as those defined in equilibrium. Thus, entropy outside of equilibrium will depend on the same state variables as in equilibrium. This can be expressed in differential form by using the Gibbs equation, which for a one-component fluid writes as

$$\theta(\mathbf{r},t)\ ds(\mathbf{r},t) = du(\mathbf{r},t) + p(\mathbf{r},t)\ d\rho^{-1}(\mathbf{r},t). \tag{1.1}$$

$\mathbf{r}$ and t denote the position vector and the time, respectively. The notation is classical : a lower case letter designates a value related to the unit mass, u is the specific internal energy,

s the specific entropy, ρ the mass density while θ stands for the absolute temperature and p for the equilibrium pressure.

For a deformable elastic body, Gibbs equation takes the form

$$\theta \, ds = du - \frac{1}{\rho} \sigma_{ij} \, d\varepsilon_{ij} \tag{1.2}$$

wherein σ_{ij} are the Cartesian components of the Cauchy stress tensor, and ε_{ij} the Cartesian components of the strain tensor

$$\varepsilon_{ij} = \frac{1}{2} (u_{i,j} + u_{j,i}) \tag{1.3}$$

with u_i the displacement vector, a comma preceding a subscript denotes partial derivation with respect to the special coordinates. Observe that (1.2) and (1.3) are only valid for small deformations; for large deformations, expression (1.2) must be replaced by a more complicated relation to be found in chapter V. In this introductory chapter and for the sake of simplicity, we limit the analysis to infinitesimally small deformations and isotropic bodies. Einstein summation convention on repeated indices will be used throughout this work.

A precise limitation on the domain of validity of the local equilibrium hypotheses can be obtained from the kinetic theory of gases. Using Chapman-Enskog's development, Prigogine has established that the hypothesis of local equilibrium is satisfactory at the condition that the distribution function expansion is limited to the first-order approximation.

b)Entropy production

The main objective of thermodynamics is to determine the changes in displacements u_i and temperature θ. The evolution of these variables is controlled by the laws of blance of momentum and energy which, in Cartesian coordinates are given by

$$\rho \, \ddot{u}_i = \sigma_{ij,j} + \rho \, F_i \tag{1.4}$$

$$\rho \, \dot{u} = - q_{i,i} + \sigma_{ij} \, \dot{\varepsilon}_{ij} \tag{1.5}$$

with F_i the external body force and q_i the heat flux; a upper dot means the material time derivative $\left(= \partial_t + v_j \, \partial_j\right)$, in a linear theory, it reduces to ∂_t. But these laws are not sufficiently restrictive, in the sense that they do not prevent the spontaneous transfer of heat from a cold to a warm body, nor the total conversion of heat into mechanical work. Moreover, these equations do not form a complete system : they contain more unknowns than equations.

To circumvent these difficulties, it is appealed to the second law of thermodynamics. It asserts that , in absence of internal heat supply, there exists a specific entropy s whose evolution is governed by

$$\rho \dot{s} = - J^s_{i,i} + \sigma^s \tag{1.6}$$

with a positive rate of production

$$\sigma^s \geq 0 \tag{1.7}$$

J^s_i stands for the entropy flux. One of the tasks of CIT is to provide an explicit expression of σ^s and J^s_i in terms of the state variables. By combining the Gibbs equation (1.1) and the law of conservation of energy (1.5), it is an easy matter to show [3-5] that σ^s may be written in a bilinear form in so-called thermodynamic fluxes J^α and forces X^α :

$$\sigma^s = \sum_\alpha J^\alpha X^\alpha \geq 0 \tag{1.8}$$

Typical thermodynamic fluxes are the heat flux, the viscous stress tensor in fluid mechanics, the electrical current, ...while the forces take generally the form of the gradient of an intensive variable like temperature, velocity, electrical potential, etc.

c)*Flux - forces relations*

The fluxes J^α and forces X^α are not independent. Within the linear approximation, they are related by expressions of the form

$$J^\alpha = \sum_\beta L^{\alpha\beta} X^\beta \tag{1.9}$$

called phenomenological or constitutive equations, $L^{\alpha\beta}$ are the phenomenological coefficients, generally functions of the variables ρ and θ. Examples of phenomenological laws are the Ohm law in electricity, the Fourier law in heat conduction and the Stokes-Newton law in hydrodynamics. It is important to realize that the phenomenological relationships (1.9) are postulated rather than derived from expression (1.8), contrary to what is sometimes claimed.

The statement that a given flux can be coupeld to any type of force is not generally correct. It is false for isotropic systems. Using representation theorems for isotropic tensors, it can be shown that in the linear approximation, fluxes and forces which have a different tensorial character cannot be coupled. Thus, the temperature gradient, a vector, will not give rise to a mechanical stress, a tensor. The law which forbids coupling between fluxes and forces of different tensorial ranks is called the Curie law or Curie principle in the CIT literature. As Truesdell observes [5], it is redundant to invoke the name Curie or the therm "principle" to establish a result which comes directly from tensor algebra.

The coefficients $L^{\alpha\beta}$ are subject to two types of restrictions. The first concerns their sign, the second is related to their symmetry. The sign of the coefficients is the consequence of the positiveness of the entropy production from which follows that

$$L^{\alpha\alpha} > 0, \; L^{\beta\beta} > 0, \; L^{\alpha\alpha} L^{\beta\beta} > \frac{1}{4} (L^{\alpha\beta} + L^{\beta\alpha})^2 \tag{1.10}$$

At the microscopic level the equations of motion are invariant with respect to a time reversal. If t is replaced by $-t$, the molecules take up their trajectory again without deviation. When the fluxes are identified as the time rates of change of the state variables, Onsager [1] has shown the symmetry of the phenomenological coefficients :

$$L^{\alpha\beta} = L^{\beta\alpha} \tag{1.11}$$

when it is assumed that there exists linear relations between fluxes and forces. Although the result (1.11) has been established on a purely microscopic theoretical basis, it is commonly accepted as correct at the macroscopic level, even when the fluxes are not time derivatives of variables. Subsequently, it has been shown by Casimir [5] that, under some circumstances, the phenomenological coefficients can be skew symmetric.

It is worth to point out that macroscopic thermodynamics does not furnish any information about the explicit form of the phenomenological coefficients $L^{\alpha\beta}$; they must either be calculated from microscopic theories (statistical mechanics or kinetic theory) or determined experimentally.

The results (1.1) to (1.11) provide the framework of CIT.

It must be realized that CIT is unable to describe materials with memory and is not adequate for studying processes taking place far from equilibrium, in particular high frequency and short wave-length phenomena. These restrictions are inherent to the local equilibrium hypothesis. Moreover, after that the phenomenological laws of CIT are substituted in the balance laws of mass, momentum and energy, one obtains partial differential equations of the parabolic type. This means that disturbances will propagate at an infinite velocity, in contradiction with the principle of causality, which demands that the cause follows the effects : in CIT, cause and effect happen simultaneously.

1.2 Rational Thermodynamics

This formalism was essentially developed by Coleman, Truesdell and Noll [2] and follows a line of thought drastically different from CIT. Its main objective is to provide a method for deriving constitutive equations.

a) *Basic hypotheses*

The basic hypothesis underlying rational thermodynamics can be summarized as follows.

1. Absolute temperature and entropy are considered as primitive concepts. They are introduced a priori in order to ensure the coherence of the theory and do not receive a precise physical interpretation.

2. It is assumed that materials have a memory : the behaviour of a system at a given instant of time is determined not only by the values of the characteristic parameters at the present time, but also by their past history. The local equilibrium hypothesis is given up since a knowledge of the values of the parameters at the present time is not sufficient to specify unambiguously the behaviour of the system.

3. The general expressions previously formulated for the balance of mass, momentum, and energy are retained. There are however two essential nuances. The first is the introduction of a specific rate of energy supply r in the balance of internal energy which is written as

$$\rho \dot{u} = -q_{i,i} + \sigma_{ij} \dot{\varepsilon}_{ij} + \rho r \qquad (1.12)$$

r is generally referred to as the power supplied or lost by radiation, per unit mass.

The second crucial point is that the body force F_i appearing in the momentum equation and the radiation term r are not given a priori as a function of space and time, but computed from the laws of momentum and energy respectively.

4. Another capital point is the mathematical formulation of the second law of thermodynamics wich serves essentially as a restriction on the form of the constitutive equations. The starting relation is the Clausius-Planck inequality stating that between two equilibrium states A and B, one has

$$\Delta S \geq \int_A^B \frac{dQ}{\theta} . \tag{1.13}$$

In rational thermodynamics, inequality (1.13) is written as

$$\frac{d}{dt}\int_{V(t)} \rho\, s\, dV \geq -\int_{\Sigma(t)} \frac{1}{\theta} q_i\, n_i\, d\Sigma + \int_{V(t)} \rho \frac{r}{\theta} dV \tag{1.14}$$

wherein Σ (t) is the surface bounding the volume V(t) of the system, n_i the unit normal pointing outwards. In local form, expression (1.14) writes as

$$\rho \dot{s} + \left(\frac{q_i}{\theta}\right)_{,i} - \rho \frac{r}{\theta} \geq 0 \tag{1.15}$$

Introducing the Helmholtz free energy $f = u - \theta s$ and eliminating r between the energy balance (1.12) and inequality (1.15) results in

$$-\rho\,(\dot{f} + s\dot{\theta}) + \sigma_{ij}\,\dot{\varepsilon}_{ij} - \frac{1}{\theta} q_i\, \theta_{,i} \geq 0 \tag{1.16}$$

This inequality is indifferently called Clausius-Duhem or the fundamental inequality.

An important problem is of course the selection of the constitutive independent variables. This choice is subordinated to the type of material that one deals with. For deformable bodies, it is usual to take as variables the strain and temperature fields. It is also known that the balance laws and Clausius-Duhem inequality introduce complementary variables, like the internal energy, the heat flux, the stress tensor and the entropy. The latter are expressed in therms of the former by means of constitutive equations. By an admissible process is meant a solution of the balance laws when the constitutive relations are taken into account and when Clausius-Duhem inequality holds.

b)*General principles*

Before deriving the constitutive equations, let us briefly examine the main principles that they must satisfy. A word of caution is required about the use and abuse of the term "principle" in rational thermodynamics. In some cases, this term is used to designate merely convenient assumptions rather than well-founded principles.

i. The principle of equipresence

This principle asserts that, if a variable is present in one constitutive equation, it will a priori be present in all the constitutive equations. The condition for the final presence or absence of a variable in a constitutive relation is however to be determined form the Clausius-Duhem inequality.

ii. The principle of memory or heredity

This principle states that past and present causes determine present effects. As a consequence the set of independent variables is no longer formed by the variables at the present time but by their whole history. If γ is an arbitrary variable, we shall denote its history up to time t by

$$\gamma^t = \gamma\ (t - t') \qquad 0 < t' < \infty\ . \tag{1.17}$$

The principles of equipresence and memory when applied to hydrodynamics assert that

$$\left\{\begin{matrix} u\ (\text{or } f) \\ s \\ q_i \\ \sigma_{ij} \end{matrix}\right\} \text{ at } \mathbf{r},t \text{ are functionals of } \left\{\begin{matrix} \varepsilon_{ij}^t \\ \\ \theta^t \end{matrix}\right\} \tag{1.18}$$

and eventually $\theta_{,i}$.

Of course, the choice of the dependent and independent variables is not unique. One could for instance permute the roles of u and θ, but since the usual attitude is to select θ as independent quantity, we shall here follow this point of view.

iii. The principle of local action

According to this principle, the behaviour of a material point should only be influenced by its immediate neighbourhood. Otherwise stated, the values of the constitutive equation at a given point are insensitive to what happens at distant points : in a first order theory, second and higher orders space derivatives should be omitted.

iv. The principle of material frame-indifference

This principle demands that the constitutive equations are independent of the frame of reference. This means that they are form invariant under arbitrary time-dependent rotations and translations of the reference frames. This statement is equivalent to require that the constitutive relations are left unaffected by the superposition of any arbitrary rigid body motion. A typical exemple is provided by Hooke's law in elasticity : the elastic constant of a spring is the same whether the spring is at a rest or on a rotating table. It is tantamount to say that the constitutive relation of the spring will not change under superposition of a rotation. For small deformations, the principle of material frame-indifference is irrelevant and therefore will not be discussed further.

c)Illustration

We proceed further with the establishment of the constitutive equations. For simplicity, we consider a very simple material namely a thermoelastic body. It is assumed to be characterized by absence of memory and described by the following set of constitutive equations.

$$\sigma_{ij} = \sigma_{ij}(\theta, \theta_{,i}, \varepsilon_{ij})$$

$$q_i = q_i(\theta, \theta_{,i}, \varepsilon_{ij})$$

$$f = f(\theta, \theta_{,i}, \varepsilon_{ij}) \tag{1.19}$$

$$s = s(\theta, \theta_{,i}, \varepsilon_{ij})$$

By formulating (1.19), the principle of equipresence was used. A stringent constraint is imposed by the second law of thermodynamics. This is achieved by substitution fo the constitutive laws (1.19) in the Clausius-Duhem inequality (1.16). Using the chain differentiation rule for calculating $\dot{f}$, inequality (1.16) reads as

$$-\rho\left(\frac{\partial f}{\partial \theta}+s\right)\dot{\theta}-\left(\rho\frac{\partial f}{\partial \varepsilon_{ij}}-\sigma_{ij}\right)\dot{\varepsilon}_{ij}-\rho\frac{\partial f}{\partial \theta_{,i}}\dot{\theta}_{,i}-\frac{1}{\theta}q_i\,\theta_{,i}\geq 0 \qquad (1.20)$$

It is noticed that inequality (1.20) is linear in $\dot{\theta}$, $\dot{\varepsilon}_{ij}$ and $\dot{\theta}_{,i}$. Now it is assumed [13] that there exists always body forces and energy supply that ensure that the balance equations of momentum and energy are identically satisfied. Therefore, the balance laws do not impose constraints on the set $\dot{\theta}$, $\dot{\varepsilon}_{ij}$, $\dot{\theta}_{,i}$ and one can give these time derivatives arbitrary and independent values. It is then clear that (1.20) could be violated, except if the coefficients of these terms are zero. This gives the following results :

$$\frac{\partial f}{\partial \theta}=-s, \qquad \frac{\partial f}{\partial \varepsilon_{ij}}=\frac{1}{\rho}\sigma_{ij}, \qquad \frac{\partial f}{\partial \theta_{,i}}=0 \qquad (1.21)$$

The two first results are classical while the third expresses that *f* cannot depend on the temperature gradient. It follows from (1.21) that

$$f=f(\theta,\sigma_{ij})$$

or in differential form, in view of (1.21),

$$df=-s\,d\theta+\frac{1}{\rho}\sigma_{ij}\,d\varepsilon_{ij} \qquad (1.22)$$

which is nothing but Gibbs equation for a thermoelastic body. Unlike CIT where the Gibbs equation is postulated at the outset, it can be said that in rational thermodynamics the Gibbs relation is derived.

Using the results (1.21), inequality (1.20) reads as

$$-\frac{1}{\theta}\, q_i\, \theta_{,i} \geq 0 \qquad (1.23)$$

In the linear approximation, the most general expression for q_i which can be constructed from (1.19 b) is simply

$$q_i = -\lambda(\theta)\, \theta_{,i} \qquad (1.24)$$

which after substitution in (1.23) leads to

$$\lambda > 0$$

expressing that the heat condutivity is a nonnegative quantity; We recognize also in (1.24) the classical Fourier law.

The steps that have led to the establishment of the constitutive relations and the constraints on the sign of the transport coefficient λ are elegant and employ a minimum of hypotheses. At no time one does call upon symmetry relations of the Onsager type. Besides, the theory is not limited to linear constitutive equations.

We do not wish to go further in the comparison of the respective merits and drawbacks of CIT and RT : these points have been widely examined and acridly discussed elsewhere. Both formalisms have greatly contributed to the advancement of our knowledge in non-equilibrium processes; nevertheless it should be kept in mind that CIT cannot cope with processes taking place at high frequencies and that RT suffers from excessive formalisation to the detriment of physical insight.

These considerations have led some people [7-13] to propose an alternative way, which under some aspects, helps to act as intermediary between CIT and RT. This new approach is nowadays generally referred to as Extended Irreversible Thermodynamics and is the subject of the next chapter.

1.3 Extended Irreversible Thermodynamics

Extended Irreversible Thermodynamics (EIT) provides a **mesoscopic** and **causal** description of non-equilibrium processes : it was born out of the double necessity to go beyond the hypothesis of local equilibrium and to avoid the paradox of propagation of disturbances with an infinite velocity. This physical unpleasant property arises as a consequence of the substitution of the Fourier, Newton and Fick laws in the balance equations of energy, momentum and mass respectively. Indeed the resulting partial differential equations are parabolic from which follows that disturbances will be felt instantaneously in the whole space.

The treatment of a thermodynamic system, say a gas of N particles ($N \approx 10^{23}$) can be performed at three different levels of description. At the **microscopic** level, the description of each particle demands six variables (three for position, three for velocity) so that in total one needs $6N$ variables. At the **macroscopic** level in hydrodynamics only five scalar variables (the mass density, the three components of the macroscopic velocity, the temperature) are necessary, the gap between both formalisms being bridged by statistical mechanics. But some systems may require more variables than five : EIT falls precisely in such an intermediate **mesoscopic** description.

The foundations of EIT can be traced back to Maxwell's celebrated paper [14] "On the dynamical theory of gases" wherein a relaxational term was introduced in the constitutive equation of gases. The presence of such a term is also found in Grad's [15] more sophisticated and elaborated treatment of the kinetic theory of gases. Generalized Fourier's laws including a time-derivative term were also proposed by Vernotte [16] and Cattaneo [17]. Early and important contributions to EIT are found in the sixties in papers by Nettleton [9] and Müller [10]. Afterwards and curiously, the theory seems to have gone quiescent with a revival of interest after the publications of some works by Lebon [11] and Lebon-Lambermont [18]. At the present time, the theory may be considered as fully developed. For a review, the reader is advised to consult the paper by Jou, Casas-Vazquez and Lebon [19]. Our objective in this chapter is to provide a brief description of the main assumptions underlying EIT

The fundamental problem in nonequilibrium thermodynamics is from one side, the identification of the new set of variables and, from the other side, the formulation of evolution equations for these variables. Moreover, one has to take into account restrictions placed by the general laws of continuum physics, like the second law of thermodynamics, the criterion of objectivity and the requirement that the entropy is a convex function of the variables; this latter property is a consequence of stability of equilibrium.

In classical irreversible thermodynamics, the set of independent variables consists of conserved quantities C with a slow decay in time so that they are easily observed and measured on ordinary time scales. These variables satisfy respectively the evolution laws of mass, momentum and energy which can be cast in the general form

$$\rho\dot{C} = -\nabla . \mathbf{J} + \sigma^c \tag{1.25}$$

wherein C designates anyone of the quantities v (volume density), **v** (velocity vector), **u** (displacement vector), a upper dot stands for the material time derivative, **J** is the flux corresponding to the quantity C and σ^c the corresponding source term; **J** and σ^c are either vectors, tensors or scalars : for instance, in the mass balance equation **J** is equal to the velocity vector while σ^c is zero, in the energy balance **J** is the heat flux vector and σ^c the energy input plus the power associated to the deformation of the system.

To close the set of balance equations, it is usual to complement them by means of phenomenological or constitutive equations relating the fluxes **J** to the gradients of the C-variables :

$$\mathbf{J} = -\alpha \nabla C \tag{1.26}$$

The quantity α designates either the diffusion coefficient, the shear viscosity, the bulk viscosity or the heat conductivity. Equations (1.26) are the familiar laws of Fick, Newton and Fourier.

This is as far as one can go with classical irreversible thermodynamics.

The procedure followed in **extended irreversible thermodynamics** is drastically different : the three main steps are outlined below.

1st step. Enlargement of the space of independent variables

Besides the classical variables mass, momentum and energy, one includes the fluxes **J** of mass, momentum, energy, plus eventually higher order fluxes among the set of variables. The space of variables V is thus formed by the union of the classical set C and the space of the fluxes **J** :

$$V = C \cup \mathbf{J} \tag{1.27}$$

The physical nature of the **J** variables is completely different from that of the C variables. The latter are conserved and slow, as their behaviour is governed by conservation laws, moreover they decay slowly in the course of time making them easily accessible to experiments. In constrast, the flux variables are non-conserved and fast : they do generally not satisfy conservation laws and their rate of decay may be very fast. In dilute gases, it is of the order of magnitude of the collision-time τ_c between the molecules, i.e 10^{-10} s. This means that for time intervals much larger than τ_c, the rate of variation on the fast variables can be ignored. This is of course no longer true at very high frequencies or for some materials (like dielectric and polymers) wherein the time of decay of the fluxes is comparable to the time-scale of usual measurements. In some polymer solutions, it is usual to observe relaxational times of the order of 1 to 10^2 seconds, in glasses up to 10^6 seconds. This explains why EIT is particularly well suited for describing these classes of materials.

2nd step. Formulation of evolution equations for the J variables

The classical variables C satisfy the usual balance laws of mass, momentum and energy : they are well known and will therefore not raise any problem. But concerning the extra flux variables **J** , we have a priori no information about their behaviour in the course of time and space. By analogy with the classical variables, it is assumed that the **J** variables obey first-order time evolution equations of the following general form

$$\rho\,\dot{\mathbf{J}} = -\,\nabla.\mathbf{J}^F + \sigma^F \tag{1.28}$$

J designates any flux (for instance the heat flux, the flux of matter, etc...), $\mathbf{J}^F$ and σ^F denote the corresponding "flux" and "source" terms respectively. At this stage of the analysis, $\mathbf{J}^F$ and σ^F are unknown quantities; they must be expressed in terms of the whole set V of variables by means of constitutive relations

$$\mathbf{J}^F = \mathbf{J}^F(V)\,, \qquad \sigma^F = \sigma^F(V)\,. \tag{1.29}$$

The explicit form of these constitutive equations will be derived from the representation theorems of tensors and can be formulated at any order of approximation. Typical examples will be analyzed in the next chapters.

3rd step. Restrictions imposed on the evolution and constitutive equations

Restrictions on the allowable forms of the evolution and constitutive equations are placed by the laws of continuum thermodynamics. In most formulations of EIT, it is assumed that there exists a specific entropy s referred per unit mass, which is function of the whole set V of variables, and subject to the following properties :

i)s is an extensive quantity;

ii)s is a convex function of the state variables;

iii) its rate of production σ^s is positive definite, σ^s being defined by

$$\sigma^s = \rho \dot{s} + \nabla . \mathbf{J}^s \geq 0 \tag{1.30}$$

with $\mathbf{J}^s$ the entropy flux which, like s, is supposed to depend on the whole set of variables.

The entropy plays a decisive role in the kinetic and statistical interpretations of the foundations of EIT : we shall come back to this point later on. From a macroscopic point of view, the notion of entropy is fundamental because it eliminates a whole class of processes that are physically not allowed. Expression (1.30) implies the absence of internal heat supply r : when the latter is present, a term of the form $\rho r / \theta$ should be added [10].

Let us now discuss the implications of the above three statements. In view of statement *i)* entropy is a field variable that is defined at each position and at each instant of time. The second statement *ii)* is introduced to recover the classical property that equilibrium is a stable state of maximum entrop; in mathematical form it can be expressed as

$$\left(\frac{\partial^2 s}{\partial V_i \partial V_j} \right)_e < 0. \tag{1.31}$$

wherein subscript e refers to the equilibrium state. This inequality (1.31) is important because it leads to the conclusion that the relaxation times are positive quantities. This ensures that the set of evolution equations is hyperbolic, allowing the disturbances to propagate with a finite velocity, even at infinite frequencies. This result is at variance with the corresponding results of classical irreversible thermodynamics wherein the momentum equation and the diffusion equation of temperature and mass are of the parabolic type, with as consequence that the distrubances will propagate at infinite velocity. Finally it can be inferred from the third statement *iii)* expressing the positiveness of the entropy production σ^s, that the transport coefficients (heat conductivity, diffusion coefficient, shear and bulk viscosities) are non-negative.

SUMMARY : EIT is made up of the three following tenets :

1. Enlargement of the space of independent variables.
2. Formulation of evolution equations in the form of first order time-rate equations.
3. Existence of a convex entropy with a positive rate of production.

EIT reduces to CIT when the relaxation times of the fluxes vanish. EIT goes beyond classical irreversible thermodynamics by including the fluxes among the set of independent variables and formulating the second law of thermodynamics in terms of a generalized entropy which depends on the fluxes, besides tne classical variables.

EIT is a general macroscopic theory based on rather broad hpotheses providing a satisfactory description of a wide variety of systems out of equilibrium. However this generality must be moderated because the equations underlying EIT contain a number of undetermined coefficients whose expression cannot be obtained from the theory itself. Just like in the classical and the rational approaches these coefficients must be derived either experimentally or computed by means of other approaches, like kinetic theory or statistical mechanics.

EIT covers a wide variety of problems; it is particularly interesting for studying high frequency and (or) short wavelength phenomena like ultrasound propagation, diffusion through membranes, light scattering in gases, neutron scattering in liquids and solids, second sound in solids, heat pulses propagation, etc... CIT can only cope with phenomena whose transport coefficients are characterized by a zero frequency and a zero wave number; EIT appears as a valuable candidate when theses coefficients are frequency and (or) wave length dependent.

Another natural field of application of EIT is provided by the class of materials with a long relaxation time; by long relaxation time is meant a duration time accessible to modern measurement devices. Polymer solutions, non-Newtonian fluids, dielectrics, superconductors and superfluids pertain to this category. An explicit derivation of the dependence of the viscometric functions, like viscosity, first and second normal stress coefficients as a function of the shear rate has been performed [48]. Non Fickean diffusion in glassy polymers and flow-induced changes in the phase diagram of polymers have also been studied by means of EIT.

A third privileged domain of application of EIT is relativity and cosmology wherein the characteristic velocities are finite and less than the speed of light : clearly a formalism like EIT which predicts that signals propagate at finite velocities is appropriate.

From a more formal point of view, it should be recognized that the impressive progress in nonequilibrium statistical mechanics has not been parallelled by similar developments in nonequilibrium thermodynamics. This gap is now filled by EIT.

EIT also deserves attention by offering a new perspective on CIT. It clarifies the definitions of some fundamental quantities like non-equilibrium temperature and entropy, it provides also a clear limitation of the local equilibrium hypothesis [19].

2. HEAT CONDUCTION IN RIGID BODIES

Our purpose is to examine the process of heat conduction by emphasizing the aspects which are the most significant and illustrative in relation with extended irreversible thermodynamics. Two main motivations underline this analysis. The first, of theoretical nature, refers to the so-called "paradox" of propagation of thermal signals with infinite speed. The second one, more closely related to experimental observations deals with second sound in solids at low temperatures.

In the classical theory of heat conduction the basic equation is parabolic which implies an infinite speed of propagation, because the influence of a signal is felt immediately at an infinite distance. A macroscopic theory with finite speed of propagation has several motivations, coming either from the kinetic theory or from experiments. From an experimental point of view, the search for generalizing the Fourier equation was launched in the 1960's by the discovery of second sound in some dielectric solids at low temperatures. This observation stimulated also the development of microscopic models of heat conduction supporting the generalized macroscopic transport equations. Such analyses are of great interest in solid-state physics.

The need for generalisation is rather evident. Consider an isolated system which is initially composed of two subsystems at different temperatures. At a given moment, they are put in thermal contact. Equilibrium thermodynamics predicts that the final equilibrium state will be that of equal temperatures. However, it does not give any information about the evolution from the initial to the final states, which could be either a pure relaxation or an oscillatory approach. Though in the oscillatory case heat would flow from cold to hot regions in several time intervals, this is not in contradiction with Clausius' principle, which is of a global nature. The problem arises when one tries to give a local formulation (in time and in space) of the second law. A local version of the Clausius principle would be in contradition with the oscillatory mode, which is observed in second sound. Two points of view may then be adopted : either one insists on the global meaning of the second law, or one tries to formulate a new local version of the second law compatible with experimental observations. The latter point of view has been undoubtedly been more fruitful than the first one, and is one of the motivations of extended irreversible thermodynamics.

2.1 The basic equations

i) *The state variables*

According to the hypotheses of EIT, an isotropic rigid solid is locally defined by the following two variables : the specific internal energy per unit mass *u* and the heat flux vector **q** .

$$V = u \cup \mathbf{q}$$

ii) *The evolution equations*

The evolution equation for *u* is the classical energy balance equation : in absence of energy supply, it writes as

$$\rho\, \dot{u} = - q_{i,i} \tag{2.1}$$

wherein a upper dot stands here for the partial time derivative.

To obtain an evolution equation for the heat flux q_i, we asssume that this flux variable obeys the following first-order time evolution equation, which is reminiscent of the balance equation (2.1) for the classical variable :

$$\rho\, \dot{q}_i = - J^q_{ij,j} + \sigma_i^q \tag{2.2}$$

J_{ij}^q denotes the "flux" of q_i, σ_i^q the corresponding "source" term. At this stage of the analysis, J^q_{ij} and σ^q_i are unknown quantities to be expressed in terms of the whole set of variables by means of constitutive relations :

$$J_{ij}^q = J_{ij}^q\,(u, q_i)$$
$$\sigma_i^q = \sigma_i^q\,(u, q_i) \tag{2.3}$$

The explicit form of these constitutive equations will be derived from the representation theorems of tensors and can be formulated at any order of approximation. For instance at the lowest order of approximation, the relevant constitutive equations are simply

$$J_{ij}^q\,(u, q_i) = a\,(u)\,\delta_{ij}\,,$$
$$\sigma_i^q\,(u, q_i) = b\,(u)\,q_i\,, \tag{2.4}$$

wherein *a(u)* and *b(u)* are undetermined functions of the energy *u* while δ_{ij} is the Kronecker symbol.

For further purpose, let us introduce the non-equilibrium temperature defined, by analogy with classical thermodynamics, as

$$\theta^{-1} = \left(\frac{\partial s}{\partial u}\right)_q \tag{2.5}$$

In contrast with classical thermodynamics, θ depends not only on *u* but also on the flux q_i :

$$\theta = \theta\,(u, q_i) \tag{2.6}$$

Resolving this relation with respect to *u* yields

$$u = u\,(\theta, q_i) \tag{2.7}$$

Substituting (2.4) in the evolution equation (2.2) for q_i and eliminating *u* by means of (2.7) results in

$$\rho\,\dot{q}_i = -\frac{\partial a}{\partial \theta}\theta_{,i} + b\,q_i \tag{2.8}$$

when terms of the second order $q_{i,j}\,q_j$ coming from $\frac{\partial a}{\partial q_j}\,q_{j,i}$ are omitted. By setting

$$b = -\frac{\rho}{\tau} \qquad \frac{\partial a}{\partial \theta} = \frac{\rho\lambda}{\tau}$$

one recovers the familiar Vernotte-Cattaneo equation expressing the time evolution of the heat flux vector, namely

$$\tau\,\dot{q}_i = -\lambda\,\theta_{,i} - q_i \tag{2.9}$$

τ is the relaxation time of the heat flux. Since τ is of the order 10^{-10} s, the non-steady term in (2.9) can be neglected for most situations of practical interest and expression (2.9) reduces then to the Fourier law

$$q_i = -\lambda\,\theta_{,i} \tag{2.10}$$

from which follows that λ can be identified with the heat conductivity.

iii)*Restrictions placed by the second law and the convexity requirement*

Restrictions on the possible forms of evolution equations will be provided by the second law of thermodynamics expressing that σ^s is a non-negative quantity :

$$\sigma^s \equiv \rho\,\dot{s} + J^s_{i,i} \geq 0 \tag{2.11}$$

The quantities s and J^s_i have to be expressed in terms of the basic variables :

$$s = s\,(u,q_i) \tag{2.12}$$

$$J^s_i = J^s_i\,(u,\,q_i) \tag{2.13}$$

As before, it is convenient to eliminate in (2.13) the variable u on behalf on θ by means of (2.7). This operation results in

$$J^s_i = J^s_i\,(\theta,\,q_i) \tag{2.14}$$

The most general form for J^s_i compatible with a linear theory is simply

$$J^s_i = \Lambda\,(\theta)\,q_i \tag{2.15}$$

wherein $\Lambda\,(\theta)$ is an unknown fonction of θ. In view of (2.10) and (2.15), expression (2.11) of the rate of entropy prodcution may be written as

$$\sigma^s \equiv \rho\frac{\partial s}{\partial u}\dot{u} + \rho\frac{\partial s}{\partial q_i}\dot{q}_i + \frac{\partial\Lambda}{\partial\theta}\theta_{,i}\,q_i + \Lambda\,q_{i,i} \geq 0 \tag{2.16}$$

The most general expression for $\dfrac{\partial s}{\partial q_i}$ reads in the linear approximation as

$$\frac{\partial s}{\partial q_i} = - \beta (\theta) q_i \tag{2.17}$$

with β an undetermined coefficient. Using the definition (2.5) for θ, the energy balance (2.1) and the evolution equation (2.9) for q_i, (2.16) takes the form

$$- \frac{1}{\theta} q_{i,i} + \frac{\rho}{\tau} \beta q_i (\lambda \theta_{,i} + q_i) + \frac{\partial \Lambda}{\partial \theta} \theta_{,i} q_i + \Lambda q_{i,i} \geq 0 \tag{2.18}$$

or, rearranging the various terms,

$$- q_{i,i} \left(\frac{1}{\theta} - \Lambda\right) + \frac{\rho \beta}{\tau} q_i q_i + q_i \theta_{,i} \left(\frac{\rho \beta \lambda}{\tau} + \frac{\partial \Lambda}{\partial \theta}\right) \geq 0 \tag{2.19}$$

The sufficient conditions guarantying the positiveness of inequality (3.19) are

$$\frac{\beta}{\tau} \geq 0 \tag{2.20}$$

$$\Lambda = \frac{1}{\theta} \tag{2.21}$$

$$\frac{\rho \beta \lambda}{\tau} = - \frac{\partial \Lambda}{\partial \theta} \tag{2.22}$$

Using (2.21), it is inferred from (2.22) that

$$\beta = \frac{\tau}{\rho \lambda \theta^2} \tag{2.23}$$

This result is important as it relates the unknown coefficient β to the physically identified quantities τ, ρ and λ.

In view of the above results, expression (2.17) of the entropy production (2.19) reduces to

$$\sigma^s = \frac{1}{\lambda\,\theta^2}\, q_i\, q_i \geq 0 \tag{2.24}$$

from which follows that

$$\lambda \geq 0 \tag{2.25}$$

which is a classical result still derived in other thermodynamic theories.

The Gibbs equation expressing s in terms of the basic variable can now be directly derived : one has

$$ds = \frac{\partial s}{\partial u}\, du + \frac{\partial s}{\partial q_i}\, dq_i$$

and, in view of the results (2.5), (2.17) and (2.23)

$$ds = \frac{1}{\theta}\, du - \frac{\tau}{\rho\,\lambda\,\theta^2}\, q_i\, dq_i \tag{2.26}$$

This is the generalized Gibbs equation predicted by E.I.T. Compared to the classical theory, it contains a supplementary term involving the heat flux. Clearly for a vanishing relaxation time of the heat flux ($\tau = 0$), one recovers the classical Gibbs equation for heat conduction.

$$ds = \frac{1}{\theta_e}\, du \tag{2.27}$$

wherein subscript recalls that the temperature refers to the local equilibrium value; the difference between θ and θ_e will be discussed in detail in section 3.

It remains to explore the consequences resulting from the convexity property of *s*. Mathematically this amounts to require that

$$\frac{\partial^2 s}{\partial q_i\, \partial q_i} < 0$$

from which follows, according to (2.26) that

$$\frac{\tau}{\lambda} > 0 \tag{2.28}$$

Since we have still shown that $\lambda > 0$, inequality (2.28) implies that the relaxation time τ is a positive quantity :

$$\tau > 0 \tag{2.29}$$

To **summarize**, the principal results issued from E.I.T. are

a) The expression of the generalized Gibbs equation

$$ds = \frac{1}{\theta} du - \frac{\tau}{\rho \lambda \theta^2} q_i \, dq_i$$

b) The expressions of the entropy flux and the entropy production

$$J_i^s = \frac{1}{\theta} q_i, \qquad \sigma^s = \frac{1}{\lambda \theta^2} q_i q_i$$

c) The generalized Fourier law :

$$\tau \dot{q}_i = - \lambda \theta_{,i} - q_i \qquad \text{(Vernotte-Cattaneo equation)}$$

d) The positiveness of λ and τ.

2.2 Finite speed of thermal waves : second sound

Consider a small thermal disturbance around an equilibrium reference state. Neglecting second order term in the fluxes, one may write for the time variation of the energy :

$$\frac{\partial u}{\partial t} = c \frac{\partial \theta}{\partial t} + 0 (q_i q_i) \tag{2.31}$$

wherein c is the specific heat per unit mass. The evolution equations for the basic variables θ and q_i are the energy and the Vernotte-Cattaneo equations :

$$\rho c \frac{\partial \theta}{\partial t} = - q_{i,i} \tag{2.32}$$

$$\tau \frac{\partial q_i}{\partial t} = - q_i - \lambda \theta_{,i} \tag{2.33}$$

When (2.33) is introduced in (2.32), one is led to a hyperbolic equation of the telegraph type, namely

$$\tau \frac{\partial^2 \theta}{\partial t^2} + \frac{\partial \theta}{\partial t} - \chi \theta_{,ii} = 0 \tag{2.34}$$

with $\chi = \lambda / \rho c$ the thermal diffusivity. By assuming plane thermal waves of the form

$$\theta = \tilde{\theta} \exp [i (kx - \omega t)] \tag{2.35}$$

where $\tilde{\theta}$ is the amplitude of the wave, ω the (real) frequency and k the (complex) wavenumber, the dispersion relation obtained by substituting (2.35) in (2.34) reads as

$$- \tau\omega^2 - i\omega + \chi k^2 = 0 \tag{2.36}$$

It follows that the phase velocity v_p and the attenuation distance α are given by

$$v_p = \frac{\omega}{\mathrm{Re}(k)} = \frac{\sqrt{2 \chi \omega}}{\sqrt{\tau \omega + \sqrt{1 + \tau^2 \omega^2}}} \tag{2.37}$$

$$\alpha = \frac{1}{\mathrm{Im}(k)} = 2 \frac{\chi}{v_p} \tag{2.38}$$

At low frequencies ($\tau\omega << 1$), $v_p = (2 \chi \omega)^{1/2}$ and $\alpha = (2 \chi / \omega)^{1/2}$, which are the results obtained from the usual theory based on Fourier's law. In the high-frequency limit ($\tau\omega >> 1$), v_p and α tend to the finite limits

$$v_{p\infty} = U = \sqrt{\frac{\chi}{\tau}}, \tag{2.39}$$

$$\alpha_\infty = 2 \sqrt{\chi \tau} \tag{2.40}$$

It is interesting to note that within the limit $\tau = 0$, which corresponds to the Fourier law, one has $v_{p\infty} = \infty$ and $\alpha_{\infty} = 0$. The solutions of equation (2.34) are usually referred to as second sound. It should be realized that the second sound is a thermal wave and should not be confused with the normal first sound, a pressure wave.

The problem of infinite speed of propagation of thermal signals was one of the first incentives for the development of extended irreversible thermodynamics. The subject of propagation of thermal signals has been dealt with by many authors. The first ones were Cattaneo [16] and Vernotte [17], who took as motivation kinetic theory arguments and added a relaxation term to the Fourier equation. This point of view has been quite common and has been used in several contexts, as in the analysis of waves in thermoelastic media, fast explosions and second sound in solids. For a wide bibliography the reader is referred to [20].

It should be observed that it is not necessary to assume relaxation terms in the heat equation to get a finite velocity for the propagation of thermals signals. Such a result is also obtained if one assumes a nonlinear diffusion equation with a thermal conductivity depending on the temperature as for instance as $\lambda \approx \theta^n$ [21]. However when $-1 < n < 0$, the parabolic nonlinear theory predicts thermal waves propagating at an infinite speed. For $-\infty < n < -1$ thermal waves do even not exist in the parabolic theory because of divergences in the energy integral [21]. Experiments and theory give as typical trends for λ in metals $\lambda \approx \theta^1$, $\lambda \approx \theta^0$, and $\lambda \approx \theta^{-2}$ at low, intermediate and high temperatures respectively, indicating that the nonlinear parabolic theory is generally not sufficient to account for thermal waves particularly at high temperatures.

2.3 A non-equilibrium temperature

While the local equilibrium temperature θ_e is only depending on the internal energy u, the non-equilibrium temperature θ depends in addition on the heat flux q_i. It is our purpose to establish the relations between these two quantities. In view of the approximation introduced in the previous sections Gibbs equation may be written as

$$ds = \theta^{-1} du - \frac{\tau}{\rho \lambda \theta_e^2} q_i \, dq_i \tag{2.41}$$

Compared to (2.26), one has in (2.41) replaced the temperature θ by θ_e in the second term of the r. h. s. The integrability condition corresponding to equation (2.41) is

$$\left(\frac{\partial \theta^{-1}}{\partial q_i}\right) = -\frac{\partial}{\partial u}\left(\frac{\tau}{\rho\,\lambda\,\theta_e^2}\right) q_i \qquad (2.42)$$

and, after integration,

$$\theta^{-1} = \theta_e^{-1}(u) - \frac{1}{2}\,\frac{\partial\,(\tau/\rho\,\lambda\,\theta_e^2)}{\partial u}\, q_i\, q_i \qquad (2.43)$$

For monoatomic gases obeying the Boltzmann equation (see section 3.3), it is found that

$$\theta^{-1} = \theta_e^{-1} + \frac{2}{5}\,\frac{\rho}{p^3\,\theta_e}\, q_i\, q_i \qquad (2.44)$$

wherein p is the pressure.

Since the coefficients of the term in $q_i\, q_i$ are very small, it turns out that in many practical situations the corrective term may be neglected. Notwithstanding that, it is interesting to examine some possible experimental consequences.

However, as repeatedly claimed, in many situations this is of no practical importance, because of the smallness of the quantity in the right hand side of (2.42).

The presence of a generalized temperature is not exclusive of extended irreversible thermodynamics. In his entropy-free formulation of non-equilibrium thermodynamics, Meixner [25] postulated the existence of a dynamical temperature depending on the interaction of the system with the outside. Müller [26] introduced a "coldness" function assumed to depend on the empirical temperature and its time derivative : in a steady state, the "coldness" reduces to the local-equilibrium temperature, in constrast with the generalized temperature appearing in EIT. More recently, Muschik [27] introduced the notion of contact temperatures and explored the conceptual difficulties of their measurement.

2.4 Heat transport in dielectric crystals at low temperature

In this section, it is shown that the basic equations proposed by Guyer and Krumhansl [22] to describe heat waves propagation in dielectric crystals at low temperature may be derived from a variational principle and are compatible with irreversible thermodynamics. The Guyer and Krumhansl equations are of primary importance because they exhibit the existence of a second sound. The presence of second sound was first detected experimentally in He II. It must however be stressed that second sound is not only detectable in He II. It was shown [20, 22] that the existence of second sound depend only on the presence of a phonon gas. Since the latter exists in any solid, second sound should be detectable in any material.

The dispersion relation for second sound in solids was derived by Guyer and Krumhansl [22] who based their analysis on Boltzmann's equation. Neglecting thermal energy transport by free electrons, which is a reasonable assumption for dielectric crystals, Guyer and Krumhansl obtained the following equations for the temperature and heat flux fields :

$$c \frac{\partial \theta}{\partial t} = - q_{i,i} \qquad (2.45)$$

$$\frac{\partial q_i}{\partial t} + \frac{1}{3} c\, c_s^2\, \theta_{,i} + \frac{1}{\tau_R} q_i = \tau_N \frac{c_s^2}{5} (q_{i,jj} + 2\, q_{j,ii}). \qquad (2.46)$$

c_s is the mean speed of the phonons, τ_R a relaxation time for the momentum non-conserving resistive processes, the so-called umklapp processes, and τ_N the relaxation time associated with the momentum preserving processes. Equation (2.45) is the energy balance while (2.46) is the evolution equation for the heat flux; it reduces to Cattaneo's relation when τ_N is zero. However, in contrast with Cattaneo's equation, expression (2.46)) does not reduce to Fourier's law in steady conditions.

Our first purpose in this section is to derive the Guyer and Krumhansl results within the framework of Extended Irreversible Thermodynamics. The second objective of this section is to show that the Guyer and Krumhansl relations may be recovered from a variational principle stating that steady states are characterized by a minimum of dissipated energy. This criterion was originally proposed Prigogine [3-5] who derived this result within the context of classical irreversible thermodynamics. Prigogine's formulation concerned the minimum entropy production and was subordinated to some conditions like linear constitutive equations of the Fourier type and was unable to cope with Guyer and Krumhansl equations. We show here that the steady Guyer-Krumhansl equations can be recovered from the principle of minimum dissipated energy at the condition to work within the framework of Extended Irreversible Thermodynamics. This result is important as it asserts the physical background of this formalism.

2.4.1 *Extended thermodynamic derivation of Guyer-Krumhansl equations for second sound*

According to the general philosophy behind EIT, the space of state variables is enlarged to include, besides the classical variable u, the heat flux vector q_i and the flux of the heat J^q_{ij}.The variables u (or Θ), q_i and J^q_{ij} are on the same footing and supposed to obey evolution equations respectively given by the energy law

$$\frac{\partial u}{\partial t} = - q_{i,i} \tag{2.47}$$

and

$$\tau \frac{\partial q_i}{\partial t} = - J^q_{ij,j} + \sigma^q_{i,} \tag{2.48}$$

$$\tau_q \frac{\partial J^q_{ij,j}}{\partial t} = - J^{qq}_{ijk,k} + \sigma^{qq}_{ij} \tag{2.49}$$

wherein τ and τ_q are relaxation times. The fluxes J^q_{ij} and J^{qq}_{ijk} are second and third order tensors respectively. The source σ^q_i is a vector and σ^{qq}_{ij} a second order tensor. To solve the problem, it is necessary to formulate constitutive equations for the unknown quantities appearing in (2.47 - 2.49), namely u , J^{qq}_{ij}, σ^q_i and σ^{qq}_{ij}. Moreover, we restrict the analysis to linear developments and assume that the quantities u, J^{qq}_{ijk}, σ^q_i and σ^{qq}_{ij} depend linearly on the variables θ, q_i and J^q_{ij}. We do not alter the generality of the formalism by supposing that J^q_{ij} is symmetric. It is also convenient to introduce the notation

$$V = \left\{\theta, q_i, J^q_{ij}\right\} \tag{2.50}$$

where V means the whole set of variables under parentheses.

General (linear) constitutive equations are given by

$$u = c\,\Theta \tag{2.51}$$

$$J^{qq}_{ijk} = 3\,A\left(q_i\delta_{ij}\right)^s \tag{2.52}$$

$$\sigma_i^q = - q_i \tag{2.53}$$

$$\sigma_{ij}^{qq} = - J_{ij}^q + \left[B_1(\theta) + B_2 J_{kk}^q\right] \delta_{ij} \tag{2.54}$$

All the coefficients are constants except B_1 which may depend on the temperature. The notation is classical : δ_{ij} designates the identity tensor and superscript s symmetrization. After substitution of (2.51)-(2.54) in the evolution equations (2.49)-(2.50), one finds :

$$c \frac{\partial \theta}{\partial t} = - q_{i,i} \tag{2.55}$$

$$\tau \frac{\partial q_i}{\partial t} = - J_{ij,j}^q - q_i \tag{2.56}$$

$$\tau_q \frac{\partial J_{ij}^q}{\partial t} = - A\left[(q_{k,k}\, \delta_{ij}) + 2\, (q_{i,j})^s\right] - J_{ij}^q + (B_1 + B_2 J_{kk}^q)\, \delta_{ij} \tag{2.57}$$

From now on, it is assumed that τ_q is negligibly small. Deriving J_{ij}^q from (2.57) and substituting in (2.56), one obtains the following equation

$$\tau \frac{\partial q_i}{\partial t} = - \lambda\, \theta_{,i} - q_i + A\, q_{i,jj} + B\, q_{k,k\,i} \tag{2.58}$$

wherein the coefficients λ and B are defined respectively by

$$\lambda = \frac{1}{1 - 3B_2} \frac{\partial B_1}{\partial \theta}, \quad B = 2A \frac{1 - \frac{1}{2} B_2}{1 - 3B_2}$$

Clearly, expressions (2.55) and (2.58) are the same as Guyer and Krumhansl's equations (2.45) and (2.46) at the condition to perform the following identifications :

$$\tau = \tau_R, \quad \lambda = \frac{1}{3} c\, c_s^2\, \tau_R, \quad A = \frac{1}{5} \tau_N\, \tau_R\, c_s^2, \quad B = 2A \tag{2.59}$$

Moreover, since we have in mind the formulation of a variational principle, the expression of the entropy production generated by the propagation of second sound in crystals is now calculated.

Assuming that s and the entropy flux J_i^s are given by the constitutive equations

$$s = s\ (V), \qquad J_i^s = J_i^s\ (V)$$

it is an easy matter to calculate the entropy production σ^s from $\sigma^s = \dot{s} + J^s_{i,i}$. Since θ has been selected as variable, it is natural to work with the Helmholtz free energy $f = u - \theta s$ as potential function. Up to second order terms in the fluxes, f and J^s_i are given by

$$f = f_1 + \frac{1}{2} f_2\, q_i\, q_i + \frac{1}{2} f_3\, J^q_{ij}\, J^q_{ij} \tag{2.60}$$

$$J_i^s = \overset{*}{\gamma}_0\, q_i + \overset{*}{\gamma}_1\, J^q_{ij}\, q_j + \overset{*}{\gamma}_2\, (J^q_{kk})\, q_i \tag{2.61}$$

wherein all the coefficients may depend on Θ. On the other side, EIT predicts that f_2 and f_3 are proportional to the relaxation times τ and τ_q as confirmed later on by equation (2.67) for the coefficient f_2. After elimination of J^q_{ij} between (2.57) written in the approximation $\tau_q = 0$ and (2.61), the entropy flux can be expressed in terms of the heat flux alone, namely,

$$J_i^s = \gamma_0\, q_i + \gamma_1\, (q_{i,j})^s\, q_j + \gamma_2\, q_{k,k}\, q_i \tag{2.62}$$

wherein γ_0, γ_1 and γ_2 stand respectively for

$$\gamma_0 = \overset{*}{\gamma}_0 + \frac{B_1}{1 - 3B_2}\, \overset{*}{\gamma}_1 + \frac{3B_1}{1 - B_2}\, \overset{*}{\gamma}_2$$

$$\gamma_1 = -\, 2A\, \overset{*}{\gamma}_1$$

$$\gamma_2 = -\frac{A\,(1 + B_2)}{1 - 3B_2}\, \overset{*}{\gamma}_1 - \frac{5A}{1 - B_2}\, \overset{*}{\gamma}_2$$

Using the energy balance, one obtains the following expression for the dissipated energy, which is equal to the entropy production time the temperature :

$$\theta\sigma^s = - s \frac{\partial\theta}{\partial t} - \frac{\partial f}{\partial t} - q_{i,i} + \theta J^s_{i,i} \geq 0 \tag{2.63}$$

Using the chain derivation rule to calculate $\frac{\partial f}{\partial t}$ and the relations (2.57), (2.60) and (2.62), expression (2.63) becomes up to second order terms in the heat flux

$$\begin{aligned}\theta\sigma^s = &- \left(\frac{\partial f}{\partial\theta} + s\right) \frac{\partial\theta}{\partial t} - \frac{1}{\tau} f_2\, q_i\, (- \lambda\, \theta_{,i} - q_i + A\, q_{i,jj} + B\, q_{k,ki}) \\ &- q_{i,i} + \theta \left[\frac{\partial \gamma_0}{\partial\theta} \theta_{,i}\, q_i + \gamma_0\, q_{k,k}\right] + \theta\, \gamma_1 \left[q_{i,j}\, q_{j,i} + q_i\, q_{i,kk}\right] \\ &+ \theta\, \gamma_2 \left[q_{i,i}\, q_{j,j} + q_i\, q_{k,ki}\right] \geq 0\end{aligned} \tag{2.64}$$

It is important to stress that (2.64) is linear in the derivative $\frac{\partial\theta}{\partial t}$. Since this quantity can be assigned arbitrary values, the positiveness of σ^s could be violated except if the coefficient of $\frac{\partial\theta}{\partial t}$ is zero, from which follows the classical result

$$\frac{\partial f}{\partial\theta} + s = 0 \tag{2.65}$$

In view of this result and after grouping the various second order terms, inequality (2.64) reduces to

$$\begin{aligned}\theta\sigma^s = &\frac{1}{\tau} f_2\, q_i\, q_i + \left[\frac{\lambda}{\tau} f_2 + \theta \frac{\partial\gamma_0}{\partial\theta}\right] q_i\, \theta_{,i} \\ &- \left[\frac{A}{\tau} f_2 - \theta\, \gamma_1\right] q_i\, q_{j,ji} - \left[\frac{B}{\tau} f_2 - \theta\gamma_2\right] q_i\, q_{k,ki} \\ &+ \theta\, \gamma_1\, q_{i,j}\, q_{j,j} + \theta\, \gamma_2\, q_{i,j}\, q_{j,j} - q_{i,i}\, (1 - \theta\gamma_0) \geq 0\end{aligned} \tag{2.66}$$

The positiveness of σ^s allows us also to identify the coefficients f_2, γ_0, γ_1 and γ_2 as

$$f_2 = \frac{\tau}{\lambda\theta}, \quad \gamma_0 = \frac{1}{\theta}, \quad \gamma_1 = \frac{A}{\lambda\theta^2}, \quad \gamma_2 = \frac{B}{\lambda\theta^2} \tag{2.67}$$

moreover, they have to comply the inequalities

$$\frac{f_2}{\tau} > 0, \quad \gamma_1 > 0, \quad \gamma_2 > 0 \tag{2.68}$$

In terms of the notation introduced by Guyer and Krumhansl [see relations (2.59)], the results (2.67) write as

$$f_2 = \frac{3}{c\, c_s^2}, \quad \gamma_1 = \frac{3}{5}\frac{\tau_N}{c\,\theta^2}, \quad \gamma_2 = \frac{6}{5}\frac{\tau_N}{c\,\theta^2} \tag{2.69}$$

In view of the inequalities (2.68) and the general result $c > 0$, it is inferred that

$$\tau_R > 0, \quad \tau_N > 0 \tag{2.70}$$

These results prove that the relaxation times τ_R and τ_N are positive. The results (2.69) are also worth to be mentioned as they indicate that the coefficients f_2, γ_1 and γ_2 are no longer undefined quantities but are related to the basic physical parameters c, c_s, τ_R and τ_N. Moreover, we were able to prove that the relaxation times τ_R and τ_N are indeed positive quantities.

The final expression of $\theta\sigma^s$ and J_i^s are thus given by

$$\theta\sigma^s = \frac{3}{\tau_R\, c\, c_s^2}\, q_i\, q_i + \frac{3}{5}\frac{\tau_N}{c}\,(q_{i,j}\, q_{j,i} + 2\, q_{i,i}\, q_{j,j}) \tag{2.71}$$

$$J_i^s = \frac{q_i}{\theta} + \frac{3}{5}\frac{\tau_N}{c\theta^2}\left[(q_{i,j})^s\, q_j + 2\, q_{j,ji}\right] \tag{2.72}$$

Compared to classical irreversible thermodynamics, $\theta\sigma^s$ and J_i^s contain terms arising from the presence of the "normal" relaxation time τ_N.

2.4.2 *A variational formulation*

Variational principles have always played a privileged role in physics [e.g. 23]. This is justified because a variational criterion presents the advantage of concision : a single equation stands for a set of differential equations, initial and boundary conditions. From a practical point of view, variational principles provide specific methods, like Rayleigh-Ritz method, for solving the differential equations governing the physical process. Besides their power of synthesis and their utility in numerical analysis, variational formulations reveal also frequently interesting from a physical point of view because in some problems, the functional submitted to variation possesses a physical meaning.

A variational principle meeting the above properties is Prigogine's minimum entropy production principle expressing that purely dissipative processes (without convection) evolves in such a way that in the steady state, the total entropy production is stationary, truly a minimum. Mathematically, this is expressed by

$$\delta \int \sigma_P^s dV = 0, \qquad (2.73)$$

wherein δ is the usual variational symbol, dV an elementary volume element and σ_P^s the entropy production derived from classical irreversible thermodynamics; for heat conduction, one has :

$$\sigma_P^s = q_i \, \theta^{-1}_{,i} \, .$$

It should however be noticed that Prigogine's principle is only applicable to processes governed by Fourier's law with a heat conductivity λ varying like θ^{-2}. In the case of λ proportional to θ^{-1}, Prigogine's principle must be changed into

$$\delta \, \phi_P (\theta) = 0$$

where ϕ_P stands for the total dissipated energy

$$\phi_P (\theta) = \int \theta \, \sigma_P^s dV$$

It is a simple exercise to check that the corresponding Euler-Lagrange equation is :

$$q_{i,i} = 0$$

It is now shown that the minimum dissipated energy principle still holds for describing the Guyer-Krumhansl steady equations. The difference with Prigogine's criterion is that the expression of the dissipated energy to be used is not that derived from classical irreversible thermodynamics but rather expression (2.71) of the dissipated energy obtained in the context of EIT. Moreover, a look on Guyer-Krumhansl steady equations

$$q_{i,i} = 0, \tag{2.74}$$

$$\frac{1}{\tau_R} q_i + \frac{1}{3} c\, c_s^2\, \theta_{,i} = \tau_N \frac{c_s^2}{5} (q_{i,jj} + 2\, q_{j,ji}) \tag{2.75}$$

indicates that the energy balance equation (2.74) may be considered as a constraint to be satisfied by the heat flux vector whose behavior is governed by equation (2.75) : equation (2.74) plays a role similar to the incompressibility condition in fluid mechanics.

It is our purpose to show that the heat flux that satisfies the Guyer-Krumhansl equation is the one corresponding to the minimum of dissipated energy :

$$\delta \int \theta\, \sigma^s_{EIT}\, dV = 0 \tag{2.76}$$

wherein σ^s_{EIT} is given by (2.71).

To prove this result, we shall introduce a Lagrange multiplier γ. Consider the integral

$$\phi(q) = \int \left(\frac{3}{2\, \tau_R\, c\, c_s^2} q_i\, q_i + \frac{3}{10} \frac{\tau_N}{c}\, q_{i,j}\, q_{j,i} + q_{i,i}\, q_{j,j} \right) dV - \int \gamma q_{i,i}\, dV \tag{2.77}$$

and determine the necessary condition for ϕ to be stationnary under arbitrary variations of q_i, when q_i takes prescribed values at the boundaries :

$$\delta\, q_i = 0 \quad \text{(at the boundaries)} \tag{2.78}$$

Calculation of the first variation of (2.77) and use of the divergence theorem lead to

$$\delta\phi = \int \left[\left(\frac{3}{\tau_R\, c\, c_s^2}\, q_i - \frac{3\tau_N}{5\,c}\, [q_{i,jj} + 2\, q_{j,ji}] + \gamma_{,i} \right) \delta q_i - q_{i,i}\, \delta\gamma \right] dV\,. \tag{2.79}$$

Clearly, $\delta\phi = 0$ under the conditions that

$$q_{i,i} = 0 \tag{2.80}$$

and

$$\frac{3}{\tau_R\, c\, c_s^2}\, q_i + \gamma_{,i} - \frac{3}{5}\frac{\tau_N}{c}\, (q_{i,jj} + 2\, q_{j,ji}) = 0 \tag{2.81}$$

which are nothing else than the Guyer-Krumhansl equations after that the Lagrange multiplier has been identified with the absolute temperature θ. The minimum property of the variational principle (2.76) is evident as the dissipated energy is a quadratic form.

The physical meaning of the functional ϕ is is thus clear as it represents the dissipated energy calculated in extended irreversible thermodynamics.

It should be observed that all above results are obtained in a straightforward way following strictly the line of thought of Extended Irreversible Thermodynamics. In particular, the basic coefficients appearing in the formalism are well defined and expressed in terms of the heat capacity c, the sound velocity c_s and the relaxation times τ_N and τ_R associated to the normal and non-conserving momentum processes : these four quantities represent the four parameters of the theory. We have also proposed a variational formulation which is a generalization of Prigogine's celebrated minimum entropy production principle, [3-5] formulated a few years ago in the context of classical irreversible thermodynamics.

3. KINETIC THEORY

Our objectif in this chapter is to propose a microscopic interpretation of extended irreversible thermodynamics. This will be accomplished via the kinetic theory of gases. It will in particular be shown that the interface between the macroscopic description and the kinetic theory of gases is much wider in extended thermodynamics than in the classical theory of irreversible processes.

Our purpose is to justify the hypotheses and the main results of extended irreversible thermodynamics. To be explicit, we have to justify :

a) the choice of the dissipative fluxes as independent variables;

b) the form of the generalized Gibbs equation;

c) the form of the entropy flux;

d) the evolution equations for the flux as well as the relations between the transport coefficicients appearing in these equations.

As before, we shall concentrate our attention on the linear terms of the evolution equations. For simplicity, only thermal effects are investigated ignoring the role of viscous effects. Moreover it is not our intention to enter into the details of the calculations for which we refer the reader to the original works.

3.1 Basic concepts

We consider ideal or highly diluted gases. The basis for the analysis is given by the microscopic definitions of entropy, entropy flux and dissipative fluxes in terms of the velocity distribution function $f(r_i, c_i, t)$, where c_i denotes the ith component of the molecular velocity, r_i the ith component of the position-vector of the particle. We recall that the distribution function accounts for the number of particles located between r_i and $r_i + dr_i$ with a velocity between c_i and $c_i + dc_i$ at time *t*.

The mass density ρ, the mean velocity v_i and the internal energy u per unit mass are defined in terms of the distribution function *f* as

$$\rho(r_i, t) = \int m f \, d\mathbf{c} \qquad (3.1)$$

$$\rho(r_i, t) \; v_i(r_i, t) = \int m \, c_i \, f \, d\mathbf{c} \qquad (i = 1,2,3) \qquad (3.2)$$

$$\rho\,(r_i, t)\ u\,(r_i, t) = \int \frac{1}{2}\ m\,(c_i - v_i)\,(c_i - v_i)\ f\ d\ c \tag{3.3}$$

wherein *m* is the mass of the particle. Denoting by $C_i = (c_i - v_i)$, the relative velocity of the particule with respect to the mean velocity v_i the equilibrium pressure *p* is given by

$$p\,(r_i, t) = \frac{1}{3}\int m\,C^2\,f\,d\,c \qquad (C^2 = C_i\,C_i) \tag{3.4}$$

It follows then from the definition (3.3) of the internal energy that

$$p = \frac{2}{3}\ \rho\ u \tag{3.5}$$

The macroscopic thermal equation of state for ideal gases leads then to the following definition of the absolute temperature θ

$$p = \frac{2}{3}\ \rho\ u = n\,k\,\theta \tag{3.6}$$

wherein n (= ρ/m) is the number of particles per unit volume and k the Boltzmann constant.

Other important expressions are these of entropy per unit mass *s*, entropy flux J_i^s and heat flux q_i : they are respectively defined as

$$\rho\,(r_i, t)\ s\,(r_i, t) = -\,k \int f\ \ln f\ dc \tag{3.7}$$

$$J_i^s\,(r_i, t) = -\,k \int C_i\,f\ \ln f\ dc \tag{3.8}$$

$$q_i\,(r_i, t) = \frac{1}{2}\int m\,C^2\,C_i\ f\ dc \tag{3.9}$$

The distribution function is obtained by solving the Boltzmann equation [5,11]

$$\partial f / \partial t + c_i\,\partial f / \partial r_i + (F_i/m)\ \partial f / \partial c_i = J\,(f, f) \tag{3.10}$$

wherein J(f,f) represents the collision integral

$$J(f, f) = \int d\tilde{c} \int d\alpha\, |c - \tilde{c}|\, \sigma\, (c_i - \tilde{c}_i, \alpha)\, [\, f(c'_i)\, f(\tilde{c}'_i) - f(c_i)\, f(\tilde{c}_i)\,] \tag{3.11}$$

σ is the collision cross section of the particles, one of them with velocity c_i, the other one with velocity $\tilde{c}_i$ before collision, and velocities c'_i and $\tilde{c}'_i$ after collision, α is the angle formed by c_i and c'_i, F_i the external force acting on the particles. We consider that the usual conditions for the validity of Boltzmann equation (only binary collisions, not too strong inhomogeneities, stossahlansatz) are fulfilled.

In equilibrium, the collision integral J(f, f) is zero and the solution of (3.10) is the equilibrium distribution function f_e given by

$$f_e = n\,(m / 2\pi k \theta)^{3/2} \exp(- m C^2 / 2 k \theta) \tag{3.12}$$

In nonequilibrium situations, Boltzmann's equation which is a nonlinear integrodifferential equation is not integrable and only approximate solutions can be derived. Among them, let us quote the approaches of Chapman-Enskog and Grad [11,29]. They consist in expanding the nonequilibrium distribution function in the form

$$f = f_e\,(1 + \phi^{(1)} + \phi^{(2)} + \dots) \tag{3.13}$$

where $\phi^{(1)}, \phi^{(2)}, \dots$ are small corrections expressed in terms of a small parameter, for instance the ratio of the relaxation time to the macroscopic time, the ratio of the mean-free path to a characteristic length of the macroscopic inhomogeneities, the higher-order moments of the velocity distribution function, etc. The function f_e represents the distribution function either in global or in local equilibrium; in the first case, ρ, v_i and θ do not depend on the position or time and in the second case they depend on these variables.

The quantities ρ, v_i and θ are determined from the first five moments of the distribution function (3.1) - (3.3). This imposes on $\phi^{(i)}$ the conditions [15,29]

$$\int f_e \phi^{(i)} dc = 0, \quad \int f_e \phi^{(i)} C_i\, dc = 0, \quad \int f_e \phi^{(i)} C^2 dc = 0 \tag{3.14}$$

When (3.13) is introduced into (3.7) and (3.8), one obtains up to second order for s and J_i^s,

$$\rho s = \rho s_e - \frac{1}{2} k \int f_e \phi^{(1)2} \, dc \tag{3.15}$$

$$J_i^s = \theta^{-1} q_i - \frac{1}{2} k \int f_e C_i \phi^{(1)2} \, dc \tag{3.16}$$

Note that, due to restrictions (3.14), $\phi^{(2)}$ does not contribute to the entropy. The first terms in the right-hand side of (3.15-16) are the classical ones. The second terms are related to the nonclassical corrections on which we shall focus our attention.

3.2 Grad's solution

Two important models, namely the Chapman-Enskog and Grad models, have been proposed to solve the Boltzmann equation. In the Chapman-Enskog approach, f is expressed in terms of the first five moments ρ, u (or θ) and v_i and their gradients. Then $\phi^{(1)}$ is proportional to $v_{i,j}$ and $\theta_{,i}$, $\phi^{(2)}$ includes terms in $\theta v_{i,kj}$, $\theta_{,ij}$ and so on [29]. Lebon [11] has compared the Chapman-Enskog development up to second order (Burnett approximation) with the full expressions derived from extended irreversible thermodynamics. We refer the reader to this work for further details. Here, we find preferable to discuss Grad's formalism for reasons that will become clear later on.

In Grad's model [15], f is developed in terms of its moments with respect to the molecular velocity. But from their very definition, the thermodynamic fluxes are directly related to the moments of the velocity distribution function (see for instance expression (3.9) of q_i). Therefore, the mean values of the thermodynamic fluxes are considered in Grad's theory as independent variables. In this way, Grad's theory is closer to extended thermodynamics than Chapman-Enskog's since both approaches use the same set of independent variables.

In Grad's method, the nonequilibrium distribution function f (r_i, c_i, t) is replaced by the infinite set of variables $\rho\,(r_i,t)$, $v_i\,(r_i,t)$, $\theta\,(r_i,t)$, $a_n\,(r_i,t)$ where a_n stands for the successive higher-order moments of the distribution function. Within the so-called thirteen-moment approximation the distribution function is given by [15]

$$f = f_e \,[1 + A_i C_i + B_i C^2 C_i] \tag{3.17}$$

in the case of heat conduction in a rigid body.
The coefficients A_i and B_i are determined by introducing (3.17) in (3.3) and (3.9). This allows to identify A_i and B_i in terms of the heat flux q_i , pressure and temperature :

$$A_i = -(m/p\,\theta)\,q_i \tag{3.18}$$

$$B_i = (\tfrac{1}{5})\,(m^2/p\,k^2\,\theta^2)\,q_i \tag{3.19}$$

As a consequence, f may be explicitly written as

$$f = f_e\left\{1 + (\tfrac{2}{5})\,(m/p\,k^2\,\theta^2)\,[\tfrac{1}{2}m\,C^2 - (\tfrac{5}{2})\,k\,\theta\,]\;C_i\,q_i\right\} \tag{3.20}$$

After substitution of this result in (3.15-16), the expressions for the entropy and the entropy flux are easily derived and given by

$$\rho s = \rho s_e - (m/5\,p\,k\,\theta^2)\,q_i\,q_i \tag{3.21}$$

$$J_i^s = \theta^{-1}\,q_i \tag{3.22}$$

These relations confirm the general results of extended irreversible thermodynamics stating that the entropy may depend on the dissipative fluxes and that for heat conduction, the entropy flux keeps its classical expression.

It remains to justify the form of the evolution equation for the heat flux. This is achieved by inserting expression (3.20) of the distribution function in Boltzmann's equation (3.10). Within the thirteen-moment approximation and in absence of viscous effects, one is led to [15]

$$\frac{3}{2\,\rho\,\gamma}\,\dot{q}_i = -\frac{15}{4}\,\frac{p\,k}{\rho\,m\,\gamma}\,\theta_{,i} - q_i \tag{3.23}$$

where the positive coefficient γ is a rather complicated expression given in terms of the collision coefficients as

$$\gamma = (\tfrac{2}{5})\,(2\pi)^{1/2}\int_0^{\infty} dx\,x^6\,e^{-x^2/2}\int(\tfrac{1}{m})\,[\sigma\,\sin^2\alpha\,\cos^2\alpha\,]\,d\alpha > 0$$

Equation (3.23) may be compared with the linearized evolution equation (2.9) for q_i obtained in the framework of extended thermodynamics and given by

$$\tau \dot{q}_i = - \lambda \, \theta_{,i} - q_i \tag{3.24}$$

By idenfication with (3.23) it is found that

$$\tau = \frac{3}{2 \rho \gamma}, \qquad \lambda = \frac{5}{2} \frac{p k}{m} \tau \tag{3.25}$$

Since γ is a positive quantity the positiveness of τ and λ follows directly from expressions (3.25).

We now turn out our attention to the Gibbs equation (2.26) which after integration yields

$$\rho s = \rho s_e - (\tau / 2 \lambda \theta^2) q_i q_i \tag{3.26}$$

Making use of the result

$$\tau / 2 \lambda = m / 5 pk .$$

obtained from (3.25), it is seen that (3.26) is strictly identical with the microscopic result (3.21). It may thus be concluded that, within the linear range, there is a complete agreement between Grad's theory and extended irreversible thermodynamics.

3.3 Relaxation time approximation

This agreement is not limited to Grad's model. It is also reached, for instance, in the relaxation-time approximation [19]. In this approach, the collision term in Boltzmann's equation is modelled as follows

$$\partial f / \partial t + c_i \partial f / \partial r_i + (F_i / m) \partial f / \partial c_i = - (1 / \tau)(f - f_e) \tag{3.27}$$

For the steady nonequilibrium solution in absence of external forces, it is found that

$$f = [1 + \tau c_i \partial_i]^{-1} f_e$$

a formal expression which may be given an operative meaning by developing it in powers of $\tau c_i \partial_i$. In the simple case of a temperature gradient but no velocity gradient, this leads to

$$f = \left(1 + \phi^{(1)}\right) f_e \tag{3.28}$$

$$\phi^{(1)} = -(\tau / k\theta^2)\,[\tfrac{1}{2} m C^2 - \tfrac{5}{2} k\theta]\, C_i\, \theta_{,i} \tag{3.29}$$

when nonlinear terms involving the products of the temperature gradients are omitted.

Introducing (3.29) in expression (3.15) of entropy, one obtains

$$\rho s = \rho s_e - (\tau^2 / 2k\theta^4) \left\{ \int [\tfrac{1}{2} m C^2 - \tfrac{5}{2} k\theta]^2\, C_j C_j f_e \, dc \right\} \theta_{,i}\, \theta_{,i} \tag{3.30}$$

and from (3.16), one finds for the entropy flux

$$J_i^s = \theta^{-1} q_i \tag{3.31}$$

Substituting (3.29) in the definition (3.9) of the heat flux and taking into account that

$$\int [\tfrac{1}{2} m C^2 - \tfrac{5}{2} k\theta]\, C_i C_i f_e \, dc = 0,$$

the expression for the heat flux may at the first order approximation, be written as

$$q_i = -(\tau / k\theta^2) \left\{ \int [\tfrac{1}{2} m C^2 - \tfrac{5}{2} k\theta]^2\, C_j C_j f_e \, dc \right\} \theta_{,i} \tag{3.32}$$

confirming the result $q_i = -\lambda\, \theta_{,i}$ obtained in extended thermodynamics for steady situations.

Furthermore, (3.30) may equivalently be written as

$$\rho s = \rho s_e - \frac{\tau\lambda}{2\,\theta^2}\,\theta_{,i}\,\theta_{,i} \tag{3.33}$$

This is nothing but the second-order approximation of (3.26) when the flux q_i is replaced by its first-order steady approximation. We take advantage of this point to recall that the entropy does not coincide with the local equilibrium entropy even in the steady case.

The good agreement between the kinetic theory and the macroscopic predictions reinforces undoubtedly the consistency of extended irreversible thermodynamics. Although the analysis was limited to the linear range and pure heat conduction, the accord between EIT and kinetic theory is much wider than the agreement of the kinetic theory with the classical local-equilibrium thermodynamics. This has not been achieved at the expenses of heavy difficulties, but in a rather straightforward way. Moreover, the same quality of agreement is obtained when viscous and relativistic effects are included [19].

4. THERMOELASTICITY

This chapter is devoted to the problem of heat conduction in elastic bodies. The hypothesis of non-deformability used in the previous chapters is now given up.

Since elasticity is a pure reversible effect, extended irreversible thermodynamics will not bring new elements in the description of elastic bodies. It is only in the formulation of thermal effects that extended thermodynamics will play a role. Since heat conduction has been treated in detail in chapter III, the present one can be considered as a simple exercice. Nevertheless, such an application is interesting as it shows how to connect extended thermodynamics with a well developed theory like elasticity. After recalling some notation and definitions we shall in a first part consider large deformations. Afterwards, we shall apply the analysis to infinitesimally small deformations.

4.1 Balance laws

Consider a deformable, homogeneous and anisotropic continuous medium B. Its configuration changes continuously under external mechanical actions and heating, from an original undeformed to a deformed state. Let Ω_0 and Ω denote the volumes, respectively in the underformed and deformed states, with surfaces A_0 and A and outwards pointing normals n_0 and n. Rectangular Cartesian coordinates will be used throughout this chapter : X_K refers to the underformed state, x_k to the deformed one.

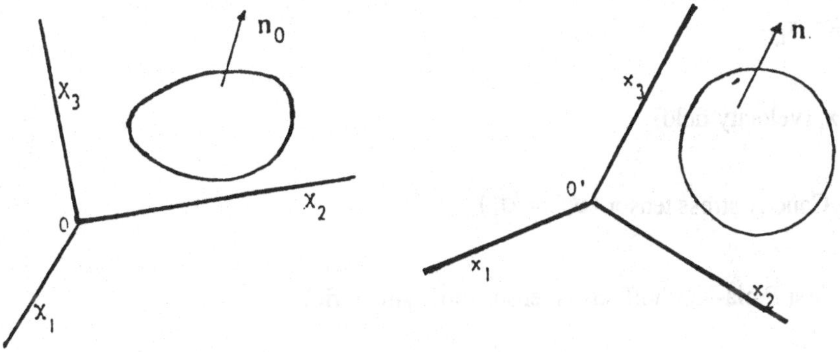

Fig. 4.1

Laws of balance : For systems undergoing large deformations at finite velocity, the balance equations can be expressed either in the material description (X_1, X_2, X_3, t as independent variables) or in the spatial description (x_1, x_2, x_3, t as independent variables). In both descriptions, the equations are given by [31]

Material description	Spatial description	
Mass balance : $\frac{\rho_0}{\rho} = J$	$\dot{\rho} = - \rho\, v_{j,j}$	(4.1)
Momentum balance : $\rho_0 \dot{v}_i = T_{Ji,J} + \rho_0 F_{oi}$	$\rho\, \dot{v}_i = \sigma_{ij,j} + \rho\, F_i$	(4.2)
Energy balance : $\rho_0 \dot{u} = - Q_{I,I} + \frac{1}{2} T_{IJ} \dot{C}_{IJ}$	$\rho\, \dot{u} = - q_{i,i} + \sigma_{ij} V_{ij}$	(4.3)

Index o refers to quantities measured in the original configuration.

A superposed dot stands as usually for the material time derivative; a comma preceeding a subscript denotes partial derivation with respect to the coordinates, in the deformed configuration if the subscript is a minuscule, in the underformed state if the subscript is a majuscule.

In (4.1)-(4.3), the following **notation** is used.

$J = \det x_{i,K}$

$v_i = \dot{x}_i$ (velocity field)

σ_{ij} = Cauchy stress tensor ($\sigma_{ij} = \sigma_{ji}$)

T_{Ij} = first Piola-Kirchoff stress tensor (not symmetric)

$$T_{Ij} = \frac{\rho_0}{\rho} X_{I,m}\, \sigma_{mj} \qquad (T_{Ij} \neq T_{jI}) \tag{4.4}$$

T_{IJ} = second Piola-Kirchoff stress tensor (symmetric)

$$T_{IJ} = X_{I,m}\, T_{Jm} \qquad (T_{IJ} = T_{JI}) \tag{4.5}$$

q_i = heat flux vector across the deformed surface A
Q_I = heat flux vector measured per unit area of A_0

$$q_i\, n_i\, dA = Q_I\, n_{Io}\, dA_o \tag{4.6}$$

$$Q_I = J\, X_{I,k}\, q_k \tag{4.7}$$

C_{IJ} = Green strain tensor
V_{ij} = symmetric rate of deformation tensor

$$V_{ij} = \frac{1}{2}\,(v_{i,j} + v_{j,i}) \tag{4.8}$$

4.2 General formulation

Consider an elastic body subject to large deformations and temperature gradients. Since the main objective in thermoelasticity is to determine the temperature and deformation fields, it is indicated to choose the Helmholtz free energy $f\,(= u - \theta s)$ as basic thermodynamical potential rather than the entropy s itself. In term of f, the rate of entropy production takes the form

$$\theta\sigma^s = \rho_0\, \dot{u} - \rho_0\, \dot{f} - \rho_0\, s\, \dot{\theta} + \theta\, J_{K,K} \geq 0 \tag{4.9}$$

To provide a complete description of the system, the field Eqs. (4.1)-(4.3) and the entropy inequality (4.9) must be supplemented by the following set of constitutive relations:

$$T_{IJ} = T_{IJ}\,(\theta, C_{KL}, Q_K), \tag{4.10 a}$$

$$f = f\,(\theta, C_{KL}, Q_K), \tag{4.10 b}$$

$$s = s\,(\theta, C_{KL}, Q_K), \tag{4.10 c}$$

$$J_I^s = J_I^s\,(\theta, C_{KL}, Q_K), \tag{4.10 d}$$

wherein the extra variables Q_I is assumed to obey an evolution equation of the Vernotte-Cattaneo type, i.e

$$\tau \dot{Q}_I = -\lambda_{IJ}\,\theta_{,J} - Q_I \tag{4.11}$$

This is justified because only heat conduction contributes to the irreversible process. It should be observed that the principle of objectivity [2,19] demands that the material time derivative in (4.11) should be replaced by an objective time derivative; nevertheless, we shall not enter into these considerations here because they do not modify the final conclusions. Neglecting nonlinear contributions the entropy flux J_I^s may be written as

$$J_I^s = \Lambda_{IJ}(\theta)\, Q_J \tag{4.12}$$

with $\Lambda(\theta)$ an unknown function of the temperature.

Elimination of $\dot{u}$ between the energy balance (4.3) and the entropy inequality (4.9) leads to

$$-\rho_0(\dot{f} + s\dot{\theta}) - Q_{I,I} + \frac{1}{2}T_{IJ}\,\dot{C}_{IJ} + \theta\frac{\partial \Lambda_{IJ}}{\partial \theta}\theta_{,I}Q_J - \theta\,\Lambda_{IJ}\,Q_{J,I} \geq 0 \tag{4.13}$$

wherein expression (4.12) has been taken into account. Using the chain differentiation rule to calculate $\dot{f}$, one obtains

$$-\rho_0\left(\frac{\partial f}{\partial \theta} + s\right)\dot{\theta} + \left(\frac{1}{2}T_{IJ} - \rho_0\frac{\partial f}{\partial C_{IJ}}\right)\dot{C}_{IJ} + Q_{J,I}(\theta\Lambda_{IJ} - \delta_{IJ}) + \frac{\rho_0}{\tau}\frac{\partial f}{\partial Q_I}(\lambda_{IJ}\,\theta_{,J} + Q_I)$$

$$+\,\theta\frac{\partial \Lambda_{IJ}}{\partial \theta}Q_J\,\theta_{,I} \geq 0 \tag{4.14}$$

after use is made of (4.11). Assume that at the lowest order of approximation

$$+\,\theta\frac{\partial \Lambda_{IJ}}{\partial \theta}Q_J\,\theta_{,I} \geq 0 \qquad (\alpha_{IJ} = \alpha_{JI}) \tag{4.15}$$

terms in C_{JK} cannot be present in (4.15) because it is required that at equilibrium $\partial f/\partial Q_I$ vanishes. With (4.15), inequality (4.14) becomes

$$-\rho_0 \left(\frac{\partial f}{\partial \theta} + s\right) \dot{\theta} + \left(\frac{1}{2} T_{IJ} - \rho_0 \frac{\partial f}{\partial C_{IJ}}\right) \dot{C}_{IJ} + \frac{\rho_0}{\tau} \alpha_{IJ} Q_I Q_J + Q_{J,I} (\theta \Lambda_{IJ} - \delta_{IJ})$$

$$+ \theta_{,J} Q_I \left(\frac{\rho_0}{\tau} \alpha_{KI} \lambda_{KJ} + \theta \frac{\partial \Lambda_{JI}}{\partial \theta}\right) \geq 0 \tag{4.16}$$

From now on the procedure is classical. Following the same reasoning as in chapter III, it is inferred form the positiveness property of (4.16) that

$$\frac{\partial f}{\partial \theta} + s = 0 \tag{4.17}$$

$$\frac{1}{2} T_{IJ} = \rho_0 \frac{\partial f}{\partial C_{IJ}} \tag{4.18}$$

$$\Lambda_{IJ} = (1/\theta)\, \delta_{IJ} \tag{4.19}$$

$$\alpha_{KI} \lambda_{KJ} = \frac{\tau}{\rho_0 \theta^2} \delta_{IJ} \tag{4.20}$$

The results (4.17) and (4.18) are typical of the classical theory of thermoelasticity while from (4.12) and (4.19), it is found that

$$J_I^s = \frac{1}{\theta} Q_I \tag{4.21}$$

which represents also a classical result. In view of (4.17-20) the rate of entropy production and the generalized Gibbs equation are given respectively by

$$\sigma^s = \frac{\rho_0}{\tau \theta} \alpha_{IJ} Q_I Q_J \tag{4.22}$$

$$df = -s\, d\theta + \frac{1}{2\rho_0} T_{IJ}\, dC_{IJ} + \alpha_{IJ} Q_I\, dQ_J \tag{4.23}$$

wherein the third term in the r.h.s of (4.23) is new compared to the classical theory of thermoelasticity.

4.3 Small deformations

For small deformation gradients, the distinction between ρ and ρ_0 disappears, the second Piola-Kirchoff and the Cauchy stress tensors are identical and the Green elastic tensor reduces to the infinitesimal strain tensor

$$\varepsilon_{ij} = \frac{1}{2}(u_{i,j} + u_{j,i}) \tag{4.24}$$

wherein u_i is the displacement vector

$$u_i = x_i - X_i \tag{4.25}$$

Moreover the material and the spatial descriptions are now identical.

Let us expand the Helmholtz free energy f in a Taylor series about the reference state assumed to be undeformed, stress free, at uniform temperature and without heat flux. Denoting the temperature by $\theta_0 + \theta$ where θ_0 is the constant reference temperature, we obtain

$$\rho f(\theta_0 + \theta, \varepsilon_{ij}, q_i) - \rho f_0(\theta_0, 0, 0) = -a\theta - b_i q_i - \frac{1}{2} d\theta^2 + \frac{1}{2} b_{ij} q_i q_j + \frac{1}{2} A_{ijkl} \varepsilon_{ij} \varepsilon_{kl} - \beta_{ij} \theta \varepsilon_{ij} + c_{ijk} \varepsilon_{ij} q_k - a_i \theta q_i \tag{4.26}$$

wherein all the coefficients a, b_i, ... are constants. From condition (4.15) requiring that $\partial f / \partial q_i$ vanishes at equilibrium, it is inferred that

$$b_i = c_{ijk} = a_i = 0 \tag{4.27}$$

It follows from the classical results (4.17) and (4.18) that the expressions of the entropy and the stress tensor can be automatically obtained by derivation of f respectively with respect to θ and ε_{ij}. In the linear approximation, it is found that

$$\rho_0 s = a + d\theta + \beta_{ij} \varepsilon_{ij} \tag{4.28}$$

$$\sigma_{ij} = A_{ijkl}\,\varepsilon_{kl} - \beta_{ij}\,\theta \tag{4.29}$$

while the Vernotte-Cattaneo equation (4.11) reads now as

$$\tau\dot{q}_i = -\lambda_{ij}\,\theta_{,j} - q_i \tag{4.30}$$

One obtains directly the linearized expression of the internal energy from (4.26) and (4.28), the final result is

$$\rho\,u = (\rho\,f_0 + \theta_0\,a) + c_v\,\theta + \theta_0\,\beta_{ij}\,\varepsilon_{ij} \tag{4.31}$$

In (4.28-31), all the coefficients are constant and defined as usual : β_{ij} is the thermal modulus, λ_{ij} the heat conductivity and c_v $(= \theta_0\, d)$ the heat capacity per unit volume.

At the linear order of approximation it is seen that the termodynamic potentials *s*, *u* and *f* as well as σ_{ij} keep their classical forms [31, 32]. The only difference is expression (4.30) which contains a non-stationary term in the heat flux. Nevertheless, this additional term influences strongly the nature of the linearized balance equation of energy which, for an isotropic body, takes the form

$$c_v\,(\tau\,\ddot{\theta} + \dot{\theta}) + \beta\,(\tau\,\ddot{\varepsilon}_{ij} + \dot{\varepsilon}_{ij}) = \lambda\,\theta_{,ii}\,, \tag{4.32}$$

this result has been obtained by combining (4.3), (4.30) et (4.31).

Compared to the classical expressions [32], equation (4.32) contains two additional terms in $\ddot{\theta}$ and $\ddot{\varepsilon}_{ij}$. The result (4.32) is the same as that obtained earlier in Lord and Schulman's paper [33], which is now put on firm thermodynamic bases.

The theory presented here differs from the Green and Lindsay approach [34] by a reduction of the number of coefficients appearing in the constitutive equations. With respect to classical thermoelasticity, equation (4.32) introduces only one extra coefficient, the relaxation time τ of the heat flux.

5. LINEAR VISCOELASTICITY

In ordinary incompressible fluids, the flow and transport phenomena are fairly well described by Newton's linear constitutive equation

$$\sigma_{ij} = -p\,\delta_{ij} + 2\eta\, V_{ij} \tag{5.1}$$

The notation is classical : σ_{ij} is the stress tensor, δ_{ij} the Kronecker symbol, p the hydrodynamic pressure, $V_{ij} = \frac{1}{2}(v_{i,j} + v_{j,i})$ the symmetric velocity gradient tensor, η the shear viscosity, which may depend on temperature and pressure but not on V_{ij}.

However, it was observed that there exists several materials, like polymers, soap solutions, asphalts, physiological fluids, ... that fail to obey the Newton law (5.1) : these materials are generally referred to as viscoelastic materials. These materials behave as fluids with a behaviour reminiscent of solids by exhibiting elastic effects. In ordinary fluids the relaxation time of the stress tensor is very short, in elastic bodies it is infinite so that no relaxation is observed : viscoelastic materials are characterized by relaxation times between these two limits.

For pedagogical reasons, we shall in this chapter only be concerned with linear viscoelasticity. Historically, the linear viscoelastic models were of pure mechanical nature and constructed by combining linear springs and dashpots. Although very simple, they have proven to be useful for describing a wide class of phenomena and for providing the first step towards more realistic non-linear descriptions. It is our purpose to show that linear viscoelasticity is easily interpreted within the framework of EIT. It si true that the description of viscoelastic materials by means of thermodynamic theories is not new. Several papers have been devoted to the subject but the majority of them is inspired either from classical irreversible thermodynamics [35-37] or rational thermodynamics [38-40]. Here we propose a derivation based on extended irreversible thermodynamics. It is seen that the classical models of Maxwell, Kelvin-Voigt, Poynting-Thomson and Jeffreys are recovered as particular cases of the formalism.

5.1 Constitutive and evolution equations of linear viscoelasticity

We take for granted the following hypotheses :

(a) the deformations are infinitesimally small,

(b) the body is at uniform temperature and heat effects are neglected,

(c) the material is isotropic.

Generalization to anisotropic systems underdoing large deformations and submitted to temperature gradients, although not a trivial matter, should not raise fundamental difficulties.

The choice of the variables is inspired by earlier results derived for viscous fluids [11-13]. Let us recall that in the latter case, the basic parameters used in EIT are, from one side the velocity field v_i and the internal energy u , from the other side an extra variable selected as being the viscous part σ^v_{ij} of the stress tensor defined through

$$\sigma_{ij} = - p\, \delta_{ij} + \sigma^v_{ij} \tag{5.2}$$

with p the hydrostatic pressure, a type-C variable.

By analogy with (5.2), we shall in the case of viscoelastic materials, decompose the stress tensor, assumed symmetric, into an elastic part σ'_{ij} and an inelastic part σ''_{ij} :

$$\sigma_{ij} = \sigma'_{ij} + \sigma''_{ij} \tag{5.3}$$

wherein the elastic part obeys Hooke's law

$$\sigma'_{ij} = 2\, G\, \varepsilon_{ij} + \lambda\, \varepsilon_{kk}\, \delta_{ij} \tag{5.4}$$

with λ and G the Lamé coefficients and ε_{ij} the symmetric strain tensor.

In parallel to the treatment of a viscous fluid, we choose as basic variables the velocity of deformation $\dot{u}_i$ (= $\partial_t\, u_i$), the internal energy u and the inelastic stress σ''_{ij}.

The behaviour of the classical variables $\dot{u}_i$ and u is governed by the usual balance laws of momentum and energy :

$$\rho\, \ddot{u}_i = \sigma_{ij,j} + \rho\, F_i \tag{5.5}$$

$$\rho\, \dot{u}_i = \sigma_{ij}\, \dot{\varepsilon}_{ij} \tag{5.6}$$

The heat flux does not appear in the energy balance (5.6) as heat effects are omitted.

For simplicity, bulk effects are ignored so that all the stress and strain tensors are traceless. A more complete treatment including bulk effects can be found in [41].

By strict analogy with the classical balance laws (5.5) and (5.6), it is assumed that the supplementary variable σ''_{ij} satisfies a balance equation of the general form

$$\dot{\sigma}''_{ij} = -J_{(ij)k,k} + S_{(ij)} \tag{5.7}$$

round brackets mean a traceless symmetrization. In order that the description be complete, it remains to express the flux J_{ijk} and source S_{ij} as functions of the basic set of variables by means of constitutive equations :

$$J_{(ij)k} = J_{ijk}(u, \dot{u}_i, \sigma''_{ij}) \tag{5.8}$$

$$S_{(ij)} = S_{ij}(u, \dot{u}_i, \sigma''_{ij}) \tag{5.9}$$

The most general constitutive relations compatible with a linear analysis are

$$J_{(ij)k} = -\frac{2\eta}{\tau_\sigma}\left[\frac{1}{2}\left(\dot{u}_i\,\delta_{jk} + \dot{u}_j\,\delta_{ik}\right) - \frac{2}{3}\,\dot{u}_k\,\delta_{ij}\right] \tag{5.10}$$

$$S_{(ij)} = -\frac{1}{\tau_\sigma}\,\sigma''_{ij} \tag{5.11}$$

where τ_σ and η are undetermined coefficients. Subsitution of (5.10) and (5.11) in (5.7) results in the following field equation for the extra variable

$$\tau_\sigma\,\dot{\sigma}''_{ij} = -\sigma''_{ij} + 2\eta\,\dot{\varepsilon}_{ij} \tag{5.12}$$

Equation (5.12) is the required linear evolution equation of the new variables σ''_{ij}.

It is possible to obtain an evolution equation for the total stress tensor by elimination of σ''_{ij} between (5.3) and (5.12). This operations leads to the following result

$$\tau_\sigma\,\dot{\sigma}_{ij} + \sigma_{ij} = 2G\,(\varepsilon_{ij} + \tau_\varepsilon\,\dot{\varepsilon}_{ij}) \tag{5.13}$$

wherein τ_ε stands for

$$\tau_\varepsilon = \frac{\eta}{G} + \tau_\sigma \tag{5.14}$$

It is worth noticing that relation (5.13) is the constitutive equation for a Poynting-Thomson body and that this relation arises naturally from extended irreversible thermodynamics.

The following particular cases are also of interest. By setting $\tau_\sigma = 0$ in (5.13), it is found

$$\sigma_{ij} = 2\,G\,\varepsilon_{ij} + 2\,\eta\,\dot{\varepsilon}_{ij} \tag{5.15}$$

Equation (5.15) is representative of a Kelvin-Voigt body. If one assumes that in equation (5.13) one has G = 0, one recovers the basic equation of Maxwell's model, namely,

$$\tau_\sigma \dot{\sigma}_{ij} + \sigma_{ij} = 2\,\eta\,\dot{\varepsilon}_{ij} \tag{5.16}$$

An interesting generalization is provided by the following model which is a coupling of Newton's viscous fluid and the material described by equation (5.13). Let us write for the stress tensor a relation of the form

$$\sigma_{ij} = \sigma^v_{ij} + \sigma^p_{ij} \tag{5.17}$$

σ^v_{ij} is the viscous stress tensor whose traceless part is given by

$$\sigma^v_{ij} = 2\,\eta^v\,\dot{\varepsilon}_{ij} \qquad \text{(Newton's law)}$$

with η^v the shear viscosity, while σ^p_{ij} is supposed to be formed of an elastic and an inelastic part :

$$\sigma^p_{ij} = 2\,G\,\varepsilon_{ij} + \sigma''_{ij} \tag{5.18}$$

The decomposition (5.17) is frequently used in rheology to describe dilute solutions of polymers : σ^v_{ij} is the contribution of the solvent (usually a Newtonian fluid), and σ^p_{ij} is the mean stress tensor of the polymer molecules. Repeating the reasoning leading to (5.12), with σ''_{ij} selected as basic variables, we obtain the relaxational equation

$$\tau\,\dot{\sigma}''_{ij} = -\,\sigma''_{ij} + 2\,\eta^p\,\dot{\varepsilon}_{ij} \tag{5.19}$$

η^p has been introduced to avoid confusion with η^v. By taking the time derivative of equation (5.17) and eliminating σ''_{ij} by means of (5.19), one obtains

$$\tau \dot{\sigma}_{ij} + \sigma_{ij} = 2 G \varepsilon_{ij} + 2 G \left(\tau + \frac{\eta^p + \eta^v}{G}\right) \dot{\varepsilon}_{ij} + 2 \tau \eta^v \ddot{\varepsilon}_{ij} . \tag{5.20}$$

Setting G = 0 in (5.20) yields

$$\tau \dot{\sigma}_{ij} + \sigma_{ij} = 2 (\eta^p + \eta^v) \dot{\varepsilon}_{ij} + 2 \tau \eta^v \ddot{\varepsilon}_{ij} \tag{5.21}$$

which is nothing but Jeffrey's model while by setting $\eta^v = 0$ in (5.20), one recovers the previous model expressed by equation (5.13). Jeffrey's equation is conveniently cast into the form

$$\tau \dot{\sigma}_{ij} + \sigma_{ij} = 2 \eta (V_{ij} + \tau_0 \dot{V}_{ij})$$

with

$$\eta = \eta^p + \eta^v , \quad \tau_0 = \frac{\tau \eta^v}{\eta^p + \eta^v} , \quad V_{ij} = \dot{\varepsilon}_{ij}$$

The material coefficients appearing in the aforementioned evolution equations for the stress tensor components are subjected to some constraints imposed by the second law of termodynamics, the criterion of objectivity, and the condition that entropy is a convex function at equilibrium. These restrictions are examined in the next subsection.

5.2 Restrictions imposed on the evolution equations

We postulate the existence fo a regular and continuous function s, the specific entropy, given by the constitutive relation

$$s = s (u, \varepsilon_{ij}, \sigma''_{ij}) ,$$

and obeying a balance equation of the form

$$\rho \dot{s} = - J^s_{i,i} + \sigma^s , \quad (\sigma^s > 0)$$

Objectivity prevents the entropy to depend on the velocity because the latter is a non-objective quantity. In absence of heat effects, the entropy flux J_i^s is zero [11-13] so that

$$\sigma^s = \rho \dot{s};$$

Moreover, it is more convenient to work with the Helmholtz free energy f (= u - θs) assumed to be given by

$$f = f(\theta, \varepsilon_{ij}, \sigma''_{ij}) \tag{5.22}$$

At fixed temperature and in terms of f, $\theta\, \sigma^s$ writes as

$$\theta\, \sigma^s = \rho \dot{u} - \rho \dot{f} \tag{5.23}$$

Using the chain differentiation rule to calculate $\dot{f}$, one obtains

$$\theta\sigma^s = \rho \left(\dot{u} - \frac{\partial f}{\partial \varepsilon_{ij}} \dot{\varepsilon}_{ij} - \frac{\partial f}{\partial \sigma''_{ij}} \dot{\sigma}''_{ij} \right) \geq 0 \tag{5.24}$$

Define as usual the components of the elastic pressure σ'_{ij} tensor by

$$\frac{\partial f}{\partial \varepsilon_{ij}} = \frac{1}{\rho} \sigma'_{ij} \tag{5.25}$$

Making use of the energy balance (5.6) and the evolution equation (5.12), inequality (5.24) reads as

$$\theta\sigma^s = \sigma''_{ij} \dot{\varepsilon}_{ij} - \frac{\rho}{\tau_\sigma} (-\sigma''_{ij} + 2\eta \dot{\varepsilon}_{ij}) \frac{\partial f}{\partial \sigma''_{ij}} \geq 0 \tag{5.26}$$

For isotropic systems, the most general form for the derivatives of s with respect to σ''_{ij} is given, in the linear approximation, by

$$\frac{\partial f}{\partial \sigma''_{ij}} = d\, \sigma''_{ij} + \wedge\, \varepsilon_{ij} \tag{5.27}$$

By taking the mixed derivative of (5.27) with respect to ε_{ij} and comparing with the mixed derivative of (5.25) with respect to σ''_{ij}, it is found that the coefficient $\wedge$ is zero. Substitution of (5.27) in (5.26) results in

$$\theta\sigma^s = \rho\,\frac{d}{\tau_\sigma}\,\sigma''_{ij}\,\sigma''_{ij} + \sigma''_{ij}\,\dot{\varepsilon}_{ij}\left(1 - 2\frac{d\eta\rho}{\tau_\sigma}\right) \geq 0 \tag{5.28}$$

Positiveness of expression (5.28) demands that the following sufficient conditions are satisfied :

$$\frac{d}{\tau_\sigma} > 0 \tag{5.29 a}$$

$$2\eta = \frac{\tau_\sigma}{\rho\, d} > 0 \tag{5.29 b}$$

the positiveness of η is a direct consequence of the result (5.29 a).

We turn now our attention to the consequences stemming from the convexity requirement of entropy. Expanding f around equilibrium, for fixed values of temperature and strain, one obtains

$$f(\theta, \varepsilon_{ij}, \sigma''_{ij}) = f_e(\theta, \varepsilon_{ij}) + \frac{1}{2}\, d\, \sigma''_{ij}\,\sigma''_{ij} \tag{5.30}$$

recalling that f is minimum at equilibrium at fixed values of temperature and strain, it is shown that $d > 0$. By combining this result with inequality (5.29 a),it is to concluded that

$$\tau_\sigma > 0 \tag{5.31}$$

The requirement that the entropy production is positive definite has thus led to the important result that the viscosity coefficient η is positive while from the concavity property of Helmholtz function f, it is inferred that the relaxation time τ_σ is positive.

5.3 The Gibbs equation for viscoelastic bodies

In classical irreversible thermodynamics the corner-stone is the Gibbs equation which is postulated a priori. In contrast, in the present formalism Gibbs' relation is derived.

One has

$$ds = \frac{\partial s}{\partial u} du + \frac{\partial s}{\partial \varepsilon_{ij}} d\varepsilon_{ij} + \frac{\partial s}{\partial \sigma''_{ij}} d\sigma''_{ij} \tag{5.32}$$

In view of the results (5.25), (5.27), (5.29 b) and $\theta ds = du - df$ (at uniform θ), expression (5.32) can be written in the form

$$\theta\, ds = du - \frac{1}{\rho}\sigma'_{ij}\, d\varepsilon_{ij} - \frac{\tau_\sigma}{2\rho\eta}\sigma''_{ij}\, d\sigma''_{ij} \tag{5.33}$$

In the liming case of a zero relaxation time ($\tau_\sigma = 0$), relation (5.33) reduces to

$$\theta\, ds = du - \frac{1}{\rho}\sigma'_{ij,}\, d\varepsilon_{ij} \tag{5.34}$$

In the case of a Poynting-Thomson body, some authors [36, 42] have proposed the following expression for the Gibbs equation which, for further purpose, is written in terms of time derivatives :

$$\theta\,\dot{s} = \dot{u} - \frac{1}{\rho}\sigma_{ij}\,\dot{\varepsilon}_{ij} + \frac{1}{\rho}\sigma''_{ij}\,\dot{\varepsilon}''_{ij} \tag{5.35}$$

This expression looks apparently different from (5.33) and introduces a new quantity, the so-called inelastic strain tensor ε''_{ij}. To recover (5.35) from (5.33), let us substitute the evolution equation (5.12) in expression (5.33), from which follows that

$$\theta\,\dot{s} = \dot{u} - \frac{1}{\rho}\sigma'_{ij}\,\dot{\varepsilon}_{ij} - \frac{1}{2\rho\eta}\sigma''_{ij}\left(-\sigma''_{ij} + 2\eta\,\dot{\varepsilon}_{ij}\right) \tag{5.36}$$

At this point, define an inelastic strain tensor ε''_{ij} through

$$\dot{\varepsilon}''_{ij} = \frac{1}{2\eta}\sigma''_{ij} \tag{5.37}$$

After introduction of (5.37) in (5.36), one is led to

$$\theta\,\dot{s} = \dot{u} - \frac{1}{\rho}\,(\sigma'_{ij} + \sigma''_{ij})\,\dot{\varepsilon}_{ij} + \frac{1}{\rho}\,\sigma''_{ij}\,\dot{\varepsilon}''_{ij} \tag{5.38}$$

which is nothing but the result (5.35) found by Kluitenberg [36] and Lambermont and Lebon [42] on completely different bases. It should also be added that the same authors obtained the above expression (5.34) for the Kelvin-Voigt body.

6. FINAL COMMENTS

At this stage of the analysis, it may be instructive to compare the merits of extended irreversible thermodynamics with the internal variable theory (IVT) which fuelled much interest during the last decades (e.g. Garcia-Colin and Uribe [8], Meixner [35], Coleman and Gurtin [43], Kestin and Bataille [44], Maugin and Drouot [45], Bampi and Morro [46] and parts I, II and IV of this book by Muschik, Haupt and Maugin).

In extended irreversible thermodynamics (EIT), the nature of the extra variables is known from the onset since these extra variables are identified as the flux of energy, flux of momentum, flux of mass or higher moments of the velocity distribution function. In IVT, the physical contents of the internal variables is generally not known from the onset but in most problems, these variables can be identified at the end of the procedure.

In IVT the internal variables do not appear in the usual balance equations of mass, momentum and energy as they are not controlled form the outside; in EIT, the fluxes of mass, momentum and energy which are introduced as variables take part in the mechanical work and the heat input.

Another difference between EIT and IVT is the form of the evolution equations; the variables introduced in EIT obey general evolution equations, which are partial differential equations involving space and time derivatives of the form

$$\tau \dot{J}^q = - \nabla . J^{qq} + \sigma^q \tag{6.1}$$

In contrast, the internal variables a^α satisfy time-differential equations like

$$\dot{a}^\alpha = f(a^\alpha, C^\alpha) \tag{6.2}$$

where C^α represents the set of classical variables : no divergence term is present in (6.2) as internal variables are not controllable through the boundaries of the system.

It should also be mentioned that EIT does not duplicate the results of IVT. In this respect, we refer to a work by Kestin and Bataille [44] whose objective is, like ours, to provide a thermodynamic description of viscoelasticity. In particular, Kestin and Bataille derive a relation between the relaxation times τ_σ and τ_ε of the form

$$\tau_\varepsilon = \frac{\gamma}{G} \tau_\sigma , \tag{6.3}$$

where γ is a constant coefficient. In EIT one obtains

$$\tau_\varepsilon = \frac{\eta}{G} + \tau_\sigma, \tag{6.4}$$

which differs from the IVT result by the property that τ_ε is non-zero even for a vanishing τ_σ.

The aim of section 5 was to show that linear viscoelasticity is easily and naturally incorporated into EIT. It was shown that EIT encompasses a wider class of materials than Maxwell's bodies; the Kelvin-Voigt, Poynting-Thomson, Jeffreys models are also recovered. The various material coefficients can be determined experimentally by measurements of wave velocity and wave attenuation [41].

It is worth repeating that the results of section 5 were obtained form very simple considerations : it was only required that σ''_{ij} be selected as independent variable and that it obeys a first-order time differential equation of the relaxation type. Of course, it is still possible to go beyond the linear models by allowing the stress tensor to satisfy non-linear evolution equations. In that respect, let us mention that EIT has been successfully used to derive a wide class of constitutive equations in rheology [47-49], like the Giesekus, Oldroyd and Reiner-Rivlin models. The structure of EIT is such that it exhibits a flexibility and a power of generalisation allowing it to cope with more and more sophisticated systems.

Every extra flux has been considered as an entity characterised by a single evolution equation. It is easily conceivable that some of them can be split into several independent contributions, each with their own evolution equation. This is not exceptional : typical examples are some viscoelastic models like Jeffreys, nonideal gases and polymers. EIT has been shown to be able to cope in a quite natural way with these situations : the results of this approach have also been widely confirmed by kinetic theories [19].

It should also be mentioned that in EIT, the space of the extra variables is not necessarily restricted to the ordinary dissipative fluxes. To describe the complexity of fast nonequilibrium phenomena, it may be necessary to introduce more variables like the fluxes of the fluxes as shown in section 2.4. The formalism of EIT leads then to a hierarchy of evolution equations displaying a wide variety of dynamical models [19].

Of course, it is required that in the particular case that the flux variables relax to their (local) equilibrium values - which means that the corresponding relaxation time tend to zero - one should recover the results of the classical theory of irreversible processes.

To **summarize**, it can be said that EIT was born of the necessity,

1. to go beyond the local equilibrium hypothesis,
2. to remove the paradox of infinite speed of propagation of signals; this physical unpleasant situation is a consequence of the parabolic nature of the evolution equations for temperature, concentration and velocity,
3. to provide an explicit way for calculating entropy and temperature outside (local) equilibrium,
4. to propose a formalism coping with high-frequency and (or) short wavelength phenomena as well as complicated systems like polymers, viscoelastic bodies, ...
5. to span the non-equilibrium microscopic theories, like non-equilibrium statistical mechanics by a macroscopic formalism.

ACKNOWLEDGMENTS

Parts of this text present results of the Belgian Interuniversity Poles of Attraction (PAI n°21 and 29) initiated by the Belgian State, Prime Minister's Office, Science Policy Programming. The scientific responsability is assumed by its author.

REFERENCES

1. Onsager, L.; Phys. Rev. **37**, 405 (1931); ibid **38**, 2265 (1931).
2. Truesdell, C., Rational Thermodynamics, Mc Graw Hill, New York 1969, 2nd ed. Springer, Berlin 1988.
3. Prigogine, I., Etude Thermodynamique des Phénomènes Irréversibles, Desoer, Liège, 1947.
4. Prigogine, I., Introduction to Thermodynamics of Irreversible Processes, Interscience, New York, 1961.
5. De Groot, S. and Mazur, P., Non-equilibrium Thermodynamics, North Holland, Amsterdam, 1962.
6. Coleman, B. and Owen, O., Arch. Rat. Mech. Anal. **54**, 1 (1974).
7. Eu, B., J. Chem. Phys. **73**, 2958 (1980) and several papers published in J. Chem. Phys.from 1980 to 1992.
8. Garcia-Colin, L., Rev. Mex. Fisica **34**, 344 (1988); Garcia-Colin, L. and Uribe, F., J. Non-Equil. Thermodyn., **16**, 89 (1991).
9. Nettleton, R., Physics of Fluids **3**, 216 (1960).
10. Müller, I., Z. Phys. **198**, 329 (1967).
11. Lebon, G., Bull. Acad. Roy. Sc. Belgique **64**, 456 (1978).
12. Jou, D., Casas-Vazquez, J. and Lebon, G., eds, Lecture Notes in Physics, Vol 199, Springer, Berlin, 1984.
13. Lebon, G., Casas-Vazquez, J. and Jou, D., J. Phys. **A13**, 275 (1980).
14. Maxwell, J.C., Phil. Trans. Roy. Soc London **157**, 49 (1867).
15. Grad, H., Principles of the Kinetic theory of gases, Enc. of Physics, Vol. XII, Springer, Berlin, 1958.
16. Vernotte, P., C.R. Acad. Sci. Paris **246**, 3154 (1958).
17. Cattaneo, C., Atti Sem. Mat. Fis. Univ. Modena **3**, 83 (1948).
18. Lambermont, J. and Lebon, G., Phys. Lett. **42A**, 499 (1973).
19. Jou, D., Casas-Vazquez, J. and Lebon, G., Rep. Prog. in Phys. **51**,1105 (1988); Lebon, G., Jou, D. and Casas-Vazquez, J., Contemporary Physics, **33**, 41 (1992); Jou, D., Casas-Vazquez, J. and Lebon, G., Extended Irreversible Thermodynamics, Springer, Berlin, 1993.
20. Joseph, O. and Preziosi, L., Rev. Mod. Phys. **61**, 41 (1989).

21. Luikov, A., Bubnov, V. and Soloviev, A., Int. J. Heat Mass Transfer **19**, 245 (1976).
22. Guyer, R. and Krumhansl, J., Phys. Rev., **133**, 1411 (1964).
23. Lebon, G. in Recent Developments in Mechanics of Solids, CISM, Courses and Lectures (Lebon, G. and Perzyna, P. eds) vol 262, 221, Springer, Berlin, 1980; Finlayson, B., The Method of Weighted Residuals and Variational Principles, Acad. Press, New-York, 1972.
24. Casas-Vazquez, J. and Jou, D., Acta Phys. Hungarica **66**, 99, (1989).
25. Meixner in Foundations of Continuum Thermodynamics (Delgado-Domingos, Nina and Whitelaw, eds) Wiley, New York, 1973.
26. Müller, I., Arch. Rat. Mech. Anal. **41**, 319 (1971).
27. Muschik, W., Arch. Rat. Mech. Anal. **4**, 379 (1977), and these lectures notes.
28. Thomson, W., Phil. Mag. **33**, 313 (1848).
29. Chapman, S. and Cowling, T., The Mathematical Theory of Nonuniform Gases, Cambridge Univ. Press, 1970.
30. Reichl, L., A Modern Course in Statistical Mechanics, Texas Univ. Press, Austin, 1982.
31. Suhubi, E., in Continuum Physics (A. C. Eringen ed.), vol. II, 173, Acad. Press, New York, 1975.
32. Chadwick, P., in Progress in Solid Mechanics (I. Sneddon ed.), Vol. I, 265, North-Holland, Amsterdam 1964.
33. Lord, H. and Shulman, Y., J. Mech. Phys. Solids **15**, 299 (1967).
34. Green, A. and Lindsay, K., J. of Elasticity **2**, 1 (1972).
35. Meixner, J., Z. Naturforschung **9**, 654 (1954).
36. Kluitenberg, G., in Non-equilibrium Therodynamics, Variational Techniques, and Stability (R. Donnelly, R. Herman, I. Prigogine, eds.), 91, University Chicago Press, Chicago, 1966.
37. Bataille, J. and Kestin, J., J. de Mécanique **14**, 365 (1975).
38. Rivlin, R. and Ericksen, J., J. Rat. Mech. Anal. **4**, 323 (1955).
39. Koh, S. and Eringen, C., Int. J. Engng. Sci. **1**, 199 (1963).
40. Huilgol, R. and Phan-Thien, N., Int. J. Engng. Sci. **24**, 161 (1986).
41. Lebon, G., Perez-Garcia, C. and Casas-Vazquez, J., J. Chem. Phys. **88**, 5068 (1988).
42. Lambermont, J. and Lebon, G., Int. J. Non-linear Mech. **9**, 55 (1974).

43. Coleman, B. and Gurtin, M., J. Chem. Phys. **47**, 597 (1964).

44. Kestin, J. and Bataille, J., In Continuum Models of Discrete Systems, Univ. of Waterloo Press, 1980.

45. Maugin, G. and Drouot, R., Int. J. Engng. Sci. **21**, 705 (1983), and these lectures notes.

46. Bampi, F. and Morro, A., in Lecture Notes in Physics, Vol 199 (Casas-Vazquez, J., Jou, D. and Lebon, G., eds.). Springer, Berlin, 1984.

47. Lebon, G. and Cloot, A., J. Non Newt. Fluid Mech. **28**, 61 (1988).

48. Lebon, G., Dauby, P., Palumbo, A. and Valenti, G., Rheol. Acta **29**, 127 (1990).

49. Lebon, G., in Advances in Thermodynamics Series, Extended Thermodynamic Systems, Vol 7, 310 (Sieniutycz, S. and Salomon, P., eds), Taylor and Francis, Bristol, 1992.

NON-EQUILIBRIUM THERMODYNAMICS OF ELECTROMAGNETIC SOLIDS

G.A. Maugin
C.N.R.S.
Université Pierre-et-Marie Curie, Paris, France
and
Institute for Advanced Study Berlin, Berlin, Germany

Contents

1. INTRODUCTION

The thermomechanics of electromagnetic continua is a branch of *energetics* which deals with a unification of continuum mechanics and electrodynamics of material media under the umbrella of general thermodynamics. This obviously goes in the direction indicated by the great P.Duhem early in this century (Duhem, 1911, 1914/1954). This ambitious , somewhat Aristotelian-like, scheme also adds one difficulty to the other. In effect, in addition to the cumbersome and rather heavy framework of nonlinear continuum mechanics (such as exposed in modern treatises, e.g., Truesdell and Toupin, 1960, Truesdell and Noll, 1965; Eringen 1980, Eringen 1971-1976), one has to consider *electromagnetism* (e.g., Jackson, 1962) and then combine them (in an *nonlinear manner* ; this is *not* a linear superposition) in the harmonious frame of thermodynamics. Some of the difficulties met have to do with the electrodynamics of *moving* bodies (writing of fields and equations in appropriate frames), while others relate to the introduction of a general deformation field ("material" writing of fields). Finally, there are difficulties connected with the inherent complexity of some of the behaviors (e.g., hysteresis), and even more so, the non-unique thermodynamical framework at the time of writing!

In spite of the many obstacles mentioned above and that one has to overcome to achieve as clean and rigorous an approach as possible, the most courageous and entrepreneur among us have not hesitated to attack such a formidable problem. Here a special tribute must be paid to the pioneers, R.A.Toupin, A.C.Eringen, H.F.Tiersten and D.F.Nelson who, in the Western World, have been so instrumental in organizing the general background. We are proud to have contributed to the most modern developments in the last two decades to the extent of our capabilities. The principal purpose of the present Lecture Notes is to present the latest developments which deal with *thermodynamically irreversible* effects in electromagnetic deformable solids. That is, afer recalling the more or less classical thermodynamical background for electromagnetic solids and the simplest irreversible behaviors (relaxation and classical conduction), the present lectures will be devoted to an organized description of more complex behaviors which necessitate the implementation of a richer thermodynamical approach, the so-called *thermodynamics with internal variables* (see Maugin and Muschik, 1992a,b, for a general introduction), as even *extended thermodynamics* in the sense of I.Müller (1985) and others (Jou *et al*, 1988) is insufficient to cope with these cases as it does not introduce sufficiently numerous thermodynamical state variables. The applications considered to illustrate our purpose are all recent and concern : *(i) the dielectric relaxation in deformable ceramics* and the consequences thereof in so far as nonlinear wave propagation is concerned, *(ii) magneto-mechanical hysteresis in ferromagnets* and its application to non-destructive techniques of measurement, and *(iii)* the construction of a rather complete phenomenological model of *elastic superconductors* which will prove useful with the forthcoming increasing implementation of high-temperature superconductors.

In the course of these lectures many analogies are drawn upon with the thermomechanics of solids exhibiting mechanical irreversible behaviors (viscosity, plasticity, locking) as exposed by others in this course and also by us in our book on the thermomechanics of plasticity (Maugin, 1991c). As a rule, we remain in the engineering context so that only classical nonrelativistic concepts are used. The background (modern continuum mechanics approach to electromagnetic bodies) is essentially found in the following books:

C.A.TRUESDELL and R.A.TOUPIN (1960), *The Classical Field Theories*, in: **Handbuch der Physik**, Bd.III/1, ed.S.Flügge, Springer-Verlag, Berlin.

A.C.ERINGEN (1971-1976; Editor),**Continuum Physics**, Vol.I to IV., Academic Press, New York (see especially contributions by R.A.Grot and G.A.Maugin in Vol.III).

H.PARKUS (1979;Editor), **Electromagnetic Interactions in Elastic Solids** (CISM Lecture Notes, 1977), Springer-Verlag, Wien.

A.C.ERINGEN (1980), **Mechanics of Continua** (revised and enlarged edition, see Chapter 10), Krieger, New York (First Edition: J.Wiley, New York, 1967).

D.F.NELSON (1979), **Electric, Optic and Acoustic of Dielectrics**,

J.Wiley-Interscience, New York

G.A.MAUGIN (1985),**Nonlinear Electromechanical Effects and Applications** (A Series of Lectures), World Scientific, Singapore.

G.A.MAUGIN (1988), **Continuum Mechanics of Electromagnetic Solids**, North-Holland, Amsterdam.

R.E.ROSENSWEIG (1989), *Thermodynamics of Electromagnetism*, in: **Thermodynamics : An Advanced Textbook for Chemical Engineers**, by G.Astarita, Plenum, New York.

A.C.ERINGEN and G.A.MAUGIN (1990), **Electrodynamics of Continua**, Two volumes, Springer-verlag, New York.

H.F.TIERSTEN (1990), **A Development of the Equations of Electromagnetism in Material Continua**, Springer-verlag, new York.

Modern developments concerning the case of dielectrics with applications to nonlinear phenomena may be found in:

G.A.MAUGIN, B.COLLET, R.DROUOT and J.POUGET (1992), **Nonlinear Electromechanical Couplings**, Manchester University Press, Manchester, U.K.

The most recent developments in the thermomechanics of solids including plasticity and some couplings with electromagnetic fields may also be found in :

G.A.MAUGIN (1991), **The Thermomechanics of Plasticity and Fracture**, Cambridge University Press, Cambridge, U.K.

G.A.MAUGIN (1992), **Material Inhomogeneities in Elasticity** (with applications to Fracture, Electrodynamics and Soliton Theory), Chapman and Hall, London.

2. REMINDER ON ELECTROMAGNETISM

Electromagnetic fields are of a different nature than mechanical fields dealt with in other lectures in this course. They are governed by a set of equations known as *Maxwell's equations*. As a matter of fact, after pioneering works by Coulomb, Gauss, Poisson, Oersted, Ampère, Faraday, Weber and others, Maxwell (1873) succeeded in formulating a coherent set of *dynamical* equations valid in a continuous material. At each regular point $\mathbf{x}$ in a material volume V at time t, this set reads[1]

$$\nabla \times \mathbf{E} + \frac{1}{c}\frac{\partial \mathbf{B}}{\partial t} = 0 \quad , \quad \nabla . \mathbf{B} = 0 \, ,$$
$$\nabla \times \mathbf{H} - \frac{1}{c}\frac{\partial \mathbf{D}}{\partial t} = \frac{1}{c}\mathbf{J} \, , \quad \nabla . \mathbf{D} = q_f \, , \tag{2.1}$$

together with

$$\mathbf{H} = \mathbf{B} - \mathbf{M} \quad , \quad \mathbf{D} = \mathbf{E} + \mathbf{P} \, . \tag{2.2}$$

In these equations the various symbols introduced bear the following significance: $\mathbf{E}$ is the electric field, $\mathbf{B}$ is the magnetic induction, $\mathbf{H}$ is the magnetic field, $\mathbf{D}$ is the electric displacement, $\mathbf{J}$ is the electric current, q_f is the volume density of free electric charges, and $\mathbf{M}$ and $\mathbf{P}$ are the magnetization and electric polarization per unit volume. The constant c is the velocity of light in vacuum. Here, $\mathbf{E}$, $\mathbf{D}$ and $\mathbf{P}$ are *polar* vectors while $\mathbf{H}$, $\mathbf{B}$ and $\mathbf{M}$ are *axial* vectors which reverse sign in time reversal. $\mathbf{H}$ and $\mathbf{D}$ differ from $\mathbf{B}$ and $\mathbf{E}$, respectively, only in matter. Thus the first two of eqns.(2.1) - Faraday's law and the equation indicating the nonexistence of magnetic monopoles - are valid everywhere including in a vacuum. The fields $\mathbf{M}$, $\mathbf{P}$, $\mathbf{J}$ and q_f relate to the presence of ponderable matter. Equations (2.1) can be deduced from a statistical average of microscopic Maxwell's equations with point-wise sources of charge and current (this is the point of view of H.A.Lorentz in his "theory of electrons", which has *not* been superseded so far). In that case $\mathbf{M}$, $\mathbf{P}$, $\mathbf{J}$ and q_f are given expressions in terms of the elementary charges and the motion (position and velocity) of these charges. These "microscopic" definitions may be helpful in establishing the invariance properties of these "material" fields, especially in so far as *objectivity* is concerned. However, in a *phenomenological* framework which is the one adopted here, q_f is a datum or the result of a computation while $\mathbf{M}$, $\mathbf{P}$ and $\mathbf{J}$ have to be given

[1] This is written here using so-called *Lorentz-Heaviside units* where neither factor 4π nor vacuum permeability μ_0 and dielectric constant ε_0 are involved; see Maugin ,1988, p.56 for other systems of units.

constitutive equations, e.g., to give an idea to the reader, *functional relations* of the type

$$\mathbf{M} = \mathbf{M}(\mathbf{H},.) \ , \ \mathbf{P} = \mathbf{P}(\mathbf{E},.) \ , \ \mathbf{J} = \mathbf{J}(\mathbf{E},.), \tag{2.3}$$

where the dot stands for some other variables such as temperature or the strain in a deformable solid. Note that the inverse relations, e.g., $\mathbf{E} = \mathbf{E}(\mathbf{P},.)$ or $\mathbf{D} = \mathbf{D}(\mathbf{E},.)$, $\mathbf{H} = \mathbf{H}(\mathbf{B},.)$, $\mathbf{E} = \mathbf{E}(\mathbf{J},.)$ may be preferable depending on where we want to place the emphasis.

It must be emphasized that eqns.(2.1) are expressed in a fixed frame $\mathcal{R}_L$ called the *laboratory frame.* It is known since Lorentz, Fitzerald, Larmor, Poincaré and Einstein that Maxwell's equations *in a vacuum* are invariant by the special-relativistic group of space-time transformations (so-called Lorentz-Poincaré group). For Maxwell's equations *in matter* the results depends on what one assumes for the transformation of the "material" fields $\mathbf{M}$, $\mathbf{P}$, $\mathbf{J}$ and q_f . being interested in material velocities much smaller, and dynamical processes must slower, than c, we may *impose* on eqns.(2.1) a restricted invariance which will be the same as the one already verified by the mechanical equations, i.e., the *Galilean invariance* of classical Newtonian mechanics (cf. de Groot and Suttorp, 1972; Maugin and Eringen, 1977; Maugin, 1988, Chapter 3). Bearing this in mind, it is found that eqns.(2.1) can be rewritten in a frame $\mathcal{R}_c(\mathbf{x},t)$ *co-moving* with the infinitesimal element of deformable matter at velocity $\mathbf{v}$ as

$$\begin{aligned} \nabla \times \mathcal{E} + \frac{1}{c}\overset{*}{\mathbf{B}} &= 0 \quad , \quad \nabla.\mathbf{B} = 0 \ , \\ \nabla \times \mathcal{H} - \frac{1}{c}\overset{*}{\mathbf{D}} &= \frac{1}{c}\mathcal{J} \ , \quad \nabla.\mathbf{D} = q_f \ , \end{aligned} \tag{2.4}$$

where a superimposed * denotes the convected time derivative such that

$$\begin{aligned} \overset{*}{\mathbf{P}} &\equiv \frac{\partial \mathbf{P}}{\partial t} + \nabla\times(\mathbf{P}\times\mathbf{v}) + \mathbf{v}(\nabla.\mathbf{P}) \\ &= \frac{d\mathbf{P}}{dt} - (\mathbf{P}.\nabla)\mathbf{v} + \mathbf{P}(\nabla.\mathbf{v}) \quad , \end{aligned} \tag{2.5}$$

and the "script upper case" fields are given by the following "Galilean transformation laws" between $\mathcal{R}_L$ and $\mathcal{R}_c(\mathbf{x},t)$:

$$\mathcal{E} = \mathbf{E} + \frac{1}{c}\mathbf{v}\times\mathbf{B} \quad , \quad \mathcal{H} = \mathbf{H} - \frac{1}{c}\mathbf{v}\times\mathbf{D} \; ,$$
$$\mathcal{B} = \mathbf{B} - \frac{1}{c}\mathbf{v}\times\mathbf{E} \quad , \quad \mathcal{H} = \mathcal{B} - \mathcal{M} \; ,$$
$$\mathcal{M} = \mathbf{M} + \frac{1}{c}\mathbf{v}\times\mathbf{P} \quad , \quad \mathcal{P} = \mathbf{P} \; , \tag{2.6}$$
$$\mathcal{J} = \mathbf{J} - q_f\mathbf{v} \; .$$

Clearly, in such transformations the symmetry between magnetization and polarization processes is lost (but this is fully justified by the microscopic definition of such fields; cf. Maugin, 1988). The field $\mathcal{E}$ is usually referred to as the *electromotive intensity* while $\mathcal{J}$ is called the *conduction current*. We shall say that a material is *nonmagnetizable* if and only if

$$\mathcal{M}(\mathbf{x},t) = 0 \; , \; \forall\, \mathbf{x}, \; \forall\, t \; , \; hence \;\; \mathbf{M} = \frac{1}{c}\mathbf{P}\times\mathbf{v} \; , \tag{2.7}$$

while for a *nonpolarizable* material

$$\mathcal{P} = \mathbf{P} = 0 \quad , \; \forall\, \mathbf{x} \; , \; \forall\, t \; . \tag{2.8}$$

The material is said to be a *conductor* of electricity if and only if $\mathcal{J}$ is nonzero; otherwise it is called an *insulator*. The material is said to be *electrically charged* if $q_f \neq 0$ and *uncharged* or *electrically neutral* if $q_f = 0$ at all points in the material. Finally, a material is called a *dielectric* when it is an insulator and $q_f = 0$. Most of electrically polarizable media (solids) are either dielectrics or semiconductors. Strongly magnetizable media may be insulators or conductors. Weakly magnetizable media may be extremely good conductors (in the limit, *perfect conductors)*. In metals, heat conduction usually accompanies electricity conduction. This is why *magneto-thermo-elasticity* is a fashionable subject of research. The range of possible behaviors thus is very wide.

Direct consequences of eqns.(2.1) are:

(i) the *law of conservation of electric charge* [by taking the divergence of $(2.1)_3$ and accounting for $(2.1)_4$]

$$\frac{\partial q_f}{\partial t} + \nabla.\mathbf{J} = 0 \; ; \tag{2.9}$$

(ii) an *energy identity* called the *"Poynting theorem"* [by multiplying scalarly $(2.1)_1$ by $\mathbf{H}$ and $(2.1)_3$ by $\mathbf{E}$ and making a combination]

$$\mathbf{H}.\frac{\partial \mathbf{B}}{\partial t} + \mathbf{E}.\frac{\partial \mathbf{D}}{\partial t} = - \mathbf{J}.\mathbf{E} - \nabla.\mathbf{S} \quad , \; \mathbf{S} \equiv c\, \mathbf{E} \times \mathbf{H} \tag{2.10}$$

or, using the fields introduced by (2.6),

$$\mathcal{H}.\overset{\star}{\mathbf{B}} + \mathcal{E}.\overset{\star}{\mathbf{D}} = - \mathcal{J}.\mathcal{E} - \nabla.\mathcal{S} \quad , \; \mathcal{S} \equiv c\, \mathcal{E} \times \mathcal{H} \; , \tag{2.11}$$

where nothing has been assumed concerning the behavior. These are mere identities and *not* statements of a law of thermodynamics ! To proceed further we need to introduce the *interactions* between electromagnetic fields and deformable matter in the fundamental *balance laws of thermomechanics*.

3. THERMOMECHANICS OF ELECTROMAGNETIC MATERIALS

Now we consider a material continuous body $\mathcal{B}$ which occupies the open , simply connected region V of Euclidean physical space $\mathbb{E}^3$ at time t in the current configuration $\mathcal{K}_t$ of continuum mechanics. Its regular boundary ∂V is equipped with unit outward normal $\mathbf{n}$. In the reference configuration $\mathcal{K}_R$ it occupies the regular region V with boundary ∂V of unit outward normal $\mathbf{N}$. The motion of the material continuum , for each t, ia a diffeomorphism of $\mathcal{K}_R$ onto $\mathcal{K}_t$ ($\mathbb{R}^3$ onto itself) such that

$$\mathbf{x} = \chi(\mathbf{X}, t) \; , \tag{3.1}$$

where $\mathbf{x}$, referred to coordinates x^i, i=1,2,3 , and $\mathbf{X}$, referred to coordinates X^K, K =1,2,3 , are positions in the Eulerian and Lagrangian descriptions, respectively.

We need the following elements of *deformation theory*. From (3.1) which is supposed to be invertible for each t, we define the physical velocity field $\mathbf{v}$ and the *direct motion* gradient $\mathbf{F}$ by

$$\mathbf{v} = \left.\frac{\partial \chi}{\partial t}\right|_{\mathbf{X}} \quad , \quad \mathbf{F} = \left.\frac{\partial \chi}{\partial \mathbf{X}}\right|_{t} = \nabla_R \chi \ . \tag{3.2}$$

Also

$$J_F = \det \mathbf{F} > 0 \ , \quad \mathbf{F}^{-1} = \left.\frac{\partial \chi^{-1}}{\partial \mathbf{x}}\right|_{t} = \nabla \chi^{-1} \ , \tag{3.3}$$

so that

$$\mathbf{F}.\mathbf{F}^{-1} = \mathbf{1} \quad , \quad \mathbf{F}^{-1}.\mathbf{F} = \mathbf{1}_R \quad , \tag{3.4}$$

and we can define the following *measures of finite strains* (T = transpose)

$$\mathbf{C} = \mathbf{F}^T.\mathbf{F} \quad , \ \mathbf{C}^{-1} = \mathbf{F}^{-1}.(\mathbf{F}^{-1})^T , \ \mathbf{E} = \frac{1}{2}(\mathbf{C} - \mathbf{1}_R) \ . \tag{3.5}$$

Then one checks that

$$\left.\frac{\partial \mathbf{E}}{\partial t}\right|_{\mathbf{X}} = \mathbf{F}^T.\mathbf{D}.\mathbf{F} \quad , \quad \mathbf{D} = \frac{1}{2}\{\nabla\mathbf{v} + (\nabla\mathbf{v})^T\} \equiv (\nabla\mathbf{v})_s \ . \tag{3.6}$$

Here, $\mathbf{C}$ is the Green strain tensor, $\mathbf{C}^{-1}$ is the Piola strain tensor, $\mathbf{E}$ is the Lagrangian strain tensor, and $\mathbf{D}$ is the rate-of-strain tensor.

Let the material body be acted on by mechanical surface tractions $\mathbf{T}^d$ at ∂V and physical forces (say, gravity) $\mathbf{f}$ in V, with a possible influx of heat $\mathbf{q}$ per unit area across ∂V and a supply of heat h per unit mass in V. If this body is a general electromagnetic body and is acted upon by electromagnetic fields, then the general balance laws of thermomechanics for that body can be written a priori in the following general form (see Maugin, 1988, Chapter 3):

* *Balance of mass:*

$$\frac{d}{dt}\int_V \rho \, dv = 0 \ ; \tag{3.7}$$

* *Balance of linear momentum:*

$$\frac{d}{dt}\int_V \rho\mathbf{v}\,dv = \int_V (\rho\mathbf{f} + \mathbf{f}^{em})\,dv + \int_{\partial V} (\mathbf{T}^d+\mathbf{T}^{em})\,da\ , \tag{3.8}$$

* *Balance of angular momentum:*

$$\frac{d}{dt}\int_V \left(\mathbf{r}\times\rho\mathbf{v}\right)dv = \int_V \left(\mathbf{r}\times\rho\mathbf{f} + \bar{\mathbf{c}}^{em}\right)dv + \int_{\partial V}\left(\mathbf{r}\times(\mathbf{T}^d+\mathbf{T}^{em})\right)da\ , \tag{3.9}$$

* *First law of thermodynamics:*

$$\frac{d}{dt}\int_V \left(\frac{1}{2}\rho\mathbf{v}^2 + \rho e\right)dv = \int_V \left(\rho\mathbf{f}.\mathbf{v} + \rho h + w^{em}\right)dv + \int_{\partial V}\left(\mathbf{T}^d.\mathbf{v} - \mathbf{q}.\mathbf{n}\right)da\ , \tag{3.10}$$

* *Second law of thermodynamics:*

$$\frac{d}{dt}\int_V \rho\eta\,dv \geq \int_V \rho h\theta^{-1}\,dv - \int_{\partial V} \theta^{-1}\mathbf{q}.\mathbf{n}\,da\ . \tag{3.11}$$

In these equations $\mathbf{r}$ is the radius vector, e is the density of internal energy, η is the density of entropy, and θ is the thermodynamical temperature ($\theta > 0$, $\inf\theta = 0$). The *interaction contributions* $\mathbf{f}^{em}$, $\mathbf{T}^{em}$, $\bar{\mathbf{c}}^{em}$, and w^{em} have to be determined by an analysis which is *foreign* to continuum mechanics , *per se*. There is *no* general agreement on their expressions as they depend either upon the taste of the scientist or on the quality and fineness of the model used to build them at a sub-macroscopic scale. A sensible approach based on an averaging procedure was proposed by Maugin and Eringen (1977) in the tradition set forth by Lorentz. We shall use their expressions. $\mathbf{T}^{em}$ here is irrelevant. For the other electromagnetic source terms we have (neglecting quadrupole contributions)

$$\mathbf{f}^{em} = q_f\mathcal{E} + \frac{1}{c}(\mathcal{J} + \overset{*}{\mathbf{P}})\times\mathbf{B} + (\mathbf{P}.\nabla)\mathcal{E} + (\nabla\mathbf{B}).\mathcal{M}\ , \tag{3.12}$$

$$\bar{\mathbf{c}}^{em} = \mathbf{r} \times \mathbf{f}^{em} + \mathbf{c}^{em} \quad , \tag{3.13}$$

$$w^{em} = \mathbf{f}^{em}.\mathbf{v} + \mathbf{c}^{em}.\Omega + \rho h^{em} \quad , \tag{3.14}$$

wherein (tr = trace)

$$\mathbf{c}^{em} = \mathbf{P} \times \mathcal{E} + \mathcal{M} \times \mathbf{B} \ , \tag{3.15}$$

$$\rho h^{em} = \mathcal{J}.\mathcal{E} + \mathcal{E}.\overset{\star}{\mathbf{P}} - \mathcal{M}.\overset{\star}{\mathbf{B}} + tr \ (\bar{\mathbf{t}}^{em}\mathbf{D}) \ , \tag{3.16}$$

and

$$\bar{\mathbf{t}}^{em} = \mathbf{P} \otimes \mathcal{E} - \mathbf{B} \otimes \mathcal{M} + (\mathcal{M}.\mathbf{B}) \ \mathbf{1} \tag{3.17}$$

together with

$$\Omega = \frac{1}{2} \ \nabla \times \mathbf{v} \ , \tag{3.18}$$

where $\otimes$ indicate the tensor product.

The local form of eqns.(3.7) through (3.11) is easily shown to be (the divergence of second-order tensors is taken on the *first* index)

$$\dot{\rho} + \rho \ \nabla.\mathbf{v} = 0 \ or \ \rho_o = \rho \ J_F \ \ in \ V \ , \tag{3.19}$$

$$\rho \ \dot{\mathbf{v}} = div \ \mathbf{t} + \rho\mathbf{f} + \mathbf{f}^{em} \quad , in \ V \ , \tag{3.20}$$

$$\mathbf{n}.\mathbf{t} = \mathbf{T}^{d} + \mathbf{T}^{em} \qquad at \ \partial V, \tag{3.21}$$

$$\rho \ \dot{e} = tr \ \{\mathbf{t}(\nabla\mathbf{v})^{T}\} - \mathbf{f}^{em}.\mathbf{v} + w^{em} - \nabla.\mathbf{q} + \rho h \quad in \ V \ , \tag{3.22}$$

$$\rho\dot{\eta} \geq \rho h\theta^{-1} - \nabla.(\mathbf{q}\theta^{-1}) \quad , \tag{3.23}$$

where ρ and ρ_o are the matter densities at $\mathcal{K}_t$ and $\mathcal{K}_R$, respectively, and a superimposed dot indicates material time differentiation for functions of $(\mathbf{x},t)$. Equation (3.20) has been used to transform the local form of (3.10).

Of special interest for further developments is the transformation of the local form of (3.22) and (3.23) of the first and second laws of thermodynamics. First we can transform the expression of w^{em} on account of the identity (2.10) or (2.11). Thus it can be shown that (prove this by way of exercise)

$$w^{em} = \mathbf{f}^{em}.\mathbf{v} + \rho\mathcal{E}.\dot{\pi} - \mathcal{M}.\dot{\mathbf{B}} + \mathcal{J}.\mathcal{E} \qquad (3.24)$$

or

$$w^{em} = \mathbf{J}.\mathbf{E} + \mathbf{E}.\frac{\partial\mathbf{P}}{\partial t} - \mathbf{M}.\frac{\partial\mathbf{B}}{\partial t} + \nabla.[\mathbf{v}(\mathbf{E}.\mathbf{P})] \;, \qquad (3.25)$$

or else

$$w^{em} = -\frac{\partial u^{em.f}}{\partial t} - \nabla.\{\mathbf{S} - \mathbf{v}(\mathbf{E}.\mathbf{P})\}, \qquad (3.26)$$

where

$$\pi = \mathbf{P}/\rho \quad , \; \mu = \mathcal{M}/\rho \quad , \quad u^{em.f} = \frac{1}{2}(\mathbf{E}^2+\mathbf{B}^2) \; . \qquad (3.27)$$

Accounting for the first of these in (3.22) yields the form

$$\rho\dot{e} = tr\,\{\mathbf{t}(\nabla\mathbf{v})^T\} - \nabla.\mathbf{q} + \rho h + \mathcal{J}.\mathcal{E} + \rho\mathcal{E}.\dot{\pi} - \mathcal{M}.\dot{\mathbf{B}} \quad . \qquad (3.28)$$

Introducing now the Helmholtz free energy density ψ by (Legendre transformation if $\theta = \partial e/\partial\eta$)

$$\psi = e - \eta\theta \;, \qquad (3.29)$$

and eliminating $\dot{e}$ between (3.28) and (3.23), we arrive at the so-called *Clausius-Duhem inequality in the form*

$$-\rho(\dot{\psi} + \eta\dot{\theta}) + tr\,\{\mathbf{t}(\nabla\mathbf{v})^T\} + \mathcal{J}.\mathcal{E} \qquad (3.30)$$
$$+ \rho\mathcal{E}.\dot{\pi} - \mathcal{M}.\dot{\mathbf{B}} + \theta\,\mathbf{q}.\nabla(\theta^{-1}) \geq 0 \; .$$

This, together with (3.28) can be further transformed by performing partial Legendre transformations and introduccing various types of time derivatives. For instance, introducing

$$\hat{e} = e + \mu.\mathbf{B} \quad , \; \hat{\psi} = \psi + \mu.\mathbf{B} \quad , \qquad (3.31)$$

eqn.(3.30) transforms to

$$-\rho(\dot{\hat{\psi}} + \eta\dot{\theta}) + tr\,\{\mathbf{t}(\nabla\mathbf{v})^T\} + \mathcal{J}.\mathcal{E} + \rho\mathcal{E}.\dot{\pi} + \rho\mathbf{B}.\dot{\mu} + \theta\,\mathbf{q}.\nabla(\theta^{-1}) \geq 0\,. \tag{3.32}$$

Alternately, defining $\bar{e}$ and $\bar{\psi}$ by

$$\bar{e} = e - \pi.\mathcal{E} = \hat{e} - \mu.\mathbf{B} - \pi.\mathcal{E}\,, \qquad \bar{\psi} = \bar{e} - \eta\theta = \hat{\psi} - \mu.\mathbf{B} - \pi.\mathcal{E}\,, \tag{3.33}$$

we can write (3.30) or (3.31) in the form

$$-\rho(\dot{\bar{\psi}} + \eta\dot{\theta}) + tr\,\{\mathbf{t}(\nabla\mathbf{v})^T\} + \mathcal{J}.\mathcal{E} - \mathbf{P}.\dot{\mathcal{E}} - \mathcal{M}.\dot{\mathbf{B}} + \theta\,\mathbf{q}.\nabla(\theta^{-1}) \geq 0\,. \tag{3.34}$$

The relevance of each of the inequalities (3.30) through (3.34) depends on the choice of independent variables to describe the interactions that take place between the Maxwellian electromagnetic fields $\mathcal{E}$ and $\mathbf{B}$ and the (material) polarization and magnetization fields. Furthermore, if we note that, e.g.,

$$\rho\dot{\pi} = \overset{\star}{\mathbf{P}} + (\mathbf{P}.\nabla)\mathbf{v}\,, \tag{3.35}$$

and define the symmetric stress $\mathbf{t}^E$ by

$$\mathbf{t}^E = (\mathbf{t} + \mathcal{E}\otimes\mathbf{P} + \mathbf{B}\otimes\mathcal{M})_S\,, \tag{3.36}$$

then, on account of the local form of (3.9),

$$\mathbf{t}^{em}_A = \mathbf{C}^{em} \quad, \mathbf{C}^{em} = -\,dual\ \mathbf{c}^{em}\,, \tag{3.37}$$

where A means symmetrization, we can rewrite (3.32) as

$$-\rho(\dot{\hat{\psi}} + \eta\dot{\theta}) + tr\,(\mathbf{t}^E\mathbf{D}) + \mathcal{J}.\mathcal{E} + \mathcal{E}.\overset{\star}{\mathbf{P}} + \mathbf{B}.\overset{\star}{\mathcal{M}} + \theta\,\mathbf{q}.\nabla(\theta^{-1}) \geq 0\,. \tag{3.38}$$

Another possibility is

$$-\rho(\dot{\bar{\psi}} + \eta\dot{\theta}) + tr\,(\bar{\mathbf{t}}^E\mathbf{D}) + \mathcal{J}.\mathcal{E} - \mathbf{P}.\overset{\star}{\mathcal{E}} - \mathcal{M}.\overset{\star}{\mathbf{B}} + \theta\,\mathbf{q}.\nabla(\theta^{-1}) \geq 0\,, \tag{3.39}$$

where

$$\bar{\mathbf{t}}^E = \mathbf{t}^E - (\mathcal{E}.\mathbf{P} + \mathcal{M}.\mathbf{B})\ \mathbf{1}\ .$$

This is not all in the case of deformable solids in finite strains where one often prefers to introduce *material* (Lagrangian) fields for both mechanical and electromagnetic entities (at points in space where this is meaningful for the latter). We remind the reader that this is achieved in the following manner for the electromagnetic fields. We look for material expressions of electromagnetic fields which allow one to express equations formally as (2.4) but in terms of Lagrangian fields which are functions of **X** and *t* only. As a matter of fact, the following "material" formulation of Maxwell 's equations in matter :

$$\nabla_R \times \mathfrak{E} + \frac{1}{c}\frac{\partial \mathfrak{B}}{\partial t}\Big|_{\mathbf{X}} = 0\ ,\quad \nabla_R . \mathfrak{B} = 0\ ,$$
$$\nabla_R \times \mathfrak{H} - \frac{1}{c}\frac{\partial \mathfrak{D}}{\partial t}\Big|_{\mathbf{X}} = \frac{1}{c}\mathfrak{J}\ ,\ \nabla_R . \mathfrak{D} = \mathfrak{Q}_f\ , \tag{3.40}$$

is obtained if we introduce the fields noted with Gothic letters by the following *operations of convection* (so-called "pull back") to the reference configuration $\mathcal{K}_R$ (see, e.g., Nelson, 1979 or Maugin, 1988):

$$\mathfrak{B} = J_F\ \mathbf{F}^{-1}.\mathbf{B}\quad ,\ \mathfrak{D} = J_F\ \mathbf{F}^{-1}.\mathbf{D}\ ,$$
$$\mathfrak{E} = \bar{\mathfrak{E}} - \frac{1}{c}\mathbf{V}\times\mathfrak{B}\quad ,\ \mathfrak{H} = \mathbf{H}.\mathbf{F} + \frac{1}{c}\mathbf{V}\times\mathfrak{D}\ , \tag{3.41}$$
$$\Pi = J_F\ \mathbf{F}^{-1}.\mathbf{P}\ ,\ \mathbb{M} = \mathcal{M}.\mathbf{F}\ ,\ \mathfrak{J} = J_F \mathbf{F}^{-1}.\mathcal{J}\ ,\ \mathfrak{Q}_f = J_F\ q_f\ ,$$

and

$$\bar{\mathfrak{E}} = \mathbf{E}.\mathbf{F}\quad ,\quad \bar{\mathfrak{B}} = \mathfrak{B} - \frac{1}{c}\mathbf{V}\times\bar{\mathfrak{E}} = J_F\ \mathbf{F}^{-1}.\mathcal{B} \tag{3.42}$$

with

$$\mathbf{V} = -\ \mathbf{F}^{-1}.\mathbf{v} = \frac{\partial \chi^{-1}}{\partial t}\Big|_{\mathbf{x}}\quad . \tag{3.43}$$

We then check that

$$\mathfrak{D} = J_F\ \mathbf{C}^{-1}.\bar{\mathfrak{E}} + \Pi\quad ,\ \mathfrak{H} = J_F^{-1}\ \mathbf{C}.\bar{\mathfrak{B}} - \mathbb{M} \tag{3.44}$$

which replace the simple (spatial) relations (2.2).

Introducing now the so-called second Piola-Kirchhoff (or thermodynamical) "elastic" stress $\mathbf{S}^E$ by

$$\mathbf{S}^E = J_F\, \mathbf{F}^{-1}.\mathbf{t}^E.(\mathbf{F}^{-1})^T \quad , \tag{3.45}$$

and the fields

$$\mathbf{\Omega} = J_F\, \mathbf{F}^{-1}.\mathbf{q} \quad , \qquad \nabla_R \mathcal{A} = (\nabla \mathcal{A}).\mathbf{F} \; , \tag{3.46}$$

by multiplication by J_F of eqns such as (3.38) and noting that

$$\overset{*}{\mathbf{P}} = J_F^{-1}\, \mathbf{F}.\left.\frac{\partial \Pi}{\partial t}\right|_{\mathbf{X}} \quad , \quad \overset{*}{\mathcal{M}} = J_F^{-1}\, \mathbf{F}.\left.\frac{\partial \bar{\mathbf{M}}}{\partial t}\right|_{\mathbf{X}} \; , \tag{3.47}$$

where

$$\bar{\mathbf{M}} = J_F\, \mathbf{C}^{-1}.\mathbf{M} = J_F\, \mathbf{F}^{-1}.\mathcal{M} \; ,$$

we readily show that eqs. (3.38) and (3.39) transform to

$$\begin{aligned} -(\dot{\hat{\Psi}} + N\dot{\theta}) + tr\,\{\mathbf{S}^E\dot{\mathbb{E}}\} + \mathfrak{J}.\mathfrak{E} \\ + \bar{\mathfrak{E}}.\dot{\Pi} + \mathfrak{B}.\dot{\bar{\mathbf{M}}} + \theta\, \mathbf{\Omega}.\nabla_R(\theta^{-1}) \geq 0 \end{aligned} \tag{3.48}$$

or

$$\begin{aligned} -(\dot{\bar{\Psi}} + N\dot{\theta}) + tr\,\{\bar{\mathbf{S}}^E\dot{\mathbb{E}}\} + \mathfrak{J}.\mathfrak{E} \\ - \Pi.\dot{\bar{\mathfrak{E}}} - \bar{\mathbf{M}}.\dot{\mathfrak{B}} + \theta\, \mathbf{\Omega}.\nabla_R(\theta^{-1}) \geq 0 \; , \end{aligned} \tag{3.49}$$

where we have set

$$N = \rho_o\, \eta \quad , \quad \hat{\Psi} = \rho_o \hat{\psi} \quad , \quad \bar{\Psi} = \rho_o \bar{\psi} \quad . \tag{3.50}$$

Any of the forms of the Clausius-Duhem inequality obtained may be considered as a constraint imposed on the formulation of constitutive equations for the fields, e.g.,

$$\Psi \; , \; N \; , \; \bar{\mathbf{S}}^E \; , \; \mathfrak{J} \; , \; \Pi \; , \; \bar{\mathbf{M}} \; , \; \mathbf{\Omega} \; , \tag{3.51}$$

or any equivalent set defined, for instance, through partial Legendre transformations. This requirement, usually referred to as the requirement of *thermodynamical admissibility* , is at the basis of further developments.

Technical remark. The above derivation does pay attention to the *tensorial variance* of various geometrical objects so that we have straightforwardly applied derivatives such as the one marked * to objects which , in fact , do not have the same variance. One could think that those are irrelevant mathematical details as we work in a Cartesian framework (at least in $\mathcal{K}_t$). But this is *not* the case because the *material manifold* used to describe the reference configuration *does make that difference* [e.g., between the elements of the pair $(\mathbf{P},\mathcal{M})$ or $(\mathcal{E},\mathbf{B})$]. A correct derivation is to be found (in the absence of dissipative processes) in Maugin, 1992b, Chapter 8.

4. CLASSICAL IRREVERSIBLE BEHAVIORS

We call *classical* irreversible behaviors of electromagnetic solids, those behaviors which can practically be "read" from the expression of the Clausius-Duhem inequality as their description does not necessitate the introduction of additional entities (e.g., extra thermodynamical variables of state). The corresponding formulation is the so-called T.I.P (*Theory of Irreversible Processes)*. From here on we shall specialize to the case of *solids* . The basic nondissipative mechanical behavior of a solid is *thermoelasticity* for which, classically,

$$\Psi = \Psi(\mathbf{E},\theta) \qquad , \ \Omega \equiv 0 \ . \tag{4.1}$$

Gibbs' equation , after Legendre transformation of the energy density, yields

$$d\Psi = tr\,\{\mathbf{S}^{E}d\mathbf{E}\} - N\,d\theta \ , \tag{4.2}$$

so that at *thermodynamical equilibrium*:

$$\mathbf{S}^{E} = \partial\Psi/\partial\mathbf{E} \qquad , \qquad N = -\,\partial\Psi/\partial\theta \ . \tag{4.3}$$

In *electromagnetoelasticity,* in the absence of dissipative processes, this generalizes to

$$\Psi = \bar{\Psi}(\mathbb{E}, \bar{\mathfrak{C}}, \mathfrak{B}, \theta) \quad , \mathfrak{Q} \equiv 0 , \mathfrak{J} \equiv 0 \quad , \tag{4.4}$$

and

$$d\Psi = tr\,\{\mathbf{S}^{E} d\mathbb{E}\} - \Pi . d\bar{\mathfrak{C}} - \bar{\mathbb{M}} . d\mathfrak{B} - N\, d\theta , \tag{4.5}$$

so that at thermodynamical equilibrium we have the following laws of state:

$$\mathbf{S}^{E} = \partial\bar{\Psi}/\partial\mathbb{E} \quad , \Pi = -\,\partial\bar{\Psi}/\partial\bar{\mathfrak{C}} \, , \quad \bar{\mathbb{M}} = -\,\partial\bar{\Psi}/\partial\mathfrak{B} \, , \quad N = -\,\partial\bar{\Psi}/\partial\theta \, . \tag{4.6}$$

Sligthly outside equilibrium, the temperature field θ and the electric potential ϕ may be spatially nonuniform, but we shall assume that entropy still assumes its thermostatic definition.This holds true if the slow thermo*dynamical* evolution is conceived as a succession of equilibria for which $(4.3)_2$ or $(4.6)_4$ is essentially valid. This is an expression of the *axiom of local (thermodynamical) state*. Then we shall consider

$$N = -\,\partial\bar{\Psi}/\partial\theta \, . \tag{4.7}$$

However, we define the following quantities outside equilibrium (i.e., *deviations* from the definition at equilibrium)

$$\mathbf{S}^{V} \equiv \mathbf{S}^{E} - \partial\bar{\Psi}/\partial\mathbb{E} \, , \tag{4.8}$$
$$\Pi^{d} \equiv \Pi + \partial\bar{\Psi}/\partial\bar{\mathfrak{C}} \, , \quad \bar{\mathbb{M}}^{d} = \bar{\mathbb{M}} + \partial\bar{\Psi}/\partial\mathfrak{B} \quad ,$$

so that the Clausius-Duhem inequality (3.49) provides the *dissipation inequality* in the form

$$\Phi \equiv tr\,\{\mathbf{S}^{V}\dot{\mathbb{E}}\} - \Pi^{d}.\dot{\bar{\mathfrak{C}}} - \bar{\mathbb{M}}^{d}.\dot{\mathfrak{B}} + \mathfrak{J}.\mathfrak{C} + \theta\, \mathfrak{Q}.\nabla_{R}(\theta^{-1}) \geq 0 \, . \tag{4.9}$$

Notice that in a more general thermodynamics , instead of (4.7) we could also introduce a "dissipative" entropy N^{d} by

$$N^{d} = N + \partial\bar{\Psi}/\partial\theta \, . \tag{4.10}$$

In any case the dissipation inequality is in the *bilinear* form

$$\sum_{\alpha} Y_\alpha \dot{X}_\alpha \geq 0 \tag{4.11}$$

favored by the tenants of *T.I.P*, with equilibrium defined by

$$\nabla_R \mathbf{v} = 0 \quad , \quad \Pi^d = 0 \ , \ \bar{\mathcal{M}}^d = 0 \ , \ \bar{\mathfrak{E}} = 0 \ , \ \nabla_R \theta = 0 \tag{4.12}$$

The simplest idea, on account of the slightness of deviations from equilibrium, is to consider the quantities dual to those appearing in (4.12) via the duality inherent in (4.9) or (4.11) as *linear* in the respective fields. Thus

$$\begin{aligned} \mathbf{S}^V &= \bar{\mathcal{L}}_S[\dot{\mathbf{E}}] \quad , \ \dot{\bar{\mathfrak{E}}} = -\bar{\mathcal{L}}_\pi[\Pi^d] \quad , \ \dot{\mathcal{B}} = -\bar{\mathcal{L}}_M[\bar{\mathcal{M}}^d] \quad , \\ \mathfrak{J} &= \bar{\mathcal{L}}_E[\mathfrak{E}] + \bar{\mathcal{L}}_{EQ}[\nabla_R \theta] \quad , \ \mathcal{Q} = \bar{\mathcal{L}}_Q[\nabla_R \theta] + \bar{\mathcal{L}}_{QE}[\mathfrak{E}] \ , \end{aligned} \tag{4.13}$$

where the $\bar{\mathcal{L}}$'s are *linear operators which are homogeneous of degree one*. It is clear that $(4.13)_{4-5}$ describe coupled electricity and heat *conductions* while $(4.13)_1$ refers to *viscoelasticity* in the manner of Kelvin and Voigt, and $(4.13)_{2-3}$ describe one type of electric and magnetic *relaxations*. Once the relaxation is achieved, both polarization and magnetization recover their equilibrium definition given in eqns.(4.6). *Onsager's relations* may be invoked to establish a necessary relationship between the operators $\bar{\mathcal{L}}_{EQ}$ and $\bar{\mathcal{L}}_{QE}$.

The above-given approach is the one to be found in classical books so that we do not dwell in greater detail in this. However, the expressions (4.13) deserve the following comments. First, in general the linear operators introduced in eqns.(4.13) may still contain the thermodynamical variables of state on which Ψ depended to start with, as this dependence is *not excluded* by any thermodynamical principle. Thus equations (4.13) are potentially rich of many coupled effects including, for instance, dependence of relaxation times on the electric field and strain, and temperature, and all the *thermo-galvanomagnetic* effects which occur in eqns.$(4.13)_{4-5}$. For these, including the Hall effect, we refer the reader to Eringen (1980, Chapter 10), Maugin (1988, Chapter 3), and Eringen and Maugin (1990, Vol.I). Second, typically, an equation such as $(4.13)_2$ will read (for *isotropy* to simplify the presentation)

$$\frac{\partial\bar{\mathfrak{E}}}{\partial t} = -\frac{\alpha(\theta)}{\tau_E(\theta)}(\Pi - \Pi^r) \quad , \quad \Pi^r \equiv -\partial\bar{\Psi}/\partial\bar{\mathfrak{E}} \ , \tag{4.14}$$

where $\tau_E > 0$ is a relaxation time and α will be chosen for convenience. For a *linear* nondissipative behavior, $\Psi \cong \bar{\mathfrak{E}}^2$ and $\Pi^r = \chi_p(\theta)\ \bar{\mathfrak{E}}$, where $\chi_p > 0$ is the electric susceptibility. We can take $\alpha = \chi_p^{-1}(\theta)$. Then eqn. $(4.14)_1$ in this case may also be written as

$$\Pi = \chi_p(\theta)\left(\bar{\mathfrak{E}} - \tau_E\frac{\partial\bar{\mathfrak{E}}}{\partial t}\right) \ . \tag{4.15}$$

Similarly, for magnetic processes, working along the same line, we would obtain the "relaxation" equation

$$\bar{\mathbb{M}} = \gamma_M(\theta)\left(\mathfrak{B} - \tau_M\frac{\partial\mathfrak{B}}{\partial t}\right) \ . \tag{4.16}$$

Had we considered the formulation (3.48) to start with, eqns. (4.11) would have been replaced by the set

$$\begin{aligned} &\dot{S}^V = \hat{\mathfrak{L}}_S[\mathbb{E}] \quad , \quad \dot{\Pi} = \hat{\mathfrak{L}}_\pi[\bar{\mathfrak{E}}^d] \quad , \quad \dot{\bar{\mathbb{M}}} = \hat{\mathfrak{L}}_M[\mathfrak{B}^d] \ , \\ &\mathfrak{J} = \hat{\mathfrak{L}}_E[\mathfrak{E}] + \hat{\mathfrak{L}}_{EQ}[\nabla_R\theta] \quad , \quad \mathfrak{Q} = \hat{\mathfrak{L}}_Q[\nabla_R\theta] + \hat{\mathfrak{L}}_{QE}[\mathfrak{E}] \ , \end{aligned} \tag{4.17}$$

where the $\hat{\mathfrak{L}}$'s are a priori different from the $\bar{\mathfrak{L}}$'s in (4.13). Typically, the relaxation equation for Π would read

$$\frac{\partial\Pi}{\partial t} = \frac{\beta(\theta)}{\tau_M(\theta)}\left(\bar{\mathfrak{E}} - \mathfrak{E}^r\right) \quad , \quad \mathfrak{E}^r \equiv \partial\hat{\Psi}/\partial\Pi \ . \tag{4.18}$$

With $\hat{\Psi} \cong \Pi^2$ and $\mathfrak{E}^r = \chi_p^{-1}(\theta)\ \Pi$, we can take $\beta(\theta) = \chi_p(\theta)$, and (4.18) will take on the typical form

$$\tau_\pi(\theta)\ \dot{\Pi} + \Pi = \chi_p(\theta)\ \bar{\mathfrak{E}} \ . \tag{4.19}$$

Obviously, eqns. (4.15) and (4.19) are different although they admit the same "equilibrium" limit. The truth, however, may be a mixture of the two formulas (4.15) and (4.19) involving *two* relaxation times. Such more complicated relaxation formulas, like those occuring for rheological models more involved than the Kelvin-Voigt and Maxwell models of viscoelasticity, require the consideration of *internal variables* to be justified on a thermodynamical basis. Similar to (4.19), for *magnetization*, we would obtain

$$\tau_M(\theta)\ \dot{\bar{M}} + \bar{M} = \gamma_M(\theta)\ \mathcal{B}\ , \tag{4.20}$$

or

$$\frac{\partial \bar{M}}{\partial t} = -\frac{1}{\tau_M(\theta)}\ (\bar{M} - M_{eq})\ , \tag{4.21}$$

where we have introduced the equilibrium value

$$M_{eq} = \tau_M(\theta)\gamma_M(\theta)\ \mathcal{B}\ . \tag{4.22}$$

Equation (4.21) is of the standard *Bloch type* in nuclear magnetism. Generalizations to the case of *deformable ferromagnets* with $|\mathcal{M}|$ = const. and nonsaturated *magnetic fluids* have been given elsewhere and allow one to produce a nice thermodynamical formulation of the *Gilbert* and *Landau-Lifshitz* types of *spin-lattice relaxation* in ferromagnets (see, Maugin, 1988, Chapter 6; also Maugin ,1979b) and so-called *ferrofluids* (Maugin and Drouot ,1983) where orientational relaxation and relaxation in magnitude (of magnetization) are separated.

5. THERMODYNAMICS WITH INTERNAL VARIABLES

The thermodynamics with internal variables (see ,e.g., Muschik, 1990a,b ; Maugin and Muschik , 1992) provides a characterization of dissipative continuous media in which, to define the thermodynamical state of a system, one needs to introduce, besides *observable* variables of state, a certain number of *internal* variables, collectively denoted by an n-Cartesian vector α, which can eventually be measured by a "gifted" experimentalist but cannot be *controlled* (e.g., through dual forces applied at the boundary). These are supposed to describe the internal structure (hidden to the macroscopic observer who sees only a black box) of the medium which gives rises to certain dissipative processes during structural rearrangements. The energy density (say, the free energy) will thus depend on α and these additional variables α will need new equations which will usually be *evolution equations* constrained to satisfy the second law of thermodynamics. It is clear that the basic idea for this finds its origin in the kinetic description of physico-chemical processes (Bridgman , 1941).

This characterization of a thermodynamical state by additional state variables is conceptually and mathematically *simple* , provided the number of internal variables introduced remains small as one has to identify them precisely through essentially physical insight. In addition, the advantage is that this characterization does not alter any of the other statements of thermomechanics and, therefore, it allows one to use the tenets of *T.I.P* since, here also, no large deviations from thermodynamical equilibrium are envisaged. This, in our opinion, is a *wise* generalization of classical *T.I.P.* Central to the development of this thermodynamics is the notion of *local* (thermodynamical) *accompanying state,* as the main problem still resides in the definition of entropy outside equilibrium. Again, generalizing (4.7) we must be confident that if all internal variables return sufficiently fast to their equilibrium values, which make zero the associated generalized thermodynamical forces, then the thermodynamical evolution may be considered as a succession of *constrained* equilibrium states in so far as entropy is concerned. This will be essentially true if the characteristic times of internal structural rearrangements, or return to equilibrium of internal variables, are small compared to the time characteristic of external loadings. That is, let τ_α be the first characteristic time such that $\tau_\alpha = \alpha/\dot{\alpha}$, and $\tau_m = a/\dot{a}$ the characteristic time of external loads a. The *axiom of local accompanying state* will be the better as the smaller will be the ratio provided by the *Deborah* number $\mathcal{D}$ such that (cf. Bataille and Kestin, 1975)

$$\mathcal{D} = \tau_\alpha/\tau_m \quad . \tag{5.1}$$

For $\mathcal{D}$ going to zero, we can use the formulas of thermostatics (thermodynamical equilibrium). According to this, only *pairs* "material-process" can be classified in thermodynamics , and *not only* materials, per se. This classification would be simple if, in general, there did not exist *several* external characteristic times. For the time being , however, we shall assume that $\tau_\alpha \ll \tau_m$ for all α, and we illustrate our concern with the case of *nonmagnetizable dielectrics*. All we need to note as a general background is that for electromagnetic media the introduction of internal variables of mechanical and electromagnetic natures is due to Kluitenberg (1973, 1977, 1981a,b) and Maugin (1979a,b; 1981a,b) but the idea goes back to works by Meixner (1961). This idea proves to be particularly fertile in ,and well adapted to, the treatment of relaxation and hysteresis effects in ceramics and hard ferromagnets.

In the case of *nonmagnetizable dielectrics*, the Clausius-Duhem inequality (3.48) reduces to

$$-(\dot{\hat{\Psi}} + N\dot{\theta}) + tr\,\{\mathbf{S}^E\dot{\mathbb{E}}\} + \bar{\mathcal{E}}.\dot{\Pi} + \theta\,\mathbf{\Omega}.\nabla_R(\theta^{-1}) \geq 0\ . \tag{5.2}$$

To make the essentials of the theory clear to the reader let us consider the following thermodynamical scheme which obviously draws on the modern formulation of *elastoplasticity with hardening* (see, for instance, previous contributions in this course, and Maugin, 1991c, Chapters 3 and 8). Assume that the total strain $\mathbb{E}$ is composed of an "elastic" part $\mathbb{E}^e$ and an "anelastic" part $\mathbb{E}^p$ (the "p" usually standing for "plastic"). In the like manner , the polarization Π is built of two contributions, one Π^r called the *reversible* part and the other Π^R called the *residual* polarization, i.e.,

$$\mathbb{E} = \mathbb{E}^e + \mathbb{E}^p\ , \qquad \Pi = \Pi^r + \Pi^R\ . \tag{5.3}$$

In addition, we know that rather complicated microscopic mechanisms are at work (irreversible slip motions in the crystal lattice, dislocations, irreversible motions of ferroelectric domain walls, Barkhausen effect,..) which imply a *dissipation* since they all have an irreversible nature. Avoiding to enter a detailed description of these mechanisms, we compensate for our ignorance of these by the a priori introduction of *internal variables*, say α *and* Π^{int}, respectively a set of variables accounting for mechanical hardening and a variable having the physical dimension of an electric polarization per unit volume. The observable variables now are the "elastic"strain $\mathbb{E}^e$, the reversible polarization Π^r and the temperature θ. Accordingly, Gibbs' equation that replaces (4.5) reads

$$d\Psi = -N\, d\theta + tr\{\mathbf{S}^e d\mathbb{E}^e\} + \mathfrak{E}^r.d\Pi^r - \mathbf{A}.d\alpha - \mathfrak{E}^{int}.d\Pi^{int}, \tag{5.4}$$

and this yields the *laws of state* as

$$\Psi = \hat{\Psi}(\theta, \mathbb{E}^e, \Pi^r | \alpha, \Pi^{int}) \ , \tag{5.5}$$
$$N = -\partial\hat{\Psi}/\partial\theta \ , \ \mathbf{S}^e = \partial\hat{\Psi}/\partial\mathbb{E}^e, \ \mathfrak{E}^r = \partial\hat{\Psi}/\partial\Pi^r \ ,$$
$$\mathbf{A} = -\partial\hat{\Psi}/\partial\alpha \ , \quad \mathfrak{E}^{int} = -\partial\hat{\Psi}/\partial\Pi^{int} \ .$$

On account of the axiom of local accompanying state, N is formally given by (5.5) even slightly outside equilibrium, so that the Clausius-Duhem inequality (5.3) takes on the form of the following *dissipation inequality* :

$$\Phi \equiv tr\ \{\mathbf{S}^V\dot{\mathbb{E}}^e\} + tr\ \{\mathbf{S}^E\dot{\mathbb{E}}^p\} + \mathbf{A}.\dot{\alpha} + \mathfrak{E}^{relax}.\dot{\Pi}^r \tag{5.6}$$
$$+ \bar{\mathfrak{E}}.\dot{\Pi}^R + \mathfrak{E}^{int}.\dot{\Pi}^{int} + \theta\, \mathbf{\Omega}.\nabla_R(\theta^{-1}) \geq 0 \ ,$$

where we have set

$$\mathbf{S}^V \equiv \mathbf{S}^E - \mathbf{S}^e \ , \ \mathfrak{E}^{relax} \equiv \bar{\mathfrak{E}} - \mathfrak{E}^r \ . \tag{5.7}$$

The inequality (5.6) is rewritten as

$$\Phi = \Phi_1 + \Phi_2 \geq 0 \tag{5.8}$$

with

$$\Phi_2 \equiv tr\ \{\mathbf{S}^V\dot{\mathbb{E}}^e\} - \mathfrak{E}^{relax}.\dot{\Pi}^r + \theta\, \mathbf{\Omega}.\nabla_R(\theta^{-1}) \ , \tag{5.9}$$
$$\Phi_1 \equiv tr\ \{\mathbf{S}^E\dot{\mathbb{E}}^p\} + \mathbf{A}.\dot{\alpha} + \bar{\mathfrak{E}}.\dot{\Pi}^R + \mathfrak{E}^{int}.\dot{\Pi}^{int} \ .$$

Equations (5.6) and (5.9) deserve the following comments. First, we see by comparison with the general case that the roles of $\mathbb{E}$ and Π are now played by $\mathbb{E}^e$ and Π^r in the part Φ_2 of the dissipation. This means that the usual viscosity , electric relaxation, and electric conduction processes are expressed, according to *T.I.P*, in terms of time rates of change of the *reversible* parts of strain and electric polarization (compare to Section 4) and the temperature gradient. In contradistinction, the *irreversible* parts of strain and electric polarization ($\mathbf{E}^P$ and Π^R) produce a dissipation in the *total* mechanical symmetric stress $\mathbf{S}^E$ and the *total* electric field $\bar{\mathcal{E}}$. Finally, the split (5.8) is explained by the fact that the thermodynamical fluxes in Φ_2 are assumed to be derivable from a *dissipation potential* which is homogeneous of degree *two* in the corresponding forces (this is equivalent to *T.I.P* - as the fluxes will then be linear, homogeneous of degree *one* in these forces) while the time rates in Φ_1 will be assumed to derive from a dissipation potential which is homogeneous of degree *one* only in the corresponding thermodynamically conjugated quantities , providing thus the needed *evolution equations* for the internal variables as well as *plasticity* and *hysteresis* evolution equations (without time scale) for the irreversible parts of strain and electric polarization. The first characterization of electromagnetic dissipative processes is rather similar to what was considered in Section 4 so that we do not repeat it in detail (see also the *magnetic case* in Maugin, 1991a,b). An illustration is given in the next section for *dielectric relaxation*. For the second class of dissipative processes, an illustration is given in Section 7 for both electric and magnetic cases.

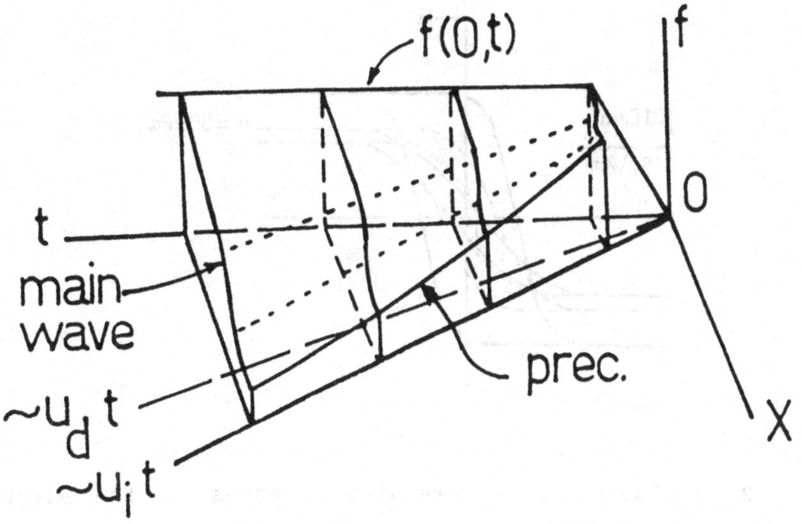

Figure 6.1. Response of a half-space with electric relaxation to a ramp f(0,*t*) in stress at the limiting plane X = 0 (after Maugin *et al* , 1992, Chapter 4).

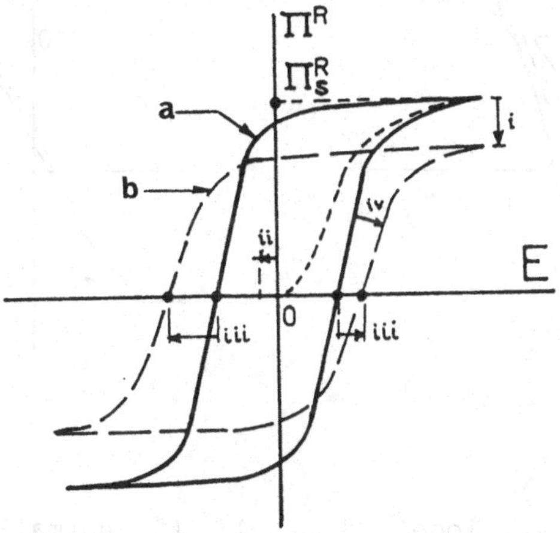

Figure 7.1. Typical hysteresis behavior of ceramics: (*a*) in the absence of applied stress and bias electric field; (*b*) in the presence of such fields.

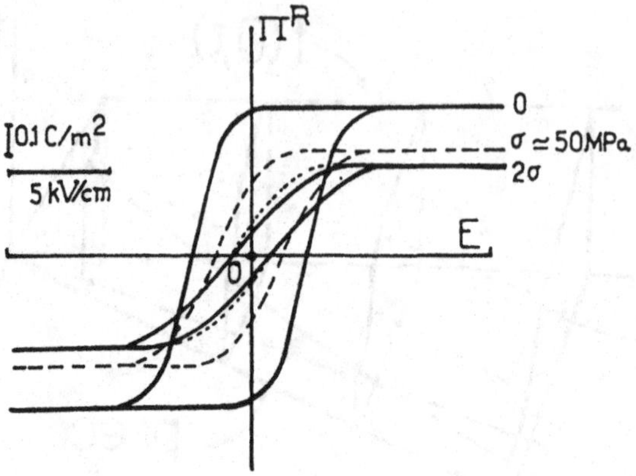

Figure 7.2. Influence of compressive stresses on the electric hysteresis loop of PLZT ceramics (after Bassiouny and Maugin, 1989a).

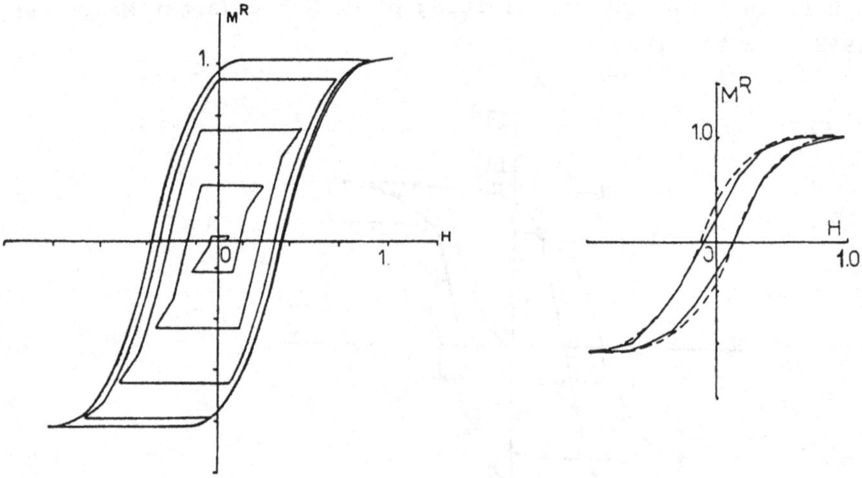

Figure 7.3. Hysteresis loop for steel (in normalized units) at zero stress: —— experimental curve after Jiles and Atherton (J.Phys.D.Appl.Phys.,**17**,1984,p.1273) and - - - "good" theoretical fit (M.Sabir, *Thesis*, Paris, 1988).

6. DIELECTRIC RELAXATION IN CERAMICS

To illustrate the viewpoint developed in Sections 4 and 5 for *dielectric relaxation* we shall consider the problem of the propagation of *transient* plane *nonlinear waves* in a one-dimensional model of ferroelectric ceramics. The problem was treated by Collet (1987) -see also Maugin, Collet *et al* (1992, Chapter 4). In this modelling we assume that

$$\alpha = 0 \quad , \mathbb{E}^{P} = 0 \quad , \Pi^{R} = 0 \ . \tag{6.1}$$

Accordingly, $\mathbb{E}^{e}$ reduces to $\mathbb{E}$ and Π^{r} to Π , so that the free energy $\hat{\Psi}$ reduces to

$$\Psi = \hat{\Psi}(\mathbb{E},\Pi,\theta \mid \Pi^{int}) \ . \tag{6.2}$$

The corresponding internal energy reads (by partial Legendre transformation)

$$\Sigma = \hat{\Psi} + N\theta = \hat{\Sigma}(\mathbb{E},\Pi,N|\Pi^{int}) \ ,$$

and the *laws of state* are

$$\theta = \partial\hat{\Sigma}/\partial N \quad , \ S^{E} = \partial\hat{\Sigma}/\partial\mathbb{E} \ , \tag{6.3}$$
$$\mathfrak{E} = \partial\hat{\Sigma}/\partial\Pi \quad , \ \mathfrak{E}^{int} = -\ \partial\hat{\Sigma}/\partial\Pi^{int} \ ,$$

where the last one merely is a definition of $\mathfrak{E}^{int}$. We have assumed that $\mathfrak{E}^{relax} = 0$, and the quasi-electrostatics framework is sufficient (hence $\bar{\mathfrak{E}}$ reduces to $\mathfrak{E}$). Then the electric relaxation, if any, is taken care of via the dependence of energy on Π^{int}only. As there is no other dissipative process (no heat conduction ; remember also that we have a *dielectric*), the *residual dissipation inequality* reads

$$\Phi_2 = \mathfrak{E}^{int}.\dot{\Pi}^{int} \geq 0 \ . \tag{6.4}$$

By way of example we may consider the *separable* case for which

$$\hat{\Sigma} = \Sigma_1(\mathbb{E},\Pi,N) + \Sigma_2(\Pi,\Pi^{int}), \tag{6.5}$$

with

$$\Sigma_2(\Pi, \Pi^{int}) = \frac{1}{2}\,\bar{a}\,\Pi^2 + \frac{1}{2}\,b\,(\Pi - \Pi^{int})^2 \,. \tag{6.6}$$

Then

$$\mathfrak{E} = (\partial\Sigma_1/\partial\Pi) + (\bar{a}+b)\Pi - b\,\Pi^{int} \tag{6.7}$$

while, applying *T.I.P.* to the inequality (6.4) we obtain a *relaxation equation* (Collet, 1987)

$$\tau_d\,\dot{\Pi}^{int} = \Pi - \Pi^{int} \quad , \tau_d \geq 0 \,, \tag{6.8}$$

where τ_d is called the dielectric (polarization) relaxation time. This shows that the equilibrium value of Π^{int} is none other than Π itself. The relaxation time τ_d is practically directly accessible to experiments in a study of shock-wave propagation in dielectrics (Yakuscov *et al* ,1968) and it obviously intervenes in transient problems which themselves exhibit a characteristic time or a characteristic length of propagation. In the present case, for a propagation process whose linear regime is characterized by the velocity c_o (say, an acoustic velocity), a characteristic length

$$L_a = c_o\,\tau_d \tag{6.9}$$

may be constructed and referred to as the *electric attenuation length*. It is clear that when the wavelengths associated with all dynamical disturbances (for small amplitude signals) are short compared to L_a , then the effect of electric relaxation is negligible and the material behaves locally as an elastic dielectric without losses. This is verified by establishing the expression that governs the behavior of an *acceleration wave* propagating through a material modelled by eqns. (6.3)-(6.6) and considering the limit where the travelled distance L is much smaller than L_a (see McCarthy, 1984b). In this limit, one-dimensional motions admit *simple-wave solutions*. We refer the reader to MacCarthy (1984a,b) and Collet (1985) for this problem. Here we prefer to deal in greater detail with a problem which directly exhibits the role played by the ratio (5.1) in nonlinear wave propagation. To that purpose we consider the effect of electric (polarization) relaxation on the transient nonlinear electroelastic motion inward a half-space $X > 0$ from the limiting plane $X = 0$. Third-order nonlinear elasticity and dielectric relaxation (6.8) are taken into account. Hence it is the *competition* between *nonlinearity* (of elastic origin) and a *dissipative mechanism* (related to another physical property, electric polarization), the two being coupled via *piezoelectricity*, on which we focus the attention.

It is shown, in one space dimension and quasi-electrostatics, that the basic equations of the problem take on the following form:

$$
\begin{aligned}
0 &= \rho_o\, \partial_t v - \partial_X T^t , \\
0 &= \partial_X \mathcal{D} , \\
\tau_d\, \partial_t \Pi^{int} &= \Pi - \Pi^{int} ,
\end{aligned}
\tag{6.10}
$$

along with the constitutive equations ($f = F-1$, where F is the only surviving component of $\mathbf{F}$)

$$
\begin{aligned}
T^t &= C\, f\, (1+ 2\gamma f) - e\, \Pi , \\
\mathcal{D} &= e\, f + a\, \Pi - b\, \Pi^{int} .
\end{aligned}
\tag{6.11}
$$

here C, γ, e , and a are, respectively, a second-order elasticity coefficient, a nondimensional third-order elasticity coefficient, a piezoelectricity coefficient, and an electric constant.

The system (6.10)-(6.11) is a *quasi-linear* system which must be solved subject to *initial* conditions at $t = 0$, *boundary* conditions at $X = 0$, and *regularity* conditions as $X \longrightarrow \infty$. These are given by

$$
\begin{aligned}
&f = 0 \ , \ \Pi = 0 \ , \ \Pi^{int} = 0 \quad \text{at } t = 0 \ , \ X > 0 \ , \\
&f = K(t) \ , \ \mathcal{D} = 0 \quad \text{at } X = 0 \ , \ t > 0 \ , \\
&f \longrightarrow 0 \ , \ \Pi \longrightarrow 0 \ , \ \Pi^{int} \longrightarrow 0 \quad \text{as } X \longrightarrow \infty, t > 0 \ .
\end{aligned}
\tag{6.12}
$$

An approximate (asymptotic) solution to this problem can be souhgt in two ways. One of them is to look at a *far-field* solution and the other one is to examine what occurs in the neighborhood of the travelling signal. These pertain to the *delayed* and *instantaneous* wave analyses, respectively.

Delayed-wave analysis:

Let τ be a characteristic time scale . We call $\eta = \max|K(t)|$ the *small* parameter related to the amplitude of the input signal. Let

$$C_d = C - (e^2/\chi_{eq}) \quad , \quad \chi_{eq} = (1+a) - b \quad , \quad V_d^2 = C_d/\rho_o \quad , \tag{6.13}$$

a piezoelectrically altered elasticity coefficient, the thermodynamical equilibrium value of the electric susceptibility ($\Pi = \Pi^{int}$ at equilibrium), and the squared *delayed* acoustic speed. We can introduce nondimensional variables (with superimposed bars) by

$$\begin{aligned} &X = V_d \tau \bar{X} \ , \quad t = \tau \bar{t} \ , \quad f = \eta \bar{f} \ , \quad v = \eta V_d \bar{v} \ , \\ &\Pi = \eta e \bar{\Pi} \ , \quad \Pi^{int} - \eta e \bar{\Pi}^{int} , \quad T = \eta V_d \bar{T} \ , \quad \mathcal{D} = \eta e \bar{\mathcal{D}} \ . \end{aligned} \tag{6.14}$$

Furthermore,

$$\varepsilon_d = \tau_d/\tau \tag{6.15}$$

is a second small parameter [this is the Deborah number of eqn.(5.1)]. Equations (6.12) are rewritten using the new variables and then overbars are discarded in order to lighten the notation. We thus obtain the following system:

$$\begin{aligned} \alpha_d\, \partial_X f + \eta \bar{\gamma}_d\, f\, \partial_X f + (e^2/C_d)\, \partial_X \Pi &= \partial_t v \ , \\ \partial_X f + \chi\, \partial_X \Pi - b\, \partial_X \Pi^{int} &= 0 \ , \\ \varepsilon_d\, \partial_t \Pi^{int} &= \Pi - \Pi^{int} \ , \end{aligned} \tag{6.16}$$

where $\alpha_d = C/C_d$, $\bar{\gamma}_d = 4\gamma\alpha_d$. Also, taking the X-derivative of

$(6.16)_1$, we get the useful equation

$$\alpha_d\, \partial_X^2 f + \eta\bar{\gamma}_d\, \partial_X (f\, \partial_X f) + (e^2/C_d)\, \partial_X^2 \Pi = \partial_t^2 f \quad . \tag{6.17}$$

The appropriate time scale for the far-field study is $O(1/\eta)$ with $(\eta/\varepsilon_d) \longrightarrow 1$ as $\varepsilon_d \longrightarrow 0$. The system (6.16) is treated by introducing *multiple strained coordinates* $X_n \equiv (\varepsilon_d)^n\, X$, $n \geq 0$. The zeroth and first-order governing systems are then deduced from (6.16). The zeroth-order solution $S_o = \{ f_o, \Pi_o, \Pi_o^{int} \}$ satisfies a linear wave equation and this provides a solution $f_o = \bar{f}_o(X_1, \zeta)$ where $\zeta = t - X_o$. The *secularity* condition for solving for the first-order solution uniformly in space provides a constraint on the functional dependence of f_o through the equation (here χ = 1+a; see details in Collet, 1987)

$$\frac{\partial f_o}{\partial X_1} + \hat{\gamma}_d\, f_o\, \frac{\partial f_o}{\partial \zeta} = \frac{1}{2}\, d\, \frac{\partial^2 f_o}{\partial\, \zeta^2} \;, \tag{6.18}$$

where we have set

$$\hat{\gamma}_d = -\,\eta\bar{\gamma}_d/2\varepsilon_d \;,\; d = \frac{\chi}{\chi_{eq}}\, \frac{C_i - C_d}{C_d} \quad , \; C_i = C - (e^2/\chi) > 0 \quad , \tag{6.19}$$

where d is a *diffusion coefficient* and C_i may be called the *instantaneous* effective second-order elasticity coefficient. Equation (6.18) is none other than a *Burgers equation* (in which the roles of time and space are played by X_1 and ζ, respectively, and that of viscosity by the diffusion coefficient. Its solution is the *delayed wave* which is valid for

$$\begin{aligned} &\tau = O(1/\eta) \gg \tau_d \quad , \\ &(X,t) \in \mathcal{D}_d = \{ X,t \mid X > L_d \gg V_d\tau_d \} \;, \\ &L_d \gg O(V_i\tau_i) \quad , \end{aligned} \tag{6.20}$$

where V_i and τ_i are defined by

$$V_i^2 = C_i/\rho_o \quad , \; \tau_i = \frac{\chi}{\chi_{eq}}\, \frac{C_i}{C_i - C_d}\, \tau_d \;. \tag{6.21}$$

The first of conditions (6.20) is none other than a condition of very small electric Deborah number (5.1), which is characteristic of the validity of the local accompanying state hypothesis in internal-variable theory.

Instantaneous wave.

The problem consists in examining what occurs in the immediate neighborhood of the wave front as the latter travels inward the half-space $X > 0$. A solution is described in this neighborhood by considering an asymptotic procedure in which the perturbation parameter is none other than the distance from the wave front of the *linear* theory. That is, we introduce the parameter $\varepsilon_i = \tau/\tau_i$ with τ_i given by (6.21) and $\tau_i > \tau_d$. In this analysis the nondimensional form of eqn.(6.10)$_3$ is

$$\partial_t \Pi^{int} = \varepsilon_i (\tau_i/\tau_d) (\Pi - \Pi^{int}) . \tag{6.22}$$

A multiple strained-coordinate ($X_n = (\varepsilon_i)^n X$) analysis in which ε_i is the small parameter with $0(\eta/\varepsilon_i) = 1$ as $\varepsilon_i \rightarrow 0$, yields a *secularity* condition on the zeroth-order solution $f_o(X_1, \zeta = t - X_o)$ in the form of the *simple* nonlinear equation (Collet, 1987)

$$\frac{\partial f_o}{\partial X_1} + \hat{\gamma}_i f_o \frac{\partial f_o}{\partial \zeta} + \frac{1}{2} f_o = 0 \tag{6.23}$$

with the boundary condition

$$f_o(t, X_o = 0 , X_1 = 0) = f_o(\zeta = t , X_1 = 0) = K(t) . \tag{6.24}$$

This solution forms a *shock* at a *breaking distance* $(X_1)_B$ given by

$$(X_1)_B = - 2 \ln \left(1 - \frac{1}{2|\hat{\gamma}_i| K'_m}\right) , \quad K'_m \equiv \max_{\xi} \left|\frac{dK(\xi)}{d\xi}\right| . \tag{6.25}$$

Here the electric losses, although small, have a *cumulative effect* on the attenuation of the wave. This is involved in (6.25) so as to prevent the formation of the shock. However, once the shock is formed, then it will evolve according to the rules of weak-shock theory. The instantaneous solution thus obtained is valid for

$$(X,t) \in \mathcal{D}_i = \{X,t|\ |V_i t - X| \ll L_i \ll V_i \tau_i\}. \tag{6.26}$$

For an input $K(t) = f(0,t)$ in the form of a ramp followed by a plateau, we obtain a space-time dynamical response as reproduced qualitatively in Figure 6.1 (after Maugin, Collet *et al*, 1992). The response consists of an exponentially damped front and of the main wave which is governed at lowest order by the Burgers equation (6.18).

Other works of a more formal nature which consider electric internal variables and nonlinear waves in electroelasticity are those of Collet (1983, 1984, 1985) and McCarthy (1984a,b). The case of *semiconducting* electroelastic bodies may be even more interesting but then the electric conduction certainly is the leading dissipative mechanism (see works by McCarthy and Tiersten; also Daher and Maugin, 1987 ; Maugin and Daher , 1986).

7. ELECTRO- AND MAGNETOMECHANICAL HYSTERESIS

Here we address the following problem. A polycrystalline multidomain dielectric material exhibits both *induced* electric polarization and *spontaneous* electric polarization. The former is *reversible* and may be modelled by Π^r while the latter is accompanied by dissipation and will be modelled by both Π^R and Π^{int}. We focus the attention on the last two fields, in particular on the first one and its relationship with the polarizing field and other stimuli such as temperature, stresses and bias electric fields. For a given temperature and fixed state of stresses and bias fields (usually zero or considered as such), the typical response $|\Pi^R|$ versus $|\mathfrak{E}|$ presents the shape of a *hystereis* loop such as *(a)* in Figure 7.1 with a *saturation* such that $|\Pi^R| \longrightarrow \Pi^R_s$ as $|\mathfrak{E}|$ goes to infinity (i.e., physically, is large enough). This hysteresis loop is obtained in an alternating polarizing field of *low frequency*. As a matter of fact, we shall assume that this loop does not, here, depend on that frequency. In other words, the electric hysteresis phenomenon is *rate-independent* (i.e., it does not depend on the time rate $\dot{\mathcal{E}}$ of the polarizing field !) and , as such, it does not involve any characteristic time (and is very much similar to *plasticity* or *locking*) in contradistinction , say, with viscosity or electric relaxation. This is probably not true for high frequencies of the polarizing field and for relatively high temperatures.

The hysteresis curve (*a*) in Figure 7.1 is essentially characterized by: (i) the level of *saturation* Π^R_s ; (ii) the value $E_c = \mathcal{E}(\Pi^R=0)$ of the *coercive field,* and (iii) the "inclination" of the hysteresis loop on the Π^R-axis (remember that perfect electric hysteresis loops of one-domain crystals have vertical, jump-like, branches ; cf. Maugin *et al*, 1992, Chapter 6). When this inclination is *not* zero - what appears to be the case for most industrial materials such as ceramics - then we say that the electric material exhibits *electric hardening* (i.e., it takes a greater value of the polarizing field to increase the polarization Π^R by a given amount). This wording is granted in analogy to *mechanical hardening.*

Under the influence of a perturbation such as a bias (dc) electric field, an applied stress or irradiation, the loop (*a*) transforms into loop (*b*) in Figure 7.1. In general, four essential effects are manifested in this transformation: (ii) the saturation level has changed; (ii) the width of the hysteresis loop at zero polarization has been altered, (iii) the coercive field has evolved, presenting different values on loading and unloading (the loop is no longer symmetrical with respect to the origin) , and (iv) the electric hardening has been modified to some extent, all facts which are made more eloquent on the *derivative* curve (instantaneous electric susceptibility). All these recognized facts may be modelled in a necessarily *nonlinear* manner. This phenomenological theory was developed by Bassiouny *et al* (1988a,b), Bassiouny and Maugin (1989a,b), Maugin (1989) and Maugin and Bassiouny (1989). It closely parodies a somewhat similar theory built precedently for *magnetomechanical hysteresis*

(Maugin and Sabir, 1990; Maugin, Sabir and Chambon, 1987) and which receives the support of a *semi-microscopic* theory of domain-wall motion in a defective material (Sabir and Maugin, 1988). We shall return to this point later on.

The above-given brief description views hysteresis as a dissipative mechanism without time scale although the past history of the electric loading of the sample clearly plays a determining role at any instant. The first point is coped with by assuming that the power dissipated in the time evolution of Π^R is homogeneous of degree *one* in $\dot{\Pi}^R$, while the second point is taken care of by the presence of the electric internal variable Π^{int}. The latter will account for *electric hardening* and will also yield an entropy production which is homogeneous of degree one in its time evolution ($\dot{\Pi}^{int}$). Accordingly, focussing on electric processes, we only keep in the thermodynamical formulation of Section 5 the effects of Π^R and Π^{int} , so that the dissipation inequality (5.8) reduces , in quasi-electrostatics , to

$$\Phi_1 \equiv \mathfrak{E}.\dot{\Pi}^R + \mathfrak{E}^{int}.\dot{\Pi}^{int} \geq 0 , \tag{7.1}$$

where Φ_1 is supposed to be homogeneous of degree *one* in elements of the set $V = \{\dot{\Pi}^R, \dot{\Pi}^{int}\}$ of generalized velocities or fluxes. The corresponding set of "forces" is $F = \{\mathfrak{E} , \mathfrak{E}^{int}\}$. Both homogeneity and non-negativity properties of Φ_1 can be satisfied automatically in the following manner. Suppose that there exists a *pseudo-potential* of dissipation $\mathcal{D}(V)$ which is positive, contains the origin in V-space, takes real values (on $\mathbb{R}^+$), is convex in the set V, and is homogeneous of degree one in the elements of V. Then $\mathfrak{E}$ and $\mathfrak{E}^{int}$ are derivable from $\mathcal{D}$ by

$$\mathfrak{E} = \partial\mathcal{D}/\partial\dot{\Pi}^R \qquad , \ \mathfrak{E}^{int} = \partial\mathcal{D}/\partial\dot{\Pi}^{int} . \tag{7.2}$$

We check that $\Phi_1 = \mathcal{D} \geq 0$. As $\mathcal{D}$ is convex we can perform a Legendre-Fenchel transformation (see Maugin, 1991c, Appendix; this transformation conserves all properties) , thus producing the conjugate or *dual* potential $\mathcal{D}^*$ by

$$\mathcal{D}^*(F) = \sup_V [\Phi_1 - \mathcal{D}(V)] , \qquad (7.3)$$

from which there follow the "inverses" of (7.2) as

$$\dot{\Pi}^R = \partial\mathcal{D}^*/\partial\mathfrak{E} \quad , \quad \dot{\Pi}^{int} = \partial\mathcal{D}^*/\partial\mathfrak{E}^{int} \quad . \qquad (7.4)$$

Like in viscoplasticity, these equations say that the time evolution of Π^R and Π^{int}, if any, takes place along the normal to the surface $\mathcal{D}^*$= const. in F-space. If such surfaces have angulous points (e.g, apices or vertices), then the concept of "normal" is generalized in that of "cone of outward normals" at a point of the surfaces. Equations (7.4) are thus referred to as *normality rules*. Now assume that the elements of F are restrained to a convex set C containg the origin in F-space, i.e.,

$$C(F) = \left\{E = (\mathfrak{E}, \mathfrak{E}^{int}) \in F \mid f(E) \le 0\right\} \quad . \qquad (7.5)$$

Here f = 0 , the boundary of the convex C in F-space, is called the *loading surface* or *flow surface* . A particular case of (7.4) follows from the condition that $\mathcal{D}^*$ is none other than the so-called *indicator of the convex set C* (this mathematically follows from the condition that Φ_1 be homogeneous of degree one; see Maugin, 1991c, Appendix 2) so that

$$\mathcal{D}^* = ind\ C = \begin{cases} 0 & if\ f \le 0 \\ +\infty & if\ f > 0 \end{cases} \quad . \qquad (7.6)$$

This physically means that dissipation occurs only when E belongs to the boundary of C. Then eqns.(7.4) are replaced by the relations (which exhibit *no* time scale)

$$\dot{\Pi}^R = \dot{\lambda}\ \partial f/\partial\mathfrak{E} \quad , \quad \dot{\Pi}^{int} = \dot{\lambda}\ \partial f/\partial\mathfrak{E}^{int} \quad , \qquad (7.7)$$

where the *unknown* multiplier λ is such that $\dot{\lambda} = 0$ if f(E) < 0 and $\dot{\lambda} \ge 0$ if f(E) = 0 . This provides a very *singular* correspondence between elements of V and E spaces which is proper to this type of rate-independent dissipative processes. The conditions pertaining to λ can be made more precise , viz : $\dot{\lambda} \geq 0$ if f = 0 *and* $\dot{f} = 0$, and $\dot{\lambda} = 0$ if f < 0 *or* f = 0 and $\dot{f} < 0$. The proof of these is elementary once the following *orthogonality* property is proved:

$$\dot{\mathfrak{E}}.\dot{\Pi}^{R} + \dot{\mathfrak{E}}^{int}.\dot{\Pi}^{int} = 0 \,, \tag{7.8}$$

i.e., symbolically, $\dot{F} \perp V$, where the derivatives are understood *to the right* . To show this one must notice that (7.7) are completely equivalent to the *variational inequality*

$$(\mathfrak{E} - \mathfrak{E}^*).\dot{\Pi}^{R} + \{\mathfrak{E}^{int} - (\mathfrak{E}^{int})^*\}.\dot{\Pi}^{int} \geq 0 \tag{7.9}$$

for any $E^* = \{\mathfrak{E}^*, (\mathfrak{E}^{int})^*\} \in C(F)$. This equation in fact means that the dissipation is *maximal* (as a matter of fact, the only case where it is *not* zero) when E belongs to the surface f = 0. This is the *electric analogue* of the principle of *maximal dissipation* of Hill and mandel in plasticity (see Maugin, 1991c, Chapter 5).

The above-described model is mathematically neat. It corresponds to the case where the electric loading surface f = 0 is identified to a surface of *equi*-pseudopotential $\mathcal{D}^*$. Several consequences and further specifications of this modelling are of particular interest (see the already quoted references for the proofs):

(i) **Local stability:**

Assume that

$$\Psi = \Psi_{e}(\Pi^{r},\theta) + \Psi_{(i)}(\Pi^{int}), \tag{7.10}$$

where Ψ_e is convex in Π^r and concave in θ, and $\Psi_{(i)}$ is convex in Π^{int}. Then (7.8) yields

$$\dot{\mathfrak{E}}.\dot{\Pi}^{R} = \dot{\Pi}^{int}.\frac{\partial^2\Psi_{(i)}}{\partial\Pi^{int}\otimes\partial\Pi^{int}}.\dot{\Pi}^{int} \geq 0 \quad . \tag{7.11}$$

Thus the vectorial increments in $\mathfrak{E}$ and $\dot{\Pi}^R$ form an accute angle or, in a one-dimensional model, this translates to

$$\partial\mathfrak{E}/\partial\Pi^R \geq 0 \tag{7.12}$$

at all points of the hysteresis loop. This, indeed, is a local stability condition (increments in $\mathfrak{E}$ and Π^R must have the same sign); thus electric hardening is positive in this model.

(ii) **Global stability:**

Another consequence of the formulation (7.9)-(7.10) is the following result:

$$\oint_{\mathfrak{C}} \mathbf{E}.d\Pi \geq 0 \tag{7.13}$$

for any closed (total) polarization cycle $\mathfrak{C}$. The proof follows the identical proof in plasticity with hardening where the roles of $\mathfrak{E}$ and Π are played by the stress and the total strain [then the inequality analogous to (7.13) is known as *Ilyushin's postulate* - when its is postulated, which is *not* the case here]. The result (7.13) means that electric hysteresis loops are always described in the *counterclockwise* direction (in a plane where $\mathfrak{E}$ is the abscissa and Π is the ordinate).

(iii) **Example of electric loading surface** $f(E) = 0$:

Let Π^{int} be a *scalar* for the sake of example, and N its thermodynamical dual such that $N = \partial\Psi/\partial\Pi^{int}$. Then $f = f(\mathfrak{E}, N)$. An example of such a function f is given by

$$f\ (\mathfrak{E},N) = (\|\mathfrak{E}\| + N)^2 - E_c^2\ . \tag{7.14}$$

It is readily shown by using eqns.(7.5) that

$$\Pi^{int}(t) = \int_0^t \{\dot{\Pi}^R(\tau).\mathbf{A}^{-1}.\dot{\Pi}^R(\tau)\}^{1/2}\, d\tau \equiv \overline{\Pi^R}\ , \tag{7.15}$$

if $\|\mathfrak{E}\|^2 = \mathfrak{E}.\mathbf{A}.\mathfrak{E}$, where the real symmetric nonsingular $\mathbf{A}$ accounts for electric anisotropy (if any). In this case Π^{int} is none other than the *cumulated* (in time) *residual polarization* at time t. It thus really accounts for the past history, up to the present, of the electric loading.

(iv) **First polarization curve:**

The above model is completed by the datum of a first polarization curve

$$\mathfrak{E} = \varphi(\Pi^R) \quad or \quad \Pi^R = \varphi^{-1}(\mathfrak{E}) \tag{7.16}$$

such that

$$\varphi(\Pi^R) = -\varphi(-\Pi^R) \ , \ hence \ \varphi(0) = 0 \ ,$$

$$\partial\varphi/\partial\Pi^R > 0 \ , \ \forall\ \mathfrak{E} \tag{7.17}$$

$$\varphi^{-1}(\mathfrak{E})\Big|_{\mathfrak{E} = \pm\infty} = \pm\ \Pi^R_s \ .$$

This allows one to construct the hysteresis curve from an alternating loading starting from a *virgin* state. Expression (7.16) is foreign to the present theory in the sense that either it is given by experiments or it is constructed from a semi-microscopic theory of elastic polarization relying on the effects of domain- wall motions. The construction of hysteresis curves on the basis of the above-developed model is described in Bassiouny *et al* (1988a,b).

(v) A correct **parametrization** of the curve (7.16) and the loading surface (7.14) in terms of stresses and temperature allows one to reproduce (at least the tendency of) the alterations to the hysteresis loop due to such fields. Figure 7.2 reproduces the result of such a parametrization for PLZT ceramics. The agreement with experimental data is not too bad.

In all, the above-given electric modelling which somewhat parodies the elastoplasticity of metals, is thermodynamically sound while reproducing physical reality in a rather satisfactory manner. It is also "comprehensible" as it adheres well to the logic that we developed in previous sections. We obviously refer the reader to the original papers for the mathematical details and many graphic illustrations. It must also be pointed out that Chen (1980, 1984) - Chen and Peercy (1979) - has also attempted to construct a model of *electric hysteresis by domain switching* which in fact amounts to a theory with electric internal variables as shown by Bassiouny *et al* (1988b). This theory reproduces well the general features of electric hysteresis and butterfly (strain vs. field) loops in PZT ceramics although it had no thermodynamical basis to start with.

Magnetomechanical hysteresis

The above-reported developments obviously admit of a magnetic analogue in the description of *ferromagnetic hysteresis* and its coupling with mechanical effects. This is especially important in view of the potential applications to non-destructive testing (*NDT*) such as the measure of residual stresses via the Barkhausen effect [cf. Rudyak (1971), Karlajainen and Moilanen (1979); Chernyi (1983), Maugin (1989, 1991a,b), Maugin and Sabir (1990)]. We refer the reader to our review (Maugin , 1991b) which, on the one hand, shows explicitly the difference between thermodynamically-based magnetic hysteresis presenting no time scale (at low frequency of the magnetizing field **H**) and *ad-hoc* mathematical models (e.g., Chua and Stromsmoe, 1971) which rather parody the rheological models with several relaxation times (these *do* involve time scales and do depend on the frequency of the magnetizing field), and so-called "physical" models (in fact pure models also) in the manner of Preisach (1935), and, on the other hand, shows the pretty good agreement which can be established between modelling and experimental data (see Figure 7.3). To conclude this point, we would like to point at the rather direct *analogy* of our approach to electric and magnetic hysteresis with the mechanical theory of *locking* where *bounded strains* occur (cf. Prager, 1957) instead of a saturation in polarization or magnetization.

8. ELASTIC SUPERCONDUCTORS

With the current development of *high-temperature superconductors*, there is need for a realistic constitutive theory of elastic *superconductors* which retains the essentials of superconductivity theory [for this, see de Gennes (1966/89) or Lynton (1969)] while admitting a proper thermodynamical framework. This can be achieved in the thermodynamic theory with internal variables when the latter can be selected to represent a *finer* level of description than the one at which the main field equations are formulated. For instance, this is achieved in a macroscopic theory of polymeric solutions in which the internal variable may be a scalar distribution function. Here we naturally select the internal variable to be the *complex-valued wave function* ψ *of Cooper's pairs* in the microscopic theory of superconductivity. This is also an *order parameter* in phase-transition theory and its spatial gradients should appear to account for weakly nonlocal effects inherent in this. Basing on other examples of theories with internal variables accounting for their gradients [e.g., in ferromagnets in Maugin (1979b), in general media and liquid crystals in Maugin (1990)], we know then that the entropy flux will no longer be the ratio of heat flux to thermodynamical temperature. Furthermore, in addition to the usual *material indifference* (objectivity) of continuum mechanics, the *gauge invariance* of superconductivity should be implemented . These two restrictions, together with the satisfaction of the second law of thermodynamics in its Clausius-Duhem form, are imposed in the present theory (Maugin, 1992a). This, in our opinion, supersedes other attempts at such a phenomenological formulation [e.g., Zhou and Miya (1991), Lebanov (1991)].

As working hypotheses we shall consider that in the starting equation (3.11) the entropy influx is not simply $\theta^{-1}\mathbf{q}$ but a more general vector $\mathbf{N} = \theta^{-1}\mathbf{q} + \mathcal{K}$, where $\mathcal{K}$ is an *extra-flux* of entropy to be determined. In addition we consider the case of *nonpolarizable media* (although this is without importance) - i.e., eqn.(2.8) applies ; equivalently $\Pi = 0$ - and we note F the free energy per unit volume in $\mathcal{K}_R$ to avoid any confusion with the complex-valued function ψ. Noting $\mathbf{N}_R$ the entropy flux in $\mathcal{K}_R$ such that

$$\mathbf{N}_R = J_F\mathbf{F}^{-1}.\mathbf{N}\,, \qquad \bar{\mathcal{K}} = J_F\mathbf{F}^{-1}.\mathcal{K}\,, \tag{8.1}$$

it is a trivial matter to show that (3.49) is replaced by the following form of the Clausius-Duhem inequality :

$$-(\dot{F}+N\dot{\theta}) + tr\,\{\mathbf{S}^E\dot{\mathbf{E}}\} + \mathfrak{J}.\mathfrak{E} - \bar{\mathbf{M}}.\dot{\mathfrak{B}} \tag{8.2}$$
$$+\nabla_R.(\theta\bar{\mathcal{K}}) - \mathbf{N}_R.\nabla_R\theta \geq 0\;.$$

Obviously, eqns.$(3.40)_{1-2}$ show that there exist [*material*, i.e., function of $(\mathbf{X},t)$] *electromagnetic potentials* φ *and* $\mathfrak{A}$ *such that*

$$\mathfrak{E} = -\left(\nabla_R\varphi + \frac{1}{c}\frac{\partial\mathfrak{A}}{\partial t}\bigg|_{\mathbf{X}}\right), \quad \mathfrak{B} = \nabla_R \times \mathfrak{A}\;. \tag{8.3}$$

The *gauge invariance* of superconductivity is such that $\mathfrak{E} = \hat{\mathfrak{E}}$ and $\mathfrak{B} = \hat{\mathfrak{B}}$ in the transformations defined by (e.g., Lynton, 1969)

$$\hat{\nabla}_R = \nabla_R - (ie^*/\hbar c)\,\mathfrak{A}\,, \quad \hat{\partial}/\partial t = \partial/\partial t + (ie^*/\hbar)\varphi\,, \tag{8.4}$$

where $\hbar$ is Planck's reduced constant, i is the unit imaginary and e^* is a characteristic electric charge (=2e for Cooper's superconducting electrons). The constitutive relations for our deformable superconductors must satisfy the inequality (8.2) and the gauge invariance (8.3). For *elastic* superconductors we naturally assume that F, to start with, is a sufficiently regular function

$$F = f(\mathbf{F},\, \mathcal{E}\,,\, \mathbf{B},\, \theta,\, \psi\,,\, \nabla_R\psi)\;, \tag{8.5}$$

where ψ is a scalar-valued complex function , and we see the first appearance of the gradient of this internal variable via $\nabla_R\psi$.

The *electric current* $\mathfrak{J}$ is composed of a *normal* contribution $\mathfrak{J}_n$ to be determined essentially by $\mathfrak{E}$ and a *superconducting* contribution $\mathfrak{J}_s$ determined by ψ and which does not contribute to the entropy growth. With this in mind we note that

$$\mathfrak{J}.\mathfrak{E} = \mathfrak{J}_n.\mathfrak{E} - \nabla_R.(\mathfrak{J}_s\ \varphi) + (\nabla_R.\mathfrak{J}_s)\ \varphi - \frac{1}{c}\ \mathfrak{J}_s.\frac{\partial\mathfrak{A}}{\partial t}\ . \tag{8.6}$$

Applying first the requirement of material indifference to *F*, we easily show that the following *F* does satisfy this invariance:

$$F = \check{F}(\mathbb{E},\ \mathfrak{E}\ ,\ \mathfrak{B}\ ,\ \theta\ ,\ \psi,\ \nabla_R\psi)\ . \tag{8.7}$$

Noting further that ψ is complex-valued while *F* is real valued, on account of gauge invariance we consider the reduced form

$$F = \hat{F}(\mathbb{E}\ ,\ \mathfrak{E}\ ,\ \mathfrak{B}\ ,\ \theta\ ,\ |\psi|\ ,\ |\hat{\nabla}_R\psi|^2)\ . \tag{8.8}$$

We now compute $\dot{F}$ while setting (Ψ^* denotes the complex conjugate of ψ):

$$\begin{aligned}\mu^* &\equiv \partial\hat{F}/\partial(\hat{\nabla}_R\psi^*)\ ,\\ \mathcal{A}_{\psi^*} &\equiv -\frac{\partial\hat{F}}{\partial\psi^*} + \hat{\nabla}_R.\mu^* + (2ie^*/\hbar c)\ \mu^*.\mathfrak{A}\ .\end{aligned} \tag{8.9}$$

On account of eqns. (8.6) and (8.9) and using (8.4) repeatedly , we transform the Clausius-Duhem inequality (8.2) to

$$-\left(\frac{\partial \hat{F}}{\partial \theta} + N\right)\dot{\theta} + tr\left\{\left(\mathbf{S}^{E} - \frac{\partial \hat{F}}{\partial \mathbb{E}}\right)\dot{\mathbb{E}}\right\} - \frac{\partial \hat{F}}{\partial \mathfrak{E}}.\dot{\mathfrak{E}}$$
$$-\left(\frac{\partial \hat{F}}{\partial \mathfrak{B}} + \bar{\mathbb{M}}\right).\dot{\mathfrak{B}} + \mathfrak{J}_{n}.\mathfrak{E} - \mathbf{N}_{R}.\nabla_{R}\theta$$
$$-\frac{1}{c}\left[\mathfrak{J}_{s} - \frac{ie^{*}}{2\hbar}(\mu^{*}\psi - \mu\psi^{*})\right].\dot{\mathfrak{A}} + 2\,\mathcal{R}e\left(\mathcal{A}_{\psi}\frac{\hat{\partial\psi^{*}}}{\partial t}\right) \tag{8.10}$$
$$-\varphi\left[\nabla_{R}.\mathfrak{J}_{s} - 2\,\mathcal{R}e\left(\frac{ie^{*}}{\hbar}\mathcal{A}_{\psi}\,\psi^{*}\right)\right]$$
$$+\nabla_{R}.\left[\theta\bar{\mathcal{K}} - \mathfrak{J}_{s}\,\varphi - \frac{1}{2}\left(\mu^{*}\frac{\hat{\partial\psi}}{\partial t} - \mu\frac{\hat{\partial\psi^{*}}}{\partial t}\right)\right] \geq 0\ .$$

Assuming that N, $\mathbf{S}^{E}$, $\bar{\mathbb{M}}$, and $\mathfrak{J}_s$ do not depend on $\dot{\theta}$, $\dot{\mathbb{E}}$, $\dot{\mathfrak{B}}$ and $\dot{\mathfrak{A}}$, respectively, we obtain that the inequality (8.16) will hold if and only if the following results hold true:

$$N = -\,\partial\hat{F}/\partial\theta\ ,\ \mathbf{S}^{E} = \partial\hat{F}/\partial\mathbb{E},\ \partial\hat{F}/\partial\mathfrak{E} = 0\ ,\ \bar{\mathbb{M}} = -\,\partial\hat{F}/\partial\mathfrak{B}\ , \tag{8.11}$$

and

$$\mathfrak{J}_{s} = \frac{ie^{*}}{2\hbar}\,(\mu^{*}\psi - \mu\psi^{*})\ . \tag{8.12}$$

Furthermore, we *select* $\bar{\mathcal{K}}$ in such a way as to avoid the appearance of a true divergence contribution in the second law of thermodynamics. Thus

$$\bar{\mathcal{K}} = \frac{1}{\theta}\left\{\mathfrak{J}_{s}\varphi + \frac{1}{2}\left(\mu^{*}\frac{\hat{\partial\psi}}{\partial t} - \mu\frac{\hat{\partial\psi^{*}}}{\partial t}\right)\right\}\ . \tag{8.13}$$

As a consequence of (8.11) through (8.13), (8.10) reduces to the following dissipation inequality:

$$\Phi \equiv \mathfrak{J}_{n}.\mathfrak{E} - \mathbf{N}_{R}.\nabla_{R}\theta + 2\,\mathcal{R}e\left(\mathcal{A}_{\psi}\frac{\hat{\partial\psi^{*}}}{\partial t}\right) \tag{8.14}$$
$$-\varphi\left[\nabla_{R}.\mathfrak{J}_{s} - 2\,Re\left(\frac{ie^{*}}{\hbar}\mathcal{A}_{\psi}\,\psi^{*}\right)\right] \geq 0\ .$$

Proceeding now like in preceding sections, i.e., exploiting *T.I.P*

-- or assuming the existence of a positive dissipation potential which is homogeneous of degree *two* in the fluxes -- for the first three contributions in (8.14) we obtain (for instance, assuming isotropy for these effects) the following *complementary laws*:

$$\mathbf{N}_R = -\frac{\kappa}{\theta}\nabla_R\theta + \nu_o\mathfrak{J}_n \quad , \quad \mathfrak{J}_n = \Sigma_n(\mathcal{E} - \pi_o\nabla_R\theta) \quad , \tag{8.15}$$

and

$$\mathcal{A}_\psi = \gamma_\psi \hbar\,(\hat{\partial}\psi/\partial t) \quad , \tag{8.16}$$

while, on account of (8.16) ,(8.12) and (8.8), we check the following *identity*

$$\nabla_R \cdot \mathfrak{J}_s \equiv 2\,Re\left(\frac{ie^*}{\hbar}\mathcal{A}_\psi\,\psi^*\right), \tag{8.17}$$

so that the last term in the left-hand side of (8.14) just vanishes. The remaining dissipation Φ is non-negative if and only if we have the following constraints on the coefficients:

$$\kappa \geq 0 \quad , \Sigma_n \geq 0 \; , \; \nu_o = \theta\,\pi_o \; , \; \gamma_\psi \geq 0 \; . \tag{8.18}$$

Equations (8.15) are *thermoelectrically* coupled *Fourier's* and *Ohm's* laws (note that the former is given directly for the entropy flux). They are of the general form of eqns.$(4.13)_{3-4}$. Equation (8.16) is the looked for gauge-invariant *evolution equation* for the internal variable ψ . The coefficient γ_ψ - which has to be positive according to $(8.18)_3$ - accounts for the *finite* relaxation time of superconducting electrons (however, this relaxation time must be sufficiently small compared to the characteristic time of external loads - see Section 5 above); $\hbar$ is introduced for notational convenience. Both $\mathfrak{J}_s$ anf $\bar{\mathcal{K}}$ are entirely expressible in terms of F and ψ on account of eqn.(8.16).

If we consider as a particular case the following free energy (clearly some kind of expansion; again $\hbar$ and the effective mass m* are introduced for convenience):

$$F = F_n(\mathbf{E},\theta) + \alpha(\mathbf{E},\theta)\,|\psi|^2 + \beta(\mathbf{E},\theta)\,|\psi|^4 + \frac{\hbar^2}{2m^*}\,|\hat{\nabla}_R\psi|^2 \quad , \tag{8.19}$$

then eqns.(8.12) and (8.16) yield

$$\mathfrak{J}_s = \frac{ie^*\hbar}{2m^*} (\psi\nabla_R\psi^* - \psi^*\nabla_R\psi) - \frac{(e^*)^2}{m^*c}|\psi|^2 \mathfrak{A} \tag{8.20}$$

and

$$\gamma_\psi\hbar \frac{\hat{\partial}\psi}{\partial t} + \alpha(\mathbf{E},\theta)\psi + \beta(\mathbf{E},\theta)|\psi|^2\psi - \frac{\hbar^2}{2m^*} \hat{\nabla}_R^2\psi = 0 , \tag{8.21}$$

respectively. Equation (8.20) is the accepted law for *supercurrent* (see , e.g., de Gennes, 1966/1989). Equation (8.21) obviously is a straightforward *dynamical* generalization of the *Ginzburg-Landau* equation of superconductivity which accounts for the coupling with strains and for dissipation by relaxing superconducting electrons (compare to Tinkham , 1964). At this point it remains to write down the *conservation-of-charge* equation deduced from (2.9) and the *heat equation* which is obtained by expanding the so-called *intrinsic-dissipation* equation (see, Maugin, 1991c, Appendix 1 for the general setting) which here reads

$$\theta\dot{N} = \left[\mathfrak{J}_n . \mathfrak{E} + 2 \, Re\left(\mathcal{A}_\psi \frac{\hat{\partial}\psi^*}{\partial t}\right)\right] + \nabla_R . (\theta \mathbf{N}_R) \; . \tag{8.22}$$

This completes the essentials of our continuum thermodynamics of thermoelastic superconductors in the Galilean and gauge-invariant form which fits well in the recently proposed general scheme for systems prone to exhibiting *dissipative structures* through strongly localized spatial variations in an internal variable (Maugin , 1990). The couplings between typically superconducting features and strains will naturally come into the picture, e.g., through the functions α and β present in the energy (8.19), possibly with drastic behaviors are phase-transition points.

9. CONCLUDING REMARKS

In electromagnetic solids, setting aside mechanical dissipative behaviors (e.g., viscosity, plasticity) and thermal ones (pure heat conduction), the main (and for the time being , the only) irreversible electromagnetic behaviors are represented by : electric polarization and magnetic relaxation, electric and magnetic (rate-dependent or rate-independent) hysteresis, classical (normal, in the limit, perfect) electric conduction , semiconductivity , and superconductivity. All these behaviors have more or less been touched upon in these lecture notes which are, *per force*, of a limited length. As the description of many classical behaviors can nowadays be found in treatises, we willingly have placed the emphasis on those behaviors which are more complex and admit a description in terms of the thermodynamics with *internal variables*. These include electric and magnetic relaxation, electric relaxation in piezoelectric ceramics and magnetomechanical hysteresis in ferromagnets, and superconductivity in deformable high-temperature superconducting magnets. Although we often pointed out at analogies between electromagnetic behaviors and purely mechanical ones [in this frame of mind, see our book (Maugin, 1991c) or Germain *et al* (1983)], we are sure that the reader, by now, can establish for himself these analogies in a rather precise manner.

At this point of conclusion we would like to mention works which study the influence of *thermomechanical irreversible* effects on electromagnetic or electromechanical properties, for instance, Bampi and Morro (1981) in MHD or Fomethe and Maugin (1982) and Maugin and Fomethe (1982) in the *elastoviscoplasticity* of ferromagnetic crystals with defects. Also relevant are those works which consider the phenomenon of *diffusion* of electric charges such as in *semiconductors* (Daher and Maugin, 1987; Maugin and Daher , 1986) or in solutions of polyelectrolytes (Morro, Drouot and Maugin, 1985; Morro, Maugin and Drouot, 1990) and which, indeed, use the concept of *internal variables* , but would require too much space for their exposition.

Finally, we would not be complete if we did not mention the frequent misunderstanding between *internal variables* and *internal degrees of freedom* as they often both pertain to the description of the *same* materials. The main difference is that where internal variables are just additional *thermodynamical state variables* satisfying *evolution equations* constrained by the second law of thermodynamics, an internal degree of freedom is governed by a *field equation* at this level of description. It thus is on an equal footing with other *observable variables* (e.g., motion) present in the theory. For example, in the electromagnetic framework, the works by Maugin and Pouget (1980; Pouget and Maugin, 1984) in *ferroelectric crystals* and Maugin and Miled (1986) in *ferromagnetic crystals*, consider electric polarization and magnetic spin as internal degrees of freedom, and the associated governing equations give rise to *solitonic structures* in the nonlinear framework, while in Sections 6 and 7 above we have electric and magnetic internal variables which are more likely, when gradients of these are introduced, to give rise to *dissipative structures*. We have emphasized these differences and also the inevitable resemblances in Maugin (1990), but confusion sometimes persists as shown by the title of a paper by Parry (1987).

10. REFERENCES

BAMPI. and MORRO A.,(1981), *Dissipative Effects and waves in Magnetofluidynamics,* **J.Non-Equilib.Thermodynam.**, **6**, 1-14.

BASSIOUNY E., GHALEB A.F. and MAUGIN G.A., (1988a), *Thermodynamical Formulation for Coupled Electromechanical Hysteresis-I-Basic Equations,* **Int.J.Engng.Sci.**, **26**, 1279-1295.

BASSIOUNY E., GHALEB A.F. and MAUGIN G.A., (1988b), *Thermodynamical Formulation for Coupled Electromechanical Hysteresis - II - Poling of Ceramics,* **Int.J.Engng.Sci.**, **26**, 1297-1306.

BASSIOUNY E. and MAUGIN G.A., (1989a), *Thermodynamical Formulation for Coupled Electromechanical Hysteresis - III - Parameter Identification ,* **Int.J.Engng.Sci.**, **27**, 975-987.

BASSIOUNY E. and MAUGIN G.A., (1989b), *Thermodynamical Formulation for Coupled Electromechanical Hysteresis - IV - Combined Electromechanical Loadings,* **Int.J.Engng.Sci.**, **27**, 989-1000.

BATAILLE J. and KESTIN J., (1975), *L'interprétation physique de la thermodynamique rationnelle,* **J.Mécanique**, **14**, 365-384.

BRIDGMAN P.W., (1941), **The Nature of Thermodynamics**, Harvard University Press, Cambridge , Mass. (New Edition, Harper and Brothers, New York, 1961).

CHEN P.J., (1980), *Three-dimensional Dynamic Electromechanical Constitutive Relations for Ferroelectric Materials,* **Int.J.Solids Struct.**, **16**, 1059-1067.

CHEN P.J., (1984), *Hysteresis Effects in Deformable Ceramics,* in: **The Mechanical Behavior of Electromagnetic Solid Materials**, ed. G.A.Maugin, pp.137-143, North-Holland , Amsterdam.

CHEN P.J. and PEERCY P.S.,(1979), *One-dimensional Dynamic Electromechanical Constitutive Equations of ferroelectric Materials,* **Acta Mechanica**, **31**, 231-241.

CHERNYI L.T., (1983), *Models of Ferromagnetic Continua with Magnetic Hysteresis,* in: **Macroscopic Theories of Matter and Fields**, ed. L.I.Sedov, pp.116-140, Mir Publishers, Moscow.

CHUA L.O. and STROMSMOE K.A.,(1971), *Mathematical Models for Dynamic Hysteresis Loops,* **Int.J.Engng.Sci.**, **9** , 564-574.

COLLET B.,(1983), *Shock Waves in Deformable Piezoelectric Materials,* in: **Proc.11th Intern. Congress of Acoustics**, Special Issue of **Revue d'Acoustique**, Vol.2, Sec.2.1, pp.125-128.

COLLET B.,(1984), *Shock Waves in Deformable Ferroelectric Materials,* in: **The Mechanical Behavior of Electromagnetic Solid Materials**, ed. G.A.Maugin, pp.157-163, North-Holland, Amsterdam.

COLLET B.,(1985), *Nonlinear Wave propagation in Elastic Dielectrics with Internal Variables*, **J.Techn.Physics**, **26**, 285-289.

COLLET B., (1987), *Transient Nonlinear Waves in Elastic Dielectric Materials*, in: **Electromagnetomechanical Interactions in Deformable Solids and Structures**, eds. Y.Yamamoto and K.Miya, pp.329-334, North-Holland, Amsterdam.

DAHER N. and MAUGIN G.A.,(1987), *Deformable Semiconductors with Interfaces: Basic Equations*, **Int.J.Engng.Sci.**, **25**, 1093-1129.

de GENNES P.G.(1966/1989), **Superconductivity of Metals and Alloys**, W.A.Benjamin , New York (New Edition, 1989).

de GROOT S.R. and SUTTORP L.G., (1972), **Foundations of Electrodynamics**, North-Holland, Amsterdam.

DROUOT R. and MAUGIN G.A.,(1985), *Continuum Modelling of Polyelectrolytes in Solution*, **Rheologica Acta**, **24**, 474-487.

DUHEM P. (1911), **Traité d'Energétique ou de Thermodynamique Générale**, Two volumes, Gauthier-Villars, Paris.

DUHEM P.,(1914/1954), **The Aim and Structure of Physical Theory**, Princeton University Press, Princeton, N.J. (1954; Translation from the Second French Edition, Rivière, Paris, 1914).

ERINGEN A.C. (1971-1976;Editor), **Continuum Physics**, Four volumes, Academic Press, new York.

ERINGEN A.C.,(1980), **Mechanics of Continua**, Revised and enlarged Edition, Krieger, new York.

ERINGEN A.C. and MAUGIN G.A., (1990), **Electrodynamics of Continua**, Two volumes, Springer-Verlag, New York.

FOMETHE A. and MAUGIN G.A., (1982), *Influence of Dislocations on Magnon-Phonon Couplings. A Phenomenological Approach*, **Int.J.Engng.Sci.**, **20**, 1125-1144.

GERMAIN P., NGUYEN QUOC SON and SUQUET P., (1983), *Continuum Thermodynamics*, **ASME Trans.J.Appl.Mech.**, **105**, 1010-1020.

JACKSON J.D. (1962), **Classical Electrodynamics**, J.Wiley, New York.

JOU D., CASAS-VASQUEZ J. and LEBON G., (1988), *Extended Irreversible Thermodynamics:A Review*, **Rep.Prog. in Physics**, **9**, 1105-1179.

KARJALAINEN L.P. and MOILAMEN M.,(1979), *Detection of Plastic Deformation during Fatigue of Mild Steel by the Measurement of the Barkhausen Noise*, **NDT International**, No.4, 51-55.

KLUITENBERG G.A.,(1973), *On Dielectric and Magnetic Relaxation Phenomena and Non-equilibrium Thermodynamics*, **Physica**, **68**,75-92.

KLUITENBERG G.A., (1977), *On Dielectric and Magnetic Relaxation Phenomena and Vectorial Integranl Degrees of Freedom in Thermodynamics,* **Physica, 87A,** 302-330.

KLUITENBERG G.A., (1981a), *On Vectorial Internal Variables and Dielectric and Magnetic Relaxation Phenomena,* **Physica, 109A, 91-122.**

KLUITENBERG G.A., (1981b), *On Transformations of Internal Variables in the Thermodynamic Theory of a...,* **Physica , 109A, 123-127.**

LOBANOV E.N., (1991), *Theory of Superconducting Composites,* **Mekh. Komposit Materialov** (in Russian), in the press.

LYNTON E.A., (1969), **Superconductivity,** Methuen and Co. (3rd Edition)- New Printing: Science Paperback, Chapman and Hall, London (1971).

MAUGIN G.A., (1979a), *Internal Variables in Ferroelectric and Ferromagnetic Continua,* in: **Recent Developments in the Theory and Application of Generalized and Oriented Media,** pp.201-204, American Academy of Mechanics, Calgary, Canada.

MAUGIN G.A., (1979b), *Vectorial Internal Variables in Magnetoelasticity,* **J.Mécanique, 18,** 541-563.

MAUGIN G.A., (1980), *The Method of Virtual Power in Continuum Mechanics: Application to Coupled Fields,* **Acta Mechanica, 35,** 1-70.

MAUGIN G.A., (1981a), *Electromagnetic Internal variables in Electromagnetic Continua,* **Arch.Mech., 33,** 927-935.

MAUGIN G.A., (1981b), *Simple Thermodynamical Model for Rigid Ferromagnets,* **Phys.Rev., B25,** 7019-7025.

MAUGIN G.A., (1985), **Nonlinear Electromechanical Effects and Applications** (A Series of Lectures), World Scientific, Singapore.

MAUGIN G.A., (1988), **Continuum Mechanics of Electromagnetic Solids,** North-Holland, Amsterdam (in Russian translation, MIR, Moskva, 1991).

MAUGIN G.A., (1989), *Coupled Magnetomechanical and Electromechanical Hysteresis Effects,* in: **Applied Electromagnetics in Materials,** eds. R.K.T.Hsieh and K.Miya, pp.5-18, Pergamon Press, London, Tokyo.

MAUGIN G.A., (1990), *Internal Variables and Dissipative Structures,* **J.Non-Equilibr.Thermodyn., 15,** 173-192.

MAUGIN G.A., (1991a), *Thermodynamics of Hysteresis,* in: **Non-equilibrium Thermodynamics,** Vol.7, "Extended Thermodynamics", pp.25-52, eds. D.Salamon and S.Sieniutycz, Taylor and Francis, New York.

MAUGIN G.A.,(1991b), *Compatibility of Magnetic Hysteresis with Thermodynamics*, **Int.J.Appl.Electrom.Mat.**, **2**, 7-19.

MAUGIN G.A., (1991c), **The Thermomechanics of Plasticity and Fracture**, Cambridge University Press (in Russian translation, MIR, Moskva, 1993).

MAUGIN G.A., (1992a), *Irreversible Thermodynamics of Elastic Superconductors*, **C.R.Acad.Sci.Paris**, **II-314**.

MAUGIN G.A., (1992b), **Material Inhomogeneities in Elasticity**, Chapman and Hall, London.

MAUGIN G.A., COLLET B., and POUGET J., (1986), *Nonlinear Wave Propagation in Coupled Electromechanical Systems*, in: **Nonlinear Wave propagation in Solids**, ed. T.W.Wright, ASME Publ.Vol.AMD-77, pp.57-84, ASME, New York.

MAUGIN G.A., COLLET B., DROUOT R. and POUGET J., (1992), **Nonlinear Electromechanical Couplings**, Manchester University Press, Manchester , U.K.

MAUGIN G.A. and BASSIOUNY E., (1989), *Continuum Thermodynamics of Electromechanical Hysteresis in Ceramics*, in: **Continuum Mechanics and its Applications**, eds. G.A.C.Graham *et al*, pp.225-235, Hemisphere Publ., New York.

MAUGIN G.A. and DAHER N., (1986), *Phenomenological Theory of Elastic Piezoelectric Semiconductors*, **Int.J.Engng.Sci.**, **21**, 703-731.

MAUGIN G.A. and DROUOT R., (1983), *Thermomagnetic Behavior of Magnetically Non-saturated Fluids*, **J.Magnetism and Magnetic Materials**, **39**, 7-10.

MAUGIN G.A. and ERINGEN A.C.,(1977), On the Equations of the Electrodynamics of Deformable Solids of Finite Extent, **J.Mécanique**, **16**, 101-147.

MAUGIN G.A. and FOMETHE A.,(1982), *On the Elastoviscoplasticity of Ferromagnetic Crystals*, **Int.J.Engng.Sci.**, **20**, 885-908.

MAUGIN G.A. amd MILED A., (1986), *Solitary Waves in Elastic Ferromagnets*, **Phys.Rev.**, **B33**, 4830-4832.

MAUGIN G.A. and MUSCHIK W.,(1992), *Thermodynamics with Internal variables: A Review, I-Generalities, II-Applications*, **J.Non-Equilibr.Thermodyn.**

MAUGIN G.A. and POUGET J., (1980), *Electroacoustic Equations for One-domain Ferroelectric Bodies*, **J.Acoust.Soc.Amer.**, **68**, 575-587.

MAUGIN G.A. and SABIR M.,(1990), *Mechanical and Magnetic Hardening of Ferromagnetic Bodies: Influence of residual Stresses and Application to Nondestructive Testing*, **Int.J.of Plasticity**, **6**, 573-589.

MAUGIN G.A., SABIR M. and CHAMBON P.,(1987), *Coupled Magnetomechanical Hysteresis in Ferromagnets: Application to Nondestructive Testing,* in: **Electromagnetomechanical Interactions in Deformable Solids and Structures**, eds.Y.Yamamoto and K.Miya, pp.255-264, North-Holland, Amsterdam.

McCARTHY M.F., (1984a), *One-Dimensional Shock Waves in Deformable Dielectrics with Internal State Variables,* **Arch.Mech.**, **35**, 97-107.

McCARTHY M.F., (1984b), *One-Dimenisonal Pulse Propagation in Deformable Dielectrics with Internal State Variables,* **Arch.Rat.Mech.Anal.**, **86**, 353-367.

MEIXNER J., (1961), *Der Drehimpulssatz in der Thermodynamik der irreversible Prozessen,* **Zeit.Phys.**, **16**, 145-155.

MORRO A., DROUOT R. and MAUGIN G.A., (1985), *Thermodynamics of Polyelectrolyte Solutions in an Electric Field,* **J.Non-Equilibr. Thermodyn.**, **10**, 131-144.

MORRO A., MAUGIN G.A. and DROUOT R., (1990) , *Diffusion in Polyelectrolyte Solutions,* **Rheologica Acta**, **29**, 215-222.

MULLER I., 91985), **Thermodynamics**, Pitman (now Longman), London.

MUSCHIK W., (1990a), *Internal Variables in Non-equilibrium Thermodynamics,* **J.Non-equilibr.Thermodyn.**, **15**, 127-137.

MUSCHIK W., (1990b), **Aspects of Non-equilibrium Thermodynamics: Six Lectures on Fundamentals and Methods**, World Scientific, Singapore.

NEEL L., (1946), *Bases d'une nouvelle théorie générale du champ coercitif,* **Ann. Univ. Grenoble**, **22**, 299-343.

NELSON D.F.,(1979), **Electric, Optic and Acoustic Interactions in Dielectrics**, J.Wiley-Interscience, New York.

PARKUS H., (1979;Editor), **Electromagnetic Interactions in Elastic Solids** (CISM Lecture Notes ,1977), Springer-verlag, Wien.

PARRY G.P., (1987), *On Internal Variable Models of Phase Transitions,* **J.of Elasticity**, **17**, 63-70.

POUGET J. and MAUGIN G.A., (1984), *Solitons and Electroacoustic Interactions in ferroelectric Crystals-I ,* **Phys.Rev.,B30**, 5306-5325.

PRAGER W.,(1957), *On Ideal Locking Materials,* **Trans.Soc.Rheology**, **1**, 169-175.

PREISACH F.,(1935), *Uber die magnetische Nachwirkung,* **Zeit. Phys.**, **94**, 277-302.

ROSENSWEIG R.E.,(1989), *Thermodynamics of Electromagnetism*, in: **Thermodynamics: An Advanced Textbook for Chemical Engineers**, by G.Astarita, Plenum , New York.

RUDYAK V.M.,(1971), *The Barkhausen Effects*, **Sov.Phys.Uspekhi**, 13, 451-479.

SABIR M. and MAUGIN G.A., (1988), *Microscopic Foundations of the barkhausen Effect*, **Arch.Mech.**, 40, 829-841.

TIERSTEN H.F., (1990), **A Development of the Equations of Electromagnetism in Material Continua**, Springer-verlag, New York.

TINKHAM M.,(1964), **Phys.Rev.Lett.**, 13, 804.

TRUESDELL C.A. and NOLL W.,(1965), *Nonlinear Field Theories of Mechanics*, in:**Handbuch der Physik**, Bd.III/3, ed. S.Flügge, Springer-verlag, Berlin.

TRUESDELL C.A. and TOUPIN R.A., (1960), *The Classical Field Theories*, in: **Handbuch der Physik**, Bd.III/1, ed. S.Flügge, Springer-verlag, Berlin.

YAKUSHCOV V.V., ROZANOV D.K. and DREMIN A.N., (1968), *On the Measurement of the Polarization relaxation Time in a Shock wave*, **Sov.Phys.JETP**, 27, 213-215.

ZHOU S.A. and MIYA K., (1991), *A Nonequilibrium Theory of Thermoelastic Superconductors* , **Int.J.Appl.Electromag.Mat.**, 2 , 21-38.

STABILITY AND CONSTITUTIVE INEQUALITIES IN PLASTICITY

H. Petryk
Polish Academy of Sciences, Warsaw, Poland

ABSTRACT

A thermodynamic theory of stability in solids with intrinsic dissipation of plastic type is developed, and various related constitutive inequalities are discussed. The thermodynamic formalism for finite strain elasto-plasticity is presented, with rate-dependent plastic behaviour and its rate-independent limit described with the help of internal variables. A general condition sufficient for stability of equilibrium in the sense of Lyapunov is formulated and then successively transformed as additional assumptions are introduced, with special attention focused on isothermal rate-independent plasticity. In the latter case the conditions for stability of a quasi-static process at varying loading are also derived and shown to differ from the respective conditions of stability of equilibrium. The stability conditions are formulated for an arbitrary continuous system with specified boundary conditions as well as for a homogeneous material element embedded in a continuum. Relation between intrinsic instability at the level of a material element and propagation of acceleration waves or strain localization in shear bands is discussed.

1. INTRODUCTION

The principal topic of these lecture notes is the stability in finite strain plasticity considered from the point of view of thermodynamics. Stability of equilibrium is examined parallely to, and distinguished from, stability of a quasi-static process. In both cases, a continuous, generally inhomogeneous material body subject to specified boundary conditions as well as a material element embedded in a continuum are investigated. Conditions for various types of stability, especially at the level of a material element, lead to a number of constitutive inequalities which form the second, and closely related to the first, topic of these lecture notes.

Stability in continua is known to be a mathematically delicate subject, and plasticity effects introduce additional complications. In the present state of theoretical development and in a general case, one has apparently to be satisfied with stability conditions which are either mathematically rigorous but difficult to be further exploited, or being simpler and verifiable but at the cost of loosing their precise connection with a mathematical definition of stability. The conditions formulated below are frequently of the second type, although an attempt has been made to give them an interpretation as either necessary or sufficient conditions for stability of some specified kind.

Fundamentals of thermodynamics of solids are not discussed here since they are presented in other Parts of this Volume. In Chapter 2 of this Part the thermodynamic constitutive formalism for finite strain elasto-plasticity is presented. Rate-dependent (visco-plastic) behaviour and its rate-independent limit are described with the help of internal variables. Two particular classes of rate-independent models are introduced as the examples which will be referred to in the next chapters when discussing detailed forms of the stability criteria. Chapter 3 is devoted to the general question of stability of thermodynamic equilibrium of an inelastic system. After certain mathematical preliminaries, a general condition for stability of a compound system is formulated. The condition is successively transformed as additional assumptions are introduced; in particular, the energy criterion for isothermal perturbations is investigated in the framework of rate-independent plasticity. In Chapter 4 the concept of stability is extended to quasi-static isothermal processes induced by slowly varying external loading. It is shown that the appropriately extended energy criterion of stability is applicable to such processes although under more restrictive assumptions. Stability at the level of a material element embedded in a continuum is investigated in Chapter 5. Under the assumption of isothermal deformations and in most cases for rate-independent plasticity, a number of conditions for intrinsic stability of the material are formulated. Connections between the thermodynamic conditions of material stability and the propagation of acceleration waves or the formation of shear bands are then investigated.

In view of a number of essential questions left open, the presentation below certainly does not provide a complete theory of thermodynamic stability in plasticity. In particular, non-isothermal perturbations receive only little attention here. However, it must be pointed out that the assumptions adopted in most previous studies on thermodynamic stability excluded, in fact, the plastic behaviour as understood in the mechanics of solids. The general line of presentation adopted here seems to be novel; also several partial results discussed below have been obtained only recently and some other appear to be new.

2. THERMODYNAMIC FORMALISM IN FINITE STRAIN PLASTICITY

2.1. Thermodynamic state of a material element

We begin with specifying the notation for basic variables. Denote by $\mathbf{x}$ and $\boldsymbol{\xi}$ the position vector of a material point in a continuum in the actual and reference configuration, respectively. A basic measure of the finite strain and rotation is the deformation gradient $\mathbf{F}=\partial\mathbf{x}/\partial\boldsymbol{\xi}$, expressed according to the polar decomposition theorem as

$$\mathbf{F} \equiv \nabla\mathbf{x} = \mathbf{R}\mathbf{U}, \quad \mathbf{R}^T\mathbf{R}=\mathbf{1}, \quad \mathbf{U} = \mathbf{U}^T, \quad \mathbf{e}=\mathbf{e}(\mathbf{U}) \ . \tag{2.1}$$

e stands for an *arbitrary* symmetric strain measure of Lagrangian type. Work-conjugate stress **t** is defined by the work differential [1]:

$$dw = \mathbf{S}\cdot d\mathbf{F} = \mathbf{t}\cdot d\mathbf{e} \quad \text{(per unit reference volume)} . \tag{2.2}$$

A dot between two quantities denotes the appropriate inner product (e.g. $\mathbf{S}\cdot d\mathbf{F} = S_{ij}dF_{ij}$ in the indical notation). **S** is the first (unsymmetric) Piola-Kirchhoff stress (the transpose of the nominal stress). Among infinitely many work-conjugate pairs (**t**,**e**), one of the most popular ones consists of the Green strain as **e** and of the second Piola-Kirchhoff stress as **t**. However, there is no need here to prefer some specific pair over another, and the reference configuration can be chosen arbitrarily. In the reference configuration any **e** vanishes and any **t** coincides with the Cauchy stress $\boldsymbol{\sigma}$.

If **e** and not **F** is used as a work variable for an anisotropic material then in transition to another reference configuration it is necessary to introduce *orientation variables*, e.g. a director triad or an orientation tensor (**Q**) [2]. We shall assume that in the reference configuration **Q=1**.

Plastic behaviour at a macroscopic scale is considered as a consequence of microstructural rearrangements [3]. They are described (in the reference configuration) by *internal variables* α_κ, $\kappa=1,2,\ldots,$ collectively denoted by $\boldsymbol{\alpha} = (\alpha_\kappa)$, whose number and character (scalar or tensorial) depend on the physical mechanisms of plastic deformation as well as on the intended degree of approximation of the real material behaviour. No balance equation for α_κ is assumed.

A general thermodynamic theory of solids is presented in the other Parts of this Volume, so that only a brief account is given here. The *material element* is regarded as a representative macroscopic sample of a heterogeneous material, sufficiently large in size to behave as if it were homogeneous and sufficiently small to be treated as a point in the continuum description. The element is regarded here as a *closed* system: chemical reactions and diffusion are disregarded so that the chemical composition is ruled out from the list of variables. In the spirit of the classical theory of irreversible processes (e.g. [4]), the material element is assumed to be in local (constrained) equilibrium when its internal energy (u), macroscopic strain ($\mathbf{e}$) and internal microstructure ($\boldsymbol{\alpha}$) are given. More generally, the local equilibrium state can be viewed as a projection of a current non-equilibrium state under the assumption that relaxation times of non-equilibrium variables (different from α_K) are very short [5][6]. The absolute temperature T, specific entropy s and other thermodynamic variables are introduced locally as in thermostatics.

It will be convenient to operate with all quantities and fields *defined in the reference configuration* (being arbitrary but fixed). To simplify the notation by eliminating the fixed reference mass density $\bar{\rho}$ from the formulae, the densities of: entropy (s), internal energy (u), Helmholtz free energy (ϕ), Gibbs function (ψ), etc., all are taken *per unit reference volume*.

The Gibbs equation is postulated in the form

$$du = \mathbf{t}\cdot d\mathbf{e} + T ds - \sum_J \mathbf{A}_J\cdot d\boldsymbol{\alpha}_J \tag{2.3}$$

where $\mathbf{A}_J$ are the thermodynamic forces (or affinities) associated with $\boldsymbol{\alpha}_J$, collectively denoted by $\mathbf{A} = (\mathbf{A}_J)$. It will be more convenient to work with the Helmholtz free energy density

$$\phi \equiv u - Ts = \phi(\mathbf{e},T,\boldsymbol{\alpha}) \equiv \phi(\Delta), \quad \Delta \equiv (\mathbf{e},T,\boldsymbol{\alpha}) \ , \tag{2.4}$$

in terms of which (2.3) becomes

$$d\phi = \mathbf{t}\cdot d\mathbf{e} - s dT - \sum_J \mathbf{A}_J\cdot d\boldsymbol{\alpha}_J \tag{2.5}$$

with

$$\mathbf{t}=\partial\phi/\partial\mathbf{e}, \qquad s = -\partial\phi/\partial T, \qquad \mathbf{A}_K = -\partial\phi/\partial\boldsymbol{\alpha}_K . \tag{2.6}$$

Viscous effects are neglected so that $\mathbf{t}$ in (2.6) is equal to that in (2.2).

The plastic strain $\hat{\mathbf{e}}$ is, by definition, the strain which remains when the current stress $\mathbf{t}$ and temperature T are reduced *at frozen* $\boldsymbol{\alpha}$ to zero and to some reference value $\hat{T}$, respectively, to produce a *relaxed state*. In defining the plastic *deformation*, there is still an indeterminacy in rotation which is removed if the relaxed state is assumed to be *isoclinic* [2], where $\mathbf{Q}=\mathbf{1}$ by definition. In an isoclinic relaxed state, we have

$$\mathbf{t}=0, \quad T=\hat{T}, \quad \mathbf{Q}=\mathbf{1} \quad \Rightarrow \quad \hat{\mathbf{e}},\ \hat{s},\ \hat{u},\ \hat{\phi},\ \hat{\mathbf{F}} = \hat{\mathbf{R}}\hat{\mathbf{U}} \quad \text{as functions of } \boldsymbol{\alpha}\,, \tag{2.7}$$

where the quantities with (^) denote in particular: plastic strain ($\hat{\mathbf{e}}$), structural entropy ($\hat{s}$), stored internal energy ($\hat{u}$), plastic deformation ($\hat{\mathbf{F}}$), plastic rotation ($\hat{\mathbf{R}}$).

The relaxed state is frequently taken as a *moving* reference state for a material element, with the free energy dependent explicitly on the elastic and not total strain. Since the elastic strain can be expressed in terms of the total strain $\mathbf{e}$ and $\boldsymbol{\alpha}$, this is contained in the description with a fixed reference state which, for our present purposes, appears to be more convenient. Various aspects of thermoplastic constitutive laws are discussed e.g. in [2,3,7,8,9].

2.2. Thermodynamic process

A thermodynamic process ρ: $t \longrightarrow \mathfrak{s}(t)$ which takes place in the representative material element in some interval of time t at varying $(\mathbf{e},T)(t)$ is viewed as a sequence of constrained equilibrium states close to each other. Transitory non-equilibrium processes *between* such states, and also all instabilities at a microscopic level which correspond to the physical nature of plastic deformation [10] are not examined here. Rapid localized structural changes in a small part of the representative volume are imagined to be "smoothed out" in space and time so that at the macroscopic level they correspond to regular variations of appropriately selected internal variables α_κ of averaging type. Consequently, the function $\mathfrak{s}(t)$ is assumed to be continuous and (at least) piecewise differentiable; a dot over a symbol will denote the right-hand time derivative at a fixed material point, e.g. $\dot{\boldsymbol{\alpha}} = (d^+/dt)\boldsymbol{\alpha}$.

The entropy production inequality (per unit reference volume) reads:

$$T\sigma \equiv \mathbf{t}\cdot\dot{\mathbf{e}} - \dot{u} + T\dot{s} - \mathbf{q}\cdot\nabla T/T \geq 0\ , \tag{2.8}$$

where $\mathbf{q}$ is the heat flux transformed to the reference configuration. The rate of entropy production (σ) is assumed to be split into two nonnegative parts, resulting in:

the thermal dissipation inequality:

$$T\sigma^{T} \equiv -\ \mathbf{q}\cdot\nabla T/T \geq 0\ , \tag{2.9}$$

the intrinsic dissipation inequality:

$$\mathcal{D} \equiv T\sigma^{P} \equiv \mathbf{t}\cdot\dot{\mathbf{e}} - (\dot{\phi} + s\dot{T}) \geq 0\ . \tag{2.10}$$

On substituting (2.5), the inequality (2.10) takes the form

$$\mathcal{D} = T\sigma^{P} = \sum_{J} \mathbf{A}_{J}\cdot\dot{\boldsymbol{\alpha}}_{J} \equiv \mathbf{A}\cdot\dot{\boldsymbol{\alpha}} \geq 0\ ; \tag{2.11}$$

in the following, the more condensed notation ($\mathcal{D} = \mathbf{A}\cdot\dot{\boldsymbol{\alpha}}$) will be used in most cases, with the obvious meaning of the dot between $\mathbf{A}$ and $\boldsymbol{\alpha}$.

Introduce the Gibbs function ψ as the partial Legendre transform of ϕ with respect to $\mathbf{e}$:

$$\psi \equiv \phi - \mathbf{e}\cdot\mathbf{t} = \psi(\mathbf{t},T,\boldsymbol{\alpha}_K), \qquad \mathbf{e} = -\ \partial\psi/\partial\mathbf{t}, \qquad \mathbf{A}_K = -\ \partial\psi/\partial\boldsymbol{\alpha}_K\ . \tag{2.12}$$

By equating the cross-derivatives of ϕ or ψ, we obtain the Maxwell relations which involve $\mathbf{A}_K$:

$$\partial\mathbf{t}(\mathbf{e},T,\boldsymbol{\alpha})/\partial\boldsymbol{\alpha}_K = -\ \partial\mathbf{A}_K(\mathbf{e},T,\boldsymbol{\alpha})/\partial\mathbf{e}\ , \tag{2.13}$$

$$\partial\mathbf{e}(\mathbf{t},T,\boldsymbol{\alpha})/\partial\boldsymbol{\alpha}_K = \partial\mathbf{A}_K(\mathbf{t},T,\boldsymbol{\alpha})/\partial\mathbf{t}\ . \tag{2.14}$$

Let the plastic part of stress-rate be defined as

$$\dot{\mathbf{t}}^p \equiv (\partial \mathbf{t}(\mathbf{e},T,\boldsymbol{\alpha})/\partial\boldsymbol{\alpha})\cdot\dot{\boldsymbol{\alpha}} = \dot{\boldsymbol{\alpha}}\cdot\partial^2\phi/\partial\boldsymbol{\alpha}\partial\mathbf{e} = -\sum_J \dot{\alpha}_J\cdot\partial\mathbf{A}_J(\mathbf{e},T,\boldsymbol{\alpha})/\partial\mathbf{e} \tag{2.15}$$

and the plastic part of strain-rate as

$$\dot{\mathbf{e}}^p \equiv (\partial \mathbf{e}(\mathbf{t},T,\boldsymbol{\alpha})/\partial\boldsymbol{\alpha})\cdot\dot{\boldsymbol{\alpha}} = -\dot{\boldsymbol{\alpha}}\cdot\partial^2\psi/\partial\boldsymbol{\alpha}\partial\mathbf{t} = \sum_J \dot{\alpha}_J\cdot\partial\mathbf{A}_J(\mathbf{t},T,\boldsymbol{\alpha})/\partial\mathbf{t} \ . \tag{2.16}$$

Then we have

$$\dot{\mathbf{e}} = \dot{\mathbf{e}}^e + \dot{\mathbf{e}}^p, \qquad \dot{\mathbf{t}} = \dot{\mathbf{t}}^e + \dot{\mathbf{t}}^p, \qquad \dot{\mathbf{t}}^p = -\ \partial^2\phi/\partial\mathbf{e}\partial\mathbf{e}\cdot\dot{\mathbf{e}}^p \ , \tag{2.17}$$

where

$$\dot{\mathbf{t}}^e = \phi_{,\mathbf{ee}}\cdot\dot{\mathbf{e}} + \phi_{,T\mathbf{e}}\,\dot{T} \ , \qquad \dot{\mathbf{e}}^e = -\psi_{,\mathbf{tt}}\cdot\dot{\mathbf{t}} - \psi_{,T\mathbf{t}}\,\dot{T} \tag{2.18}$$

are the thermoelastic stress-rate and strain-rate, respectively. The usual short-hand notation has been introduced, where a comma as a lower index denotes partial derivation with respect to the arguments which follow the comma.

In general, we have

$$\dot{\mathbf{e}}^p \neq (\hat{\mathbf{e}})^{\boldsymbol{\cdot}} \tag{2.19}$$

but in many practical cases the difference can be neglected. For further discussion on the structure of incremental relations in plasticity, see e.g. [11].

2.3. Kinetics for rate-dependent plastic solid

Evolution equations for internal variables in rate-dependent plasticity are assumed in the form

$$\dot{\alpha}_K = \dot{\alpha}_K(\mathbf{A},T,\boldsymbol{\alpha}) \ , \tag{2.20}$$

upon which the restriction (2.11) is imposed. Since under the assumption of local equilibrium $\mathbf{A}_K$ are functions of Δ by $(2.6)_3$, we can also rewrite (2.20) as $\dot{\alpha}_K = \dot{\alpha}_K(\Delta)$.

Our special attention will be directed to the cases where the equations (2.20) admit a potential with respect to the thermodynamic forces, viz.

$$\dot{\alpha}_K = \partial\omega(\mathbf{A}, T, \alpha)/\partial \mathbf{A}_K \tag{2.21}$$

which is (formally) equivalent to assuming the following generalization of the Onsager reciprocal relations to the present nonlinear kinetic law:

$$\partial\dot{\alpha}_K/\partial \mathbf{A}_J = \partial\dot{\alpha}_J/\partial \mathbf{A}_K . \tag{2.22}$$

This is, of course, an approximation to reality and not a law of nature. The potential form (2.21) can be derived from the assumption [12, 3] that $\mathbf{A}, \alpha$ can be represented by a number of scalars A_K, α_K such that each $\dot{\alpha}_K$ is dependent on the *associated* force A_K and not on other A_J, $J \neq K$, viz.

$$\dot{\alpha}_K = \dot{\alpha}_K(A_K, T, \alpha) . \tag{2.23}$$

(2.23) can evidently be recast in the form

$$\dot{\alpha}_K = \partial\omega(\mathbf{A}, T, \alpha)/\partial A_K, \qquad \omega = \sum_K \int_0^{A_K} \dot{\alpha}_K(B_K, T, \alpha)\, dB_K .$$

If A_K are expressed through (2.12) as functions of $(\mathbf{t}, T, \alpha)$ then from (2.16) we obtain [12, 3] that the plastic part of the macroscopic strain rate is also derivable from a potential, viz.

$$\dot{\mathbf{e}}^P = \partial\omega(\mathbf{t}, T, \alpha)/\partial\mathbf{t}, \qquad \omega(\mathbf{t}, T, \alpha) \equiv \omega(\mathbf{A}(\mathbf{t}, T, \alpha), T, \alpha) . \tag{2.24}$$

Under the assumption that $\dot{\alpha}_K$ is a non-decreasing function of A_K at fixed T and α, the potential ω is a convex function of A_K.

Analogously, the equation (2.24) results also from a more general assumption (2.21). If ω in (2.21) is a strictly convex (and everywhere differentiable) function of $\mathbf{A}_K$ at fixed $T,\boldsymbol{\alpha}$ then we can introduce its partial Legendre transform with respect to $\mathbf{A}_K$ to obtain an inverse form of (2.21):

$$\mathbf{A}_K = \partial d(\dot{\boldsymbol{\alpha}},T,\boldsymbol{\alpha})/\partial\dot{\alpha}_K \ , \qquad d(\dot{\boldsymbol{\alpha}},T,\boldsymbol{\alpha}) \equiv \mathbf{A}\cdot\dot{\boldsymbol{\alpha}} - \omega \ . \tag{2.25}$$

The intrinsic dissipation inequality then reads

$$\mathcal{D} = d + \omega \geq 0, \qquad (\ d + \omega = 0 \ \Leftrightarrow \ \text{all } \dot{\alpha}_K = 0 \ ) \tag{2.26}$$

while the compatibility between $\mathbf{A}_K$ in (2.25) and (2.6) requires that

$$\partial\phi/\partial\alpha_K + \partial d/\partial\dot{\alpha}_K = 0 \ . \tag{2.27}$$

Note that the potential d does *not* coincide with the rate $\mathcal{D}$ of intrinsic dissipation.

In a scale of time such that creep can be neglected, experiments on metals show that there are no macroscopically observable changes in the state of the material if the stress is kept within some (elastic) domain at T = const. Having assumed that changes of α_K are responsible for the plastic strains, it follows from (2.14) that $\partial\mathbf{A}_K/\partial t$ cannot vanish identically; hence, the elastic domain in stress space corresponds to some elastic domain in $\mathbf{A}_K$-space. Accordingly, we assume that

$$\dot{\alpha}_K = 0 \ \text{ if } \ \mathbf{A}_K \in C_K(T,\boldsymbol{\alpha}) \quad , \ 0 \in C_K \ ; \tag{2.28}$$

the fact that $\dot{\alpha}_K = 0$ need not correspond to the zero point in $\mathbf{A}_K$-space but can take place in a finite domain C_K is crucial in plasticity. The elastic domain in $\mathbf{A}$-space is obtained as a common part of all C_K:

$$C = C(T,\boldsymbol{\alpha}) \equiv \bigcap_K C_K(T,\boldsymbol{\alpha}) \ . \tag{2.29}$$

Note that even if the boundary ∂C_K of each C_K is smooth, the elastic domain C has in general a boundary ∂C which is only piecewise smooth.

If $\mathbf{A}(\mathbf{e},T,\boldsymbol{\alpha}(t)) \notin \overline{C}(T,\boldsymbol{\alpha}(t))$ where $\overline{C} \equiv C \cup \partial C$ then $\dot{\boldsymbol{\alpha}}(t) \neq 0$; we assume that if $(\mathbf{e},T)$ are kept constant then an asymptotic state Δ^{∞} is eventually approached, viz.

$$\Delta^{\infty}(\mathbf{e},T,\boldsymbol{\alpha}(0)) = (\mathbf{e},T,\boldsymbol{\alpha}^{\infty}) : \quad \alpha_K^{\infty} = \alpha_K(0) + \int_0^{\infty} \dot{\alpha}_K(\mathbf{e},T,\boldsymbol{\alpha}(t))\, dt,$$

$$\dot{\alpha}_K(\Delta^{\infty})=0, \qquad \mathbf{A}_K(\Delta^{\infty}) \in \overline{C}_K(T,\boldsymbol{\alpha}^{\infty}) \tag{2.30}$$

It is essential that $\boldsymbol{\alpha}^{\infty}$ depends in general on the initial value $\boldsymbol{\alpha}(0)$ and that $\mathbf{A}(\Delta^{\infty})$ need *not* vanish in the asymptotic state. The latter property is of fundamental importance in formulating the stability conditions in plasticity. A state Δ^{∞} satisfying (2.30) is, by definition, an *equilibrium* state of the material element at the prescribed strain $\mathbf{e}$ and temperature T.

The loading functions $f_K(\mathbf{A}_K,T,\boldsymbol{\alpha})$ are defined such that

$$f_K<0 \ \text{ if } \mathbf{A}_K \in C_K(T,\boldsymbol{\alpha}), \quad f_K=0 \ \text{ if } \mathbf{A}_K \in \partial C_K(T,\boldsymbol{\alpha}), \quad f_K>0 \text{ otherwise.} \tag{2.31}$$

2.4. Rate-independent plasticity

Experiments performed on metals show that the stresses during plastic flow often only slightly depend on the strain-rate in a certain range of deformation rates and temperatures. In that range, we can consider a rate-independent limit of plastic behaviour as relaxation times for internal structural rearrangements tend either to zero or to infinity. That limit is not uniquely defined and can depend on the time scale of changes of external loading and on the split of internal variables into completely frozen and possibly active. Suppose that the former have been eliminated from consideration. According to (2.30), in the rate independent limit we have

$$f_K \le 0, \qquad \Delta \longrightarrow \Delta^\infty \tag{2.32}$$

at every stage of a process. Alternatively, we can take

$$\omega(\mathbf{A},T,\alpha) = 0 \;\text{ if } \mathbf{A} \in \bar{C}(T,\alpha), \qquad \omega(\mathbf{A},T,\alpha) = +\infty \;\text{ if } \mathbf{A} \notin \bar{C}(T,\alpha); \tag{2.33}$$

and rewrite (2.21) and (2.25) in terms of subgradients [13,14] rather than partial derivatives.

If the internal variables α_K describe *local* rearrangements in the representative material element then the resulting limit process $t \longrightarrow \Delta^\infty(t)$ need not be continuous. For, the attractor α^∞ can suffer a discontinuous jump that corresponds to an instability at the micro-level. As indicated above, such instabilities are beyond the scope of this presentation. Rather, the variables α_K are understood as *averaging* internal variables which vary continuously in time. Consequently, only continuous processes p will be considered.

A complementary assumption to (2.32) is that $\dot{\alpha}_K$ can be non-zero only if the respective yield condition in A-space is satisfied:

$$\dot{\alpha}_K \neq 0 \quad \Rightarrow \quad f_K(\mathbf{A}_K, T, \alpha) = 0 \quad \Leftrightarrow \quad \mathbf{A}_K \in \partial C_K(T,\alpha) \; . \tag{2.34}$$

Transformed to e-space[1)], this reads:

$$g_K(\Delta) \equiv g_K(\mathbf{e},T,\alpha) \le 0, \qquad g_K\dot{\alpha}_K = 0, \qquad g_K(\Delta) \equiv f_K(\mathbf{A}_K(\Delta),T,\alpha) \; . \tag{2.35}$$

Contrary to (2.20), the actual values of $\dot{\alpha}_K$ depend now not only on the state Δ but also on $\dot{\mathbf{e}}$ and $\dot{T}$.

As examples, two particular cases of the time-independent constitutive framework are considered below. Note that evolution

---

1) It is common to consider the yield condition in stress-space rather than in strain-space. However, the description adopted here is applicable under more general assumptions.

equations for $\dot{\alpha}_K$ are contained in an *indirect* form.

Example (1)

Consider the dissipation function of the form:

$$\mathcal{D} = \sum_J \mathbf{A}_J \cdot \dot{\alpha}_J = \mathcal{D}(T, \alpha, \dot{\alpha}) \geq 0, \qquad \mathcal{D}(T, \alpha, r\dot{\alpha}) = r\, \mathcal{D}(T, \alpha, \dot{\alpha}) \quad \text{for every } r>0 , \tag{2.36}$$

as a convex (and homogeneous of degree one as indicated) function of $\dot{\alpha}_J$. In place of (2.21) and (2.25), the normality law which follows is written as:

$$\dot{\alpha}_K \in \partial\omega(\mathbf{A}_K), \qquad \mathbf{A}_K \in \partial\mathcal{D}(\dot{\alpha}_K), \qquad T, \alpha \text{ fixed} \tag{2.37}$$

where, on account of non-differentiability of $\mathcal{D}$ at $\dot{\alpha}=\mathbf{0}$ (cf. 2.36) and of ω at $\mathbf{A}\in\partial C$ (cf. (2.33), the subdifferential notation [14] has been used in place of partial derivatives. Alternatively, the above convexity and normality assumptions can be replaced by the postulated principle of maximum dissipation rate:

$$(\mathbf{A} - \mathbf{A}^*)\cdot\dot{\alpha} \geq 0 \qquad \text{for every } \mathbf{A}^* \in \bar{C}(T,\alpha) \tag{2.38}$$

where $\mathbf{A}$ is associated with $\dot{\alpha}$. (2.38) implies that the domain C in $\mathbf{A}$-space is convex and that $\dot{\alpha}$ lies within the cone generated by outward normals to the boundary ∂C at given $\mathbf{A}(\diamond)$. For our purposes, validity of (2.38) will be required at the *actual* $\mathbf{A}$ for $|\mathbf{A}-\mathbf{A}^*| < \varepsilon$ only, where ε is an arbitrarily small positive number. It will be shown later how that hypothesis is related to material stability with respect to internal structural rearrangements.

The compatibility between (2.36) and $(2.6)_3$ requires that

$$\mathcal{D}(T,\alpha,\dot{\alpha}^*) = \mathbf{A}(\diamond)\cdot\dot{\alpha}^* \quad \Leftrightarrow \quad \sum_K (\partial\phi/\partial\alpha_K + \partial\mathcal{D}/\partial\dot{\alpha}^*_K)\cdot\dot{\alpha}^*_K = 0 \tag{2.39}$$

for all *virtual* rates $\dot{\alpha}^*$ consistent with the actual thermodynamic forces

$\mathbf{A}(\Delta)$; in obtaining the right-hand form, the Euler theorem on homogeneous functions has been used. In a process ρ, (2.38) holds for $|\mathbf{A}-\mathbf{A}^*| < \varepsilon$ at every instant, by assumption. Hence, among all virtual $\dot{\alpha}^*$ which satisfy (2.39), the actual rate $\dot{\alpha}$ in ρ must satisfy the following incremental compatibility conditions

$$
\begin{aligned}
&(\frac{d^+}{dt}\mathbf{A}(\Delta))\cdot\dot{\alpha} = \mathcal{D}_{,\alpha}(T,\alpha,\dot{\alpha})\cdot\dot{\alpha} + \mathcal{D}_{,T}(T,\alpha,\dot{\alpha})\,\dot{T}\,, \qquad \dot{\alpha} = \frac{d^+}{dt}\alpha \quad \text{in } \rho\,,\\
&(\frac{d^+}{dt}\mathbf{A}(\Delta))\cdot\dot{\alpha}^* \le \mathcal{D}_{,\alpha}(T,\alpha,\dot{\alpha}^*)\cdot\dot{\alpha} + \mathcal{D}_{,T}(T,\alpha,\dot{\alpha}^*)\,\dot{T} \qquad \text{for all } \dot{\alpha}^* \text{ which satisfy (2.39),}
\end{aligned}
\tag{2.40}
$$

where the dependence of $\mathcal{D}$ on α and T has been assumed smooth (Gateaux differentiable).

By using (2.40) and the expression for $\dot{\mathbf{A}}$ obtained from $(2.6)_3$ it can be shown that $\dot{\alpha}$ at given Δ is uniquely defined by $\dot{T}$ and $\dot{\mathbf{e}}$ if

$$
(\dot{\alpha}-\dot{\alpha}^*)\cdot\phi_{,\alpha\alpha}\cdot(\dot{\alpha}-\dot{\alpha}^*) + (\dot{\alpha}-\dot{\alpha}^*)\cdot(\mathcal{D}_{,\alpha}(T,\alpha,\dot{\alpha}) - \mathcal{D}_{,\alpha}(T,\alpha,\dot{\alpha}^*)) > 0 \quad \text{for all } \dot{\alpha}\neq\dot{\alpha}^* \text{ which satisfy (2.39).} \tag{2.41}
$$

We will make use of the additional symmetry assumption

$$
\mathcal{D}_{,\alpha}(T,\alpha,\dot{\alpha})\cdot\dot{\alpha}^* = \mathcal{D}_{,\alpha}(T,\alpha,\dot{\alpha}^*)\cdot\dot{\alpha} \tag{2.42}
$$

for every $\dot{\alpha}$, $\dot{\alpha}^*$ which satisfy (2.39). The meaning of the mathematical condition (2.42) will be seen later (cf. the property (3.56)).

The incremental framework with $\mathcal{D}$ dependent on α has been recently developed in [15,16] where references to the related earlier work can be found; the above extension to a variable temperature is straightforward.

Example (2)

The direction (but not the magnitude) of each $\dot{\alpha}_K$ is assumed to be restricted by an individual normality rule in the form

$$
\dot{\alpha}_K = \dot{\gamma}_K \mathbf{P}_K(\mathbf{A}_K, T, \alpha), \quad \mathbf{P}_K \equiv \partial f_K(\mathbf{A}_K, T, \alpha)/\partial \mathbf{A}_K\,, \quad \dot{\gamma}_K \ge 0, \quad f_K\dot{\gamma}_K = 0. \tag{2.43}
$$

The related dissipation function can be written as

$$\mathcal{D} = \mathcal{D}(\mathbf{A}_J, T, \alpha;\ \dot{\gamma}_J) = \sum_J \pi^c_J(\mathbf{A}_J, T, \alpha)\ \dot{\gamma}_J \tag{2.44}$$

while f_K can be identified with the difference between the thermodynamic "driving force" π_K and its threshold value π^c_K, viz.

$$f_K \equiv \pi_K - \pi^c_K\ , \qquad \pi_K = \pi_K(\mathbf{A}_K, T, \alpha) \equiv \mathbf{A}_K\cdot\mathbf{P}_K\ . \tag{2.45}$$

Note that this implies

$$\partial\pi^c_K/\partial\mathbf{A}_K = \mathbf{A}_K\cdot\partial^2 f_K/\partial\mathbf{A}_K\partial\mathbf{A}_K = \mathbf{A}_K\cdot\partial\mathbf{P}_K/\partial\mathbf{A}_K\ . \tag{2.46}$$

The yield functions g_K in strain space are given by (2.35), with the compatibility condition

$$g_K(\Delta)\ \overset{\bullet\bullet}{\gamma}_K = 0 \tag{2.47}$$

satisfied by all virtual $\overset{\bullet\bullet}{\gamma}_K$. The actual $\dot{\gamma}_K$ must satisfy the incremental compatibility condition

$$\dot{g}_K = \boldsymbol{\lambda}_K\cdot\dot{\mathbf{e}} + g_{KT}\ \dot{T} - \sum_J g_{KJ}\ \dot{\gamma}_J \le 0 \quad \text{if } g_K{=}0, \quad \text{with } \dot{g}_K\dot{\gamma}_K = 0\ , \tag{2.48}$$

where, on account of (2.46),

$$\boldsymbol{\lambda}_K \equiv \partial g_K/\partial\mathbf{e} = \mathbf{P}_K\cdot\mathbf{A}_{K,\mathbf{e}}\ ,$$

$$g_{KT} \equiv \partial g_K/\partial T = \mathbf{P}_K\cdot\mathbf{A}_{K,T} + \mathbf{A}_K\cdot\mathbf{P}_{K,T} - \pi^c_{K,T}\ , \tag{2.49}$$

$$g_{KJ} \equiv -\ \partial g_K/\partial\alpha_J\ \cdot\mathbf{P}_J = (\ \mathbf{P}_K\cdot\phi_{,\alpha_K\alpha_J} - \mathbf{A}_K\cdot\mathbf{P}_{K,\alpha_J} + \pi^c_{K,\alpha_J}\ )\cdot\mathbf{P}_J\ .$$

It can be shown that $\dot{\gamma}_K$ at given Δ is uniquely defined by $\dot{T}$ and $\dot{\mathbf{e}}$ if

$$(g_{KJ}) \text{ is positive definite}, \quad K,J:\ g_K=g_J=0\ . \tag{2.50}$$

An analog to the symmetry property (2.42) is

$$g_{KJ} = g_{JK}\ , \qquad K,J:\ g_K=g_J=0\ ; \tag{2.51}$$

the meaning of (2.51) will be seen later; cf. the property (3.60). Of course, all the above quantities without a dot depend on the state Δ.

The above constitutive framework is analogous to that developed in [2,17,18].

In both Examples, from the incremental compatibility condition (2.40) or (2.48) the actual values of $\dot{\alpha}_K$ can be determined for given $\dot{\mathbf{e}}, \dot{T}$. Uniqueness of such $\dot{\alpha}_K$ will be assumed as granted, except in Section 5.2. Note that this does not ensure uniqueness of $\dot{\alpha}_K$ when $\dot{\mathbf{t}}$ is given in place of $\dot{\mathbf{e}}$. If $\dot{\alpha}_K$ have been found then the incremental constitutive law (the constitutive rate equations) follows from (2.17).

For given Δ and $\dot{T}$, one can distinguish in $\dot{\mathbf{e}}$-space three constitutive cones: the total loading cone, the total unloading cone and the transitory range of partial unloading. The total loading cone contains $\dot{\mathbf{e}}$ such that the associated $\dot{\alpha}_K$ are nonzero for all K such that $g_K(\Delta)=0$. The total unloading condition in Example (2) is

$$\boldsymbol{\lambda}_K \cdot \dot{\mathbf{e}} + g_{KT}\,\dot{T} < 0 \qquad \text{if } g_K(\Delta) = 0\ . \tag{2.52}$$

We will use later (**S**, **F**) as a pair of stress and deformation measures particularly useful for examining stability at given boundary conditions. In analogy to $(2.17)_2$ and (2.15), we have

$$\dot{\mathbf{S}} = \dot{\mathbf{S}}(\Delta, \dot{\mathbf{F}}, \dot{T}) = \phi_{,\mathbf{FF}} \cdot \dot{\mathbf{F}} + \phi_{,T\mathbf{F}}\,\dot{T} + \dot{\boldsymbol{\alpha}} \cdot \phi_{,\boldsymbol{\alpha}\mathbf{F}}, \qquad \phi=\phi(\Delta), \quad \Delta = (\mathbf{F}, T, \boldsymbol{\alpha}). \tag{2.53}$$

For isothermal processes and at the assumed constitutive uniqueness in determining $\dot{\boldsymbol{\alpha}}$, each of the symmetry conditions: (2.42) or (2.51) implies existence of the velocity-gradient potential introduced by Hill [19]

$$\dot{\mathbf{S}} = \partial U/\partial\dot{\mathbf{F}}, \qquad U = U(\mathfrak{a},\dot{\mathbf{F}}) = \frac{1}{2}\,\dot{\mathbf{S}}\cdot\dot{\mathbf{F}}, \qquad T = \text{const.} \tag{2.54}$$

This property is of fundamental importance in the isothermal theory of bifurcation and stability in rate-independent plasticity.

3. STABILITY OF EQUILIBRIUM OF ELASTIC-PLASTIC SYSTEMS

3.1. General concept of stability. Lyapunov functional

Any system in an equilibrium state is subject in reality to small disturbing influences (≡ disturbances) which induce a non-equilibrium process in the system. If the process deviates only slightly from (fluctuates about) the equilibrium state then the latter is called stable. If, on the contrary, the deviation becomes finite no matter how small the disturbing influences are then the equilibrium state is called unstable and regarded as unattainable in a physical system.

That intuitive concept of stability of equilibrium has to be made precise by defining (i) the object of stability analysis, (ii) the class of disturbances considered, (iii) the measure of "strength" of disturbance, (iv) the measure of distance from the equilibrium state. Since arbitrariness in formulating such definitions cannot be avoided, the concept of stability is not an absolute one. We begin with the following conventions.

(i) We shall examine stability of an *equilibrium state* $\mathcal{S}^0$ of a given *isolated* system $\mathcal{A}$, in general defined as a compound system. (Stability of a *process* will be discussed in Chapter 4).

(ii) Small changes of the *initial* values (at an instant τ, say) of selected parameters of a (non-equilibrium) state $\mathcal{S}$ are taken as initial disturbances; the *equations* governing the subsequent non-equilibrium process $\mathcal{P}$: $t \longrightarrow \mathcal{S}(t)$, $t \geq \tau$, are regarded as unperturbed.

(iii) and (iv) The measures: (ρ^0) of an initial disturbance and (ρ) of a distance between $\mathcal{S}$ and the equilibrium state $\mathcal{S}^0$ in the process $\mathcal{P}$, called *metrics*[2], are not specified universally; they may depend on the system and on the examined type of stability. We merely impose the obvious restrictions:

$$\rho^0 = \rho^0(\mathcal{S},\mathcal{S}^0) \geq 0, \quad \rho^0(\mathcal{S}^0,\mathcal{S}^0) = 0\,, \tag{3.1}$$
$$\rho = \rho(\mathcal{S},\mathcal{S}^0) \geq 0, \quad \rho(\mathcal{S}^0,\mathcal{S}^0) = 0\,, \tag{3.2}$$

and assume that:

$$\rho(t) \equiv \rho(\mathcal{S}(t),\mathcal{S}^0) \text{ is continuous with respect to } t \text{ in any process } \mathcal{P}, \tag{3.3}$$

$$\rho \text{ is continuous at } (\mathcal{S}^0,\mathcal{S}^0) \text{ with respect to } \rho^0. \tag{3.4}$$

As discussed below, one can use "natural" measures generated by a Lyapunov functional. In particular, ρ^0 can coincide with ρ.

Following the Lyapunov-Movchan theory of stability [20][29], the definition of stability for infinitesimal initial disturbances with respect to metrics ρ^0 and ρ is now adopted.

Definition 1. An equilibrium state $\mathcal{S}^0$ is called stable for initial disturbances with respect to two metrics ρ^0, ρ if and only if for every $\varepsilon > 0$ there is $\delta(\varepsilon) > 0$ such that $\rho^0(\tau) < \delta(\varepsilon)$ implies $\rho(t) < \varepsilon$ at every instant $t > \tau$.

The *existence* of a (perturbed) process $\mathcal{P}$ in a time interval $[\tau, +\infty)$ initiated by any initial disturbance is taken here as granted. While this is a natural assumption from a physical point of view, serious mathematical difficulties can be met when the existence is attempted to be proved for a continuous nonlinear system.

The initial state of the process $\mathcal{P}$ is regarded as attainable from the examined equilibrium state in some transitory process, e.g. due to

---

[2] They need not possess all metric properties assumed in topology.

imposing appropriate time-dependent perturbations of body forces, surface tractions or temperature in some time interval $[t^0,\tau)$, with $\mathcal{S}(t^0)=\mathcal{S}^0$. (In particular, the initial disturbances must be consistent with the assumed continuity restrictions.) Since any theoretical description is approximate with respect to reality, it is natural to examine stability for such time-dependent perturbations of the governing equations (i.e. for so-called *persistent* disturbances) rather than only for initial disturbances. However, as a first step we restrict ourselves here to considering only the latter.

A powerful tool for establishing stability of equilibrium (or, more generally, of a process) is the Lyapunov direct method. Its essence lies in constructing a functional V^L (the Lyapunov functional)[3] such that

(a) V^L vanishes at the equilibrium state,
(b) V^L is positive definite with respect to the distance ρ,
(c) V^L is non-increasing in any process following an initial disturbance.

The properties (b) and (c) need to hold only in some neighborhood of $\mathcal{S}^0$, e.g. for $\rho < R$, say. If, additionally,

(d) the initial value of V^L at τ is continuous with respect to ρ^0,

then stability of equilibrium in the sense of Definition 1 can be inferred. In fact, one can prove the following theorem (Movchan [20] gave a rigorous proof under more general assumptions) on stability with respect to two metrics.

Theorem 1. Let a functional $V^L(\mathcal{S})$ exists such that

---

[3] In the following, we will use the notation $V^L(\mathcal{S})$ to indicate that the value of V^L depends on the current state. However, V^L may also depend on the process which leads from $\mathcal{S}^0$ to $\mathcal{S}$.

(a) $V^L(\mathcal{S}^0) = 0$;

(b) $V^L(\mathcal{S}) \geq a(\rho(\mathcal{S},\mathcal{S}^0))$ if $\rho(\mathcal{S},\mathcal{S}^0) < R$, R = const > 0,

$a(\cdot)$ is a continuous and monotonically increasing function, $a(0)=0$;

(c) $V^L(t) \equiv V^L(\mathcal{S}(t))$ is non-increasing with respect to t in any process $\mathcal{P}$ so long as $\rho(\mathcal{S}(t),\mathcal{S}^0) < R$,

(d) $V^L(\mathcal{S})$ continuous at $\mathcal{S}^0$ with respect to $\rho^0(\mathcal{S})$.

Then the equilibrium state $\mathcal{S}^0$ is stable with respect to two metrics ρ^0, ρ in the sense of Definition 1.

In proof, suppose that the conditions (a)÷(d) are satisfied by some $V^L(\mathcal{S})$, and take any ε such that $R > \varepsilon > 0$. From (d) it follows that $V^L(\tau) < a(\varepsilon)$ provided ρ^0 is sufficiently small with respect to $a(\varepsilon)$, i.e. if $\rho^0(\mathcal{S}(\tau),\mathcal{S}^0) < \hat{\delta}(a(\varepsilon)) \equiv \delta(\varepsilon)$, say. Then from (b) we obtain that $\rho(\tau) \leq a^{-1}(V^L(\tau)) < \varepsilon$. We find also that $\rho(t)$ cannot reach ε at any $t_\varepsilon > \tau$ since by condition (b) this would imply $V^L(t_\varepsilon) \geq a(\varepsilon) > V^L(\tau)$, in contradiction with (c). From the continuity of $\rho(t)$ we now obtain that $\rho(t) < \varepsilon$ for any $t > \tau$, implying stability according to Definition 1.

From the above shortened proof one can see how the assumed properties of V^L, ρ and ρ^0 interact leading to stability with respects to two metrics.

An important advantage of the Lyapunov direct method is that the solutions which describe perturbed processes induced by initial disturbances need not be determined. The property (c) can in principle be established from the general form of the governing equations. If a functional $V(\mathcal{S})$ is found such that its value is continuous and non-increasing in time in any process $\mathcal{P}$ and, moreover,

$$V(\mathcal{S}) - V(\mathcal{S}^0) > 0 \qquad \text{for every} \quad \mathcal{S} \neq \mathcal{S}^0 \tag{3.5}$$

then we may identify the difference $V(\mathcal{S}) - V(\mathcal{S}^0)$ with the Lyapunov functional V^L discussed above. The pseudo-metric generated by V:

$$\rho_V(\mathcal{S},\mathcal{S}^0) \equiv |V(\mathcal{S}) - V(\mathcal{S}^0)| \tag{3.6}$$

can serve as a particular measure ρ of the distance from the unperturbed equilibrium state, the so-called *Lyapunov* (pseudo-)*metric*. Moreover, we can also take $\rho^0 = \rho_V$ as a particular measure for initial disturbances. Then all the conditions in Theorem 1 are evidently satisfied. Thus, we have:

Corollary 1. An equilibrium state $\mathcal{S}^0$ is stable for initial disturbances with respect to the Lyapunov metric generated by a functional V if :

(i) $V(\mathcal{S}) > V(\mathcal{S}^0)$ for every $\mathcal{S} \neq \mathcal{S}^0$,

(ii) $V(t) \equiv V(\mathcal{S}(t))$ is continuous with respect to t in any process $\mathcal{P}$,

(iii) $d^+V/dt \leq 0$ in any process $\mathcal{P}$.

If stability in the sense of Corollary 1 has been established for a given system then one can search for other metrics ρ^0, ρ such that stability with respect to them could be concluded from Theorem 1.

3.2. Thermodynamic condition of stability of equilibrium

We restrict ourselves to considering a continuous elastic-plastic body $\mathcal{B}$ subject to *conservative* external loads (dependent on a loading parameter λ) at a fixed environmental temperature $\bar{T}$. In a state $\mathcal{S}^0$ of *thermodynamic equilibrium* of the system, all time derivatives of parameters of state vanish, by definition, at every point of the system. According to the assumptions specified in the previous chapter, in particular to the assumption of a local state, an equilibrium state $\mathcal{S}^0$ in the body is defined by the triple $(\tilde{\mathbf{x}}^0, \bar{T}, \tilde{\alpha}^0)$, or alternatively by $(\tilde{\mathbf{x}}^0, \tilde{s}^0, \tilde{\alpha}^0)$, where a tilde over a symbol is used to denote a spatial field over a domain G. To define a *non-equilibrium* state $\mathcal{S}$ of the body (but still under the assumption of local equilibrium within each material element separately), we must add the velocity field $\tilde{\mathbf{v}}$, $\mathbf{v} \equiv \dot{\mathbf{x}}$, so that $\mathcal{S}$ is defined by the quadruple $(\tilde{\mathbf{x}}, \tilde{T}, \tilde{\alpha}, \tilde{\mathbf{v}})$, or alternatively by $(\tilde{\mathbf{x}}, \tilde{s}, \tilde{\alpha}, \tilde{\mathbf{v}})$. A compatible strain field $\tilde{\mathbf{e}}$ is determined from $\tilde{\mathbf{F}} = \nabla\tilde{\mathbf{x}}$; from now on it will be convenient to use $\mathbf{F}$ rather than $\mathbf{e}$ as the external variable for a material element. For simplicity, we shall assume, if not stated otherwise, that not only $\tilde{\mathbf{x}}$ but also $\tilde{\mathbf{v}}$ and $\tilde{T}$ are continuous fields on $\bar{G}$

(including continuity of $\mathbf{v}$ and T across the body surface) and that $\mathbf{F}, T, \boldsymbol{\alpha}$ are continuous and at least right-hand differentiable with respect to t.

The internal energy $\mathbb{U}$, macroscopic kinetic energy $\mathbb{K}$, entropy $\mathbb{S}$ and Helmholtz free energy Φ of the body $\mathcal{B}$ are defined in the usual manner as volume integrals of the respective local quantities u, $\bar{\rho}\mathbf{v}\cdot\mathbf{v}/2$, s and ϕ over a fixed domain $G \subset \mathbb{R}^3$ occupied by the body in the *reference* state. The assumed conservative loads acting on the body $\mathcal{B}$ are imagined to result from mechanical interaction between the body and a *loading device* which, by definition, can exchange work and not entropy with $\mathcal{B}$. The assumed constant temperature $\bar{T}$ over the body surface ∂G is imagined to result from the thermal contact with an ideal heat reservoir of temperature $\bar{T}$, which can exchange entropy and not work with $\mathcal{B}$. It is assumed that the loading device and heat reservoir undergo only reversible processes, with no entropy production also at the interfaces with the body $\mathcal{B}$.

The global quantities for the body $\mathcal{B}$ are complemented with the potential energy $\Omega = \Omega(\tilde{\mathbf{x}}, \lambda)$ of the mechanical loading device and with the internal energy $\mathbb{E}_h$ and entropy $\mathbb{S}_h$ of the heat reservoir. Since the entropy of the mechanical loading device is unessential, it is set equal to zero for simplicity and Ω is interpreted as the internal energy of the device. We shall consider the compound system $\mathcal{A}$ which consists of the body $\mathcal{B}$, the heat reservoir and the loading device, whose total (internal and kinetic) energy $\mathbb{E}_A$ and total entropy $\mathbb{S}_A$ read

$$\mathbb{E}_A = \mathbb{U} + \mathbb{K} + \mathbb{E}_h + \Omega\,, \qquad \mathbb{S}_A = \mathbb{S} + \mathbb{S}_h\,. \tag{3.7}$$

The compound system $\mathcal{A}$ is regarded as adiabatically insulated if $\lambda = \lambda(t)$ and as fully isolated if λ = const. These restrictions can only be violated in a transitory perturbed process due to additional disturbing influences acting independently of the assumed external conditions.

An admissible thermodynamic process $\mathcal{P}$ is subject to the restrictions which follow from the basic laws of thermodynamics. From the energy balance (the first law) for the body $\mathcal{B}$ at the absence of additional supply terms, in a process $\mathcal{P}$ we have

$$\dot{E}_A = \dot{W} + \dot{K} + \dot{\Omega} = 0 \quad \text{at } \lambda\text{=const}, \qquad \dot{W} = \dot{U} - \dot{Q} \tag{3.8}$$

where $\dot{W} = \int_G \mathbf{S}\cdot\dot{\mathbf{F}}\, d\xi$ is the rate of deformation work in $\mathcal{B}$ and $\dot{Q} = -\dot{E}_h = -\bar{T}\dot{S}_h$ is the rate of heat supply to $\mathcal{B}$ from the heat reservoir. From the assumed reversibility of the surroundings and from the second law we obtain

$$\dot{S}_A = \dot{S}^i \equiv \dot{S} - \dot{Q}/\bar{T} \geq 0 \quad \text{in } \mathcal{P} . \tag{3.9}$$

Under the assumptions introduced, the entropy production rate $\dot{S}^i$ in the body $\mathcal{B}$ consists of the plastic part $\dot{S}^p$ and the thermal part $\dot{S}^T$, being the volume integrals over G of the local *nonnegative* contributions σ^p and σ^T, respectively. In general, there may be more non-negative contributions to $\dot{S}^i$, e.g. due to viscosity, strong discontinuities in $\mathbf{F}$ or T, etc. In any case, we can *assume* that (cf. (2.9) and (2.10))

$$\dot{S}^i - \dot{S}^p \geq 0, \qquad \dot{S}^p \equiv \int_G \sigma^p \, d\xi \geq 0 . \tag{3.10}$$

The former inequality will play a fundamental role in establishing a general condition of thermodynamic stability of equilibrium of systems with plastic dissipation.

Let the prefix Δ denote the difference of corresponding quantities in states $\mathcal{S}$ and $\mathcal{S}^0$, e.g. $S_A(\mathcal{S}) - S_A(\mathcal{S}^0) \equiv \Delta S_A$, $E_A(\mathcal{S}) - E_A(\mathcal{S}^0) \equiv \Delta E_A$. Suppose first that $\tilde{\alpha}$ is frozen, so that $\dot{S}^p = 0$. The classical Gibbs condition of stability of equilibrium reads

$$\Delta S_A < 0 \quad \text{whenever} \quad \Delta E_A = 0, \quad \dot{S}^p \equiv 0 . \tag{3.11}$$

The equality constraint can be eliminated in the usual way by introducing a Lagrangian multiplier, which allows the (thermo-elastic) stability condition to be written in the equivalent[4] form

$$V^{L} \equiv \Delta V_{e} > 0 \quad \text{for every } \mathcal{S} \neq \mathcal{S}^{0}, \quad \dot{\mathbb{S}}^{p} \equiv 0 \,, \tag{3.12}$$

where

$$V_{e} \equiv \mathbb{E}_{A} - \bar{T}\, \mathbb{S}_{A} \,. \tag{3.13}$$

The value of V_e varies continuously in time since the energy and entropy do, and from (3.8) and (3.9) we obtain that

$$\dot{V}_{e} \equiv \dot{\mathbb{E}}_{A} - \bar{T}\, \dot{\mathbb{S}}^{i} \leq 0 \tag{3.15}$$

in any process $\mathcal{P}$ which follows an initial disturbance and is free of further (persistent) disturbances. From Corollary 1 we arrive at the conclusion that (3.12) ensures stability of $\mathcal{S}^0$ with respect to the Lyapunov metric generated by V_e.

The conclusion would be valid also for variable $\tilde{\alpha}$. However, as it will be shown later, the inequality in (3.12) or (3.11) is generally *not* satisfied for solids with intrinsic plastic dissipation. As a consequence, if $\dot{\mathbb{S}}^{p} \neq 0$ then the Gibbs condition ((3.11) or its equivalent) does not provide a satisfactory basis for stability investigations within the present thermodynamic formalism of plasticity[5]. The hypothesis that

---

[4] Implication (3.12) $\Longrightarrow$ (3.11) is obvious. The converse implication follows from the possibility of varying $\Delta\mathbb{E}_A$ at *constant* V_e by supplying heat from external sources to the heat reservoir.

[5] This conclusion is not surprising if the macroscopic plastic behaviour is thought of as being associated with a large number of non-equilibrium jumps at a microscopic level between (metastable) equilibrium states in an elastic body. In view of entropy production during any such jump, the relation between the occurrence of macroscopic plastic dissipation and the violation of (3.11) becomes comprehensible.

equilibrium is unstable unless all modes of departure from it contradict the second law must be abandoned. The attempts to derive a stability criterion in plasticity on the basis of that hypothesis [21] refer, in effect, to a hypothetical "equivalent elastic structure" rather than to the actual inelastic solid.

The main idea developed below is to extend Gibbs' stability condition to the systems which exhibit plastic dissipation, by taking as a basis the inequality $(3.10)_1$ rather than that in (3.9). To construct the corresponding Lyapunov functional for an isolated system $\mathcal{A}$, in place of (3.13) we define

$$V \equiv \mathbb{E}_A - \bar{T}\,(\mathbb{S}_A - \Delta\mathbb{S}^p) \tag{3.16}$$

where

$$\Delta\mathbb{S}^p \equiv \int_{t^0}^{t} \dot{\mathbb{S}}^p\, dt \tag{3.17}$$

is the *intrinsic* (plastic) dissipation in the body $\mathcal{B}$ in a process starting from the equilibrium state $\mathcal{S}^0 = \mathcal{S}(t^0)$ and leading to a current state $\mathcal{S} = \mathcal{S}(t)$.

By using (3.8), $(3.9)_1$ and $(3.10)_1$ we obtain that

$$\dot{V} = \dot{\mathbb{E}}_A - \bar{T}\,(\dot{\mathbb{S}}^i - \dot{\mathbb{S}}^p) \le 0 \tag{3.18}$$

in any process $\mathcal{P}$ free of persistent disturbances. A condition of stability of the equilibrium state $\mathcal{S}^0$ can thus be formulated as

$$\Delta V > 0 \qquad \text{for every } \mathcal{S} \neq \mathcal{S}^0 \tag{3.19}$$

and for every transitory process leading from $\mathcal{S}^0$ to $\mathcal{S}$. From Corollary 1 we arrive at the following conclusion.

Corollary 2. If the condition (3.19) is satisfied then the equilibrium state $\mathcal{S}^0$ is stable for initial disturbances with respect to the Lyapunov metric generated by the functional V.

Note that the pseudo-metric (3.6) depends now not only on $\mathcal{S}$ but also on the process leading to $\mathcal{S}$ from $\mathcal{S}^0$.

If $\tilde{\alpha}$ is fixed then V reduces to V_e, so that the conditions for thermoelastic stability do not change. If $\tilde{\alpha}$ is varying then $\Delta S^p > 0$ and $V > V_e$ so that the condition (3.19) is less restrictive than (3.12) and, as it will be shown below, is applicable to plastic stability problems. The integral in (3.17) is in general path-dependent, i.e. the actual value of V depends not only on the current state but also on the whole process from $\mathcal{S}^0$ to $\mathcal{S}$, no matter how small the final variations of the state fields are. In order not to complicate the notation, this dependence is not indicated in a formal way.

The condition (3.19) is in general not necessary for stability of equilibrium; note that in Theorem 1 only a proper *local* minimum of a Lyapunov functional with respect to a metric ρ is required and not the absolute minimum property (3.19) in the whole state space. In the absence of plastic dissipation, the condition (3.11) or (3.12) but weakened to define a *local* minimum can be regarded as a necessary condition for *asymptotic* stability. For, if V_e vanishes asymptotically and is decreasing in any real (dissipative) process then its initial value must always be positive. In plasticity, the property of asymptotic stability cannot in general be expected.

3.3. **Thermoelastic stability**

Before proceeding to our main topic - stability in plasticity - it is instructive to discuss first the *relatively* simpler case when the internal variables α are frozen. Substitution of (3.7) into (3.13) yields

$$\Delta V_e = \Delta(\mathbb{K} + \mathbb{U} - \bar{T}\, \mathbf{S} + \Omega) \tag{3.20}$$

where the energy and entropy of the heat reservoir have been eliminated.

To begin with, suppose that the body is artificially constrained to undergo no displacements ($\tilde{\mathbf{x}}$ fixed, so that $\tilde{\mathbf{v}} \equiv \tilde{0}$), and that the entropy field $\tilde{s}$ is treated as the independent variable. The condition (3.12) reduces to

$$\Delta V_e\big|_{\tilde{\mathbf{x}},\tilde{\alpha}} = \Delta V_e(\tilde{s}) = \int_G (\Delta u - \bar{T}\,\Delta s)\,d\xi > 0 \qquad \text{for } \tilde{s} \neq \tilde{s}^0. \tag{3.21}$$

Assume that $u = u(\mathbf{F}, s, \boldsymbol{\alpha})$ is continuously twice differentiable with respect to s. For (3.21) to hold, we must have the stationarity property

$$\delta V_e(\tilde{s}) = \int_G (u_{,s} - \bar{T})\,\delta s\,d\xi = 0 \qquad \text{in } \mathcal{S}^0 \tag{3.22}$$

which means that in the equilibrium state $\mathcal{S}^0$ we have

$$u_{,s} \equiv T = \bar{T} \qquad \text{in } G \tag{3.23}$$

as the expected condition of *thermal equilibrium*. Moreover, (3.21) implies non-negativeness of the second variation

$$\delta^2 V_e(\tilde{s}) = \int_G u_{,ss}\,(\delta s)^2\,d\xi \geq 0 \qquad \text{in } \mathcal{S}^0$$

equivalent to

$$u_{,ss} \equiv \bar{\rho}\,\bar{T}/c \geq 0 \qquad \text{in } G \text{ in } \mathcal{S}^0\,. \tag{3.24}$$

In turn, fulfilment of (3.21) in *any* equilibrium state is ensured if c, the specific heat at constant strain, is always positive

$$c > 0 \tag{3.25}$$

which we assume to be the case. For, the functional in (3.21) is then strictly convex with respect to $\tilde{s}$, so that its stationary point is automatically an absolute minimum point. The inequality (3.25) is the (classical) condition of *thermal stability* of all constrained equilibrium states (where displacements and internal variables are treated as fixed).

Suppose now that the fields $\tilde{\mathbf{x}}$, $\tilde{\mathbf{v}}$ and $\tilde{s}$ can undergo simultaneous perturbations. On substituting (3.20) the condition (3.12) becomes

$$\Delta V_e(\tilde{\mathbf{x}},\tilde{s},\tilde{\mathbf{v}}) = \Delta K + \int_G (\Delta u - \overline{T}\,\Delta s)\,d\xi + \Delta\Omega > 0 \qquad \text{for } \mathcal{S} \neq \mathcal{S}^0. \tag{3.26}$$

We make now use of the identity (with (3.23) accounted for)

$$u(\mathbf{F},s,\boldsymbol{\alpha})-u(\mathbf{F}^0,s^0,\boldsymbol{\alpha}) - \overline{T}(s-s^0) = \phi(\mathbf{F},\overline{T},\boldsymbol{\alpha}) - \phi(\mathbf{F}^0,\overline{T},\boldsymbol{\alpha}) + \mathcal{E}_u\big|_{\mathbf{F},\boldsymbol{\alpha}}, \tag{3.27}$$

where

$$\mathcal{E}_u = \mathcal{E}_u(s,\bar{s}) \equiv u(s)-u(\bar{s})-u_{,s}(\bar{s})(s-\bar{s}), \qquad \bar{s} = -\phi_{,T}(\overline{T}), \qquad \mathbf{F},\boldsymbol{\alpha} \text{ fixed} \tag{3.28}$$

is the Weierstrass function associated with $u(s)$ at fixed $\mathbf{F}$ and $\boldsymbol{\alpha}$, which represents the excess of $u(s)$ over its linear approximation at $\bar{s}$ (with the tangent $u_{,s}(\bar{s}) = \overline{T}$). Substitution of the identity into (3.26) yields

$$\Delta V_e(\tilde{\mathbf{x}},\tilde{s},\tilde{\mathbf{v}}) = \Delta K + \int_G \mathcal{E}_u\big|_{\mathbf{F},\boldsymbol{\alpha}}\,d\xi + \Delta\Phi\big|_{\overline{T}} + \Delta\Omega\,, \tag{3.29}$$

where in the term $\Delta\Phi|_{\overline{T}}$ the free energy density is evaluated at the environmental (equilibrium) temperature $\overline{T}$ and *not* at the current local temperature T. From the assumed thermal stability condition (3.25) we obtain that $u(s)$ is strictly convex so that $\mathcal{E}_u(s,\,\bar{s}) > 0$ if $s\neq\bar{s}$. The first two terms on the right-hand side of (3.29) are thus non-negative and vanish only if $\tilde{\mathbf{v}} = \tilde{\mathbf{0}}$ and $\tilde{T} = \overline{T}$, respectively. The thermodynamic stability condition (3.26) reduces, in effect, to the quasi-static isothermal condition [22]

$$\Delta V_e\big|_{\overline{T},\tilde{\mathbf{v}}=\tilde{\mathbf{0}}} = \Delta V_e(\tilde{\mathbf{x}}) = \Delta\Phi\big|_{\overline{T}} + \Delta\Omega > 0 \qquad \text{for } \tilde{\mathbf{x}} \neq \tilde{\mathbf{x}}^0\,. \tag{3.30}$$

The inequality (3.30) means that the potential energy of the *system* which consists of the elastic body and the loading device attains under isothermal conditions an absolute minimum in the equilibrium state. As already mentioned, this is not necessary for a physically meaningful stability of equilibrium. If (3.25) is granted then (3.30) *is sufficient for thermodynamic stability of the equilibrium state* $\mathcal{S}^0$ *with respect to the Lyapunov metric generated by* V_e (at frozen $\tilde{\alpha}$ and variable $\tilde{T}$). For further discussion and possible extensions, see [23].

Examine now prerequisites for (3.30). Assuming that there are no *unilateral* constraints, $V_e(\tilde{\mathbf{x}})$ must be stationary (in the sense of the usual *weak* variations, i.e. of the identically vanishing Gateaux differential) in the equilibrium state:

$$\delta(\Phi + \Omega) = 0 \qquad \text{in } \mathcal{S}^0 , \qquad T = \text{const} \tag{3.31}$$

which gives the conditions of *mechanical equilibrium.* For instance, suppose that the displacements vanish on a part $S_u \subset \partial G$ of the body surface, the *nominal* surface tractions $\mathbf{T}$ (per unit surface area in the *reference* configuration) are prescribed on the complementary part S_T of ∂G, and the *nominal* body forces $\mathbf{b}$ (per unit reference volume) are prescribed in G . Then

$$\Delta\Omega = - \int_G \mathbf{b}\cdot\mathbf{u}\, d\xi - \int_{S_T} \mathbf{T}\cdot\mathbf{u}\, dS \tag{3.32}$$

where $\mathbf{u} = \Delta\mathbf{x}$ are displacements from the equilibrium configuration $\tilde{\mathbf{x}}^0$. The condition (3.31) has the form of the virtual work principle written in the reference configuration :

$$\delta(\Phi+\Omega) = \int_G (\mathbf{S}\cdot\nabla(\delta\mathbf{x}) - \mathbf{b}\cdot\delta\mathbf{x})\, d\xi - \int_{S_T} \mathbf{T}\cdot\delta\mathbf{x}\, dS = 0 \quad \text{if } \delta\mathbf{x} = 0 \text{ on } S_u . \tag{3.33}$$

Under appropriate regularity conditions, this can be reduced in the standard manner by employing the Green theorem to the local conditions of mechanical equilibrium in the Lagrangian formulation:

$$\mathrm{Div}\,\mathbf{S} + \mathbf{b} = 0 \ \text{ in } G\backslash S_D, \qquad [\mathbf{S}]\cdot\mathbf{n} = 0 \ \text{ on } S_D\,, \qquad \mathbf{S}\cdot\mathbf{n} = \mathbf{T} \ \text{ on } S_T\,,$$

where S_D is a possible surface of the stress discontinuity $[\mathbf{S}]$ and $\mathbf{n}$ is the unit normal to a surface (all in the reference configuration).

As another implication of (3.30), we obtain

$$\delta^2(\Phi+\Omega) \geq 0 \qquad \text{in } \mathcal{S}^0\,, \quad T = \text{const}\,, \tag{3.34}$$

where, under the assumed linear dependence of Ω on the displacement field,

$$\delta^2(\Phi+\Omega) = \delta^2\Phi = \int_G \nabla(\delta\mathbf{x})\cdot\phi_{,FF}\cdot\nabla(\delta\mathbf{x})\, d\xi\,, \qquad \delta\mathbf{x} = \mathbf{0} \text{ on } S_u\,, \tag{3.35}$$

is the second weak variation of the isothermal Helmholtz free energy (whose density is assumed to be continuously twice differentiable with respect to its arguments). The condition (3.34) may be regarded as a *necessary* condition for *mechanical stability of equilibrium*, although more in a physical than mathematically rigorous sense. Evidently, (3.34) is necessary for *asymptotic* stability.

In general, positiveness of the second *weak* variation of the total potential energy $(\Phi+\Omega)$ (i.e. when only trivial equality in (3.34) is allowed) need not imply (3.30) and thus need not ensure elastic stability. This question will be discussed later when examining material stability aspects.

3.4. Stability of equilibrium of systems with plastic dissipation

The above constraint $\tilde{\alpha}$ = const is relaxed from now on. To justify the statement about inapplicability of the condition (3.11) or (3.12) to systems with plastic dissipation, consider the first weak variation of V_e in an equilibrium state. On account of (3.22) and (3.31), it reads

$$\delta V_e = \delta V_e\big|_{\tilde{\mathbf{x}},\bar{T}} = \delta\Phi(\tilde{\alpha}) = -\int_G \mathbf{A}\cdot\delta\alpha \; d\xi \; . \tag{3.36}$$

Since the thermodynamic forces need not vanish in an equilibrium state of the elastic-plastic solid as discussed in Chapter 2, this expression is generally non-zero, in contrast to the stationarity properties (3.22) and (3.31). If the actual stresses satisfy the yield condition ($g_K=0$) within a finite part of the body $\mathcal{B}$ then any disturbance $\delta\alpha_K$ compatible with the actual thermodynamic forces makes the expression (3.36) negative. This means that the condition (3.12) or (3.11) can never be satisfied if $\tilde{\alpha}$ is variable, with the exception of the particular equilibrium states where within the body either the current stress point lies everywhere inside the elastic domain (so that $\delta\alpha_K\equiv 0$) or all thermodynamic forces $\mathbf{A}_K$ happen to vanish.

As stated in Section 3.4, the condition (3.19) and not (3.12) is taken as a basis for the stability investigations in solids which exhibit plastic dissipation. Substitution of (3.7) into (3.16) yields

$$\Delta V = \Delta\mathbb{K} + \Delta U - \bar{T}\,(\Delta S - \Delta S^{P}) + \Delta\Omega \; . \tag{3.37}$$

In the following we shall assume (except in Section 5.2) that the disturbing agency does not penetrate "inside" a material element so that (cf. (2.11))

$$\mathcal{D}(\mathfrak{s},\dot{\alpha}) = \mathbf{A}(\mathfrak{s})\cdot\dot{\alpha} \tag{3.38}$$

also in a transitory perturbed process. From (3.36) and (3.38) it follows that the first one-sided *weak* variation of V vanishes in any equilibrium state $\mathcal{S}^0$ provided the variation of $\tilde{\alpha}$ is compatible, in the sense of (3.38), with the actual thermodynamic forces in state $\mathcal{S}^0$.

Besides Corollary 2, other interpretations of (3.19) as a sufficient condition for stability can be obtained with the help of Theorem 1. For instance, let the class of admissible disturbances be restricted such that during a transitory process in $[t^0,\tau)$ there is no entropy supply (of neither sign) to the compound system $\mathcal{A}$ from its environment (e.g.

transitory processes induced by mechanical excitation with $\mathcal{A}$ thermally insulated), which is denoted by $\Delta^{ext}S_A=0$. Then $\Delta S_A = \Delta S^i$ also for $t<\tau$ so that an increment of the term in parentheses in (3.16) is non-negative by $(3.10)_1$. Suppose that (3.19) holds, and define the metric ρ^0 for *initial* disturbances at τ as the *maximal* amount of energy supplied to $\mathcal{A}$ by a disturbing agency within the time interval $[t^0,\tau]$ in order to reach the state $\mathcal{S}(\tau)$, viz.

$$\rho^0_E = \sup_{t^0\le t\le\tau} \Delta E_A(t), \qquad \Delta^{ext}S_A \equiv 0 . \tag{3.39}$$

Then

$$\rho^0_E \ge \Delta E_A(\tau) \ge \Delta V(\tau) > 0 \quad \text{in } \mathcal{S}_\tau \ne \mathcal{S}^0 , \tag{3.40}$$

so that V and ρ^0_E satisfy the condition (d) in Theorem 1. Hence, (3.19) is sufficient for stability of equilibrium with respect to ρ^0_E and the metric ρ^L generated by the functional (3.16). If the restriction $\Delta^{ext}S_A \equiv 0$ is relaxed then ΔE_A in the above formulae should be replaced by $(\Delta E_A - \bar{T}\Delta^{ext}S_A)$; this quantity is not smaller that ΔV, on account of $(3.10)_1$. As discussed below, the problem of determining a physically acceptable metric ρ for a continuum, which could satisfy the condition (b) in Theorem 1, is more complicated.

In analogy to (3.29), we can write

$$\Delta V = \Delta K + \int_G \mathcal{E}_u|_{F,\alpha}\, d\xi + \Delta\Phi|_{\bar{T}} + \Delta\Omega + \bar{T}\,\Delta S^p . \tag{3.41}$$

Since the first two terms on the right-hand side in (3.41) are non-negative, the remaining quantity

$$\Delta E \equiv \Delta\Phi|_{\bar{T}} + \Delta\Omega + \bar{T}\,\Delta S^p \tag{3.42}$$

is not greater than ΔV. On account of $(3.10)_1$, ΔE represents also a lower bound to the amount of energy which has to be supplied to $\mathcal{A}$ by a disturbing agency such that $\Delta^{ext}\underset{\wedge}{S}=0$ in order to carry the system from the equilibrium state $\mathcal{S}^0$ to another (quasi-equilibrium) state $\mathcal{S}^{qe}$ of *uniform* temperature $\bar{T}$ and *zero* velocities. It follows that (3.19) is ensured if

$$\Delta\mathbb{E} > 0 \qquad \text{in every } \mathcal{S}^{qe} \neq \mathcal{S}^0 \tag{3.43}$$

and for every process leading from $\mathcal{S}^0$ to $\mathcal{S}^{qe}$. The first two terms in (3.42) are defined as functions of $(\tilde{\mathbf{x}}, \tilde{\boldsymbol{\alpha}})$ under isothermal conditions. However, contrary to the case of thermoelastic stability, we cannot conclude that the isothermal stability implies stability also in a non-isothermal case since the last term in (3.41), previously absent, depends in general on the temperature history in the process from $\mathcal{S}^0$ to $\mathcal{S}^{qe}$.

Isothermal case

From now on we restrict ourselves to examining isothermal processes only[6]. Then $\sigma^T \equiv 0$ (cf. (2.9)) and we have $\dot{\mathbb{S}}^i = \dot{\mathbb{S}}^p$ since other sources of dissipation have been disregarded[7]. Thus (cf. (3.8))

$$\dot{V} = \dot{\mathbb{K}} + \dot{\mathbb{W}} + \dot{\Omega} = 0 \qquad \text{at } T = \text{const}, \ \lambda = \text{const}, \tag{3.44}$$

in a process $\mathcal{P}$ free of persisting disturbances; it is recalled that $\dot{\mathbb{W}} = \int_G \mathbf{S}\cdot\dot{\mathbf{F}}\, d\xi$. The stability condition (3.43) reduces to

$$\Delta\mathbb{E} = \Delta V - \Delta\mathbb{K} = \Delta\mathbb{W} + \Delta\Omega > 0 \qquad \text{in every } \mathcal{S}^{qe} \neq \mathcal{S}^0 \tag{3.45}$$

---

[6] An analogous discussion is possible for *adiabatic* processes but it is omitted here.

[7] If moving strong discontinuities of $\mathbf{F}$ were allowed then we could have $\dot{\mathbb{S}}^i > \dot{\mathbb{S}}^p$ and $\dot{V} < 0$ even in an isothermal process (cf. the discussion in Section 5.4).

and for every isothermal process from $\mathcal{S}^0$ to $\mathcal{S}^{qe}$. The functional $\mathbb{E}$ can be identified with the energy functional introduced in [24] where an increment of $\mathbb{E}$ in an isothermal quasi-static deformation process was interpreted as the amount of energy supplied from external sources to the *mechanical* system consisting of the deformed body and the loading device. Here, for isothermal processes such that $\Delta^{ext}S_{A}=0$, $\mathbb{E}$ can be identified with the internal energy of the compound thermodynamic system $\mathcal{A}$ which includes additionally the heat reservoir.

The discussion from now on will be restricted to rate-independent plasticity. Consider a kinematically admissible process of departure from the equilibrium state $\mathcal{S}^0$ with a *continuous* velocity field $\tilde{\mathbf{v}}$ such that $\mathbf{v}=\mathbf{v}^0$ on S_u and $\|\nabla\mathbf{v}\| < \infty$, where $\|\cdot\| \equiv (\int_G|\cdot|^2 d\xi)^{1/2}$. As discussed above, the first time derivative $\dot{\mathbb{E}}$ of $\mathbb{E}$ (representing the first one-sided variation for a *given* direction of departure from $\mathcal{S}^0$) vanishes identically in $\mathcal{S}^0$. For (3.45) to hold it is necessary that the following *second-order energy condition* is satisfied:

$$\ddot{\mathbb{E}} \geq 0 \quad \text{in } \mathcal{S}^0 \tag{3.46}$$

for every kinematically admissible mode of departure from $\mathcal{S}^0$, with the restriction on $\tilde{\mathbf{v}}$ mentioned above. If $\tilde{\alpha}$ is frozen then (3.46) reduces to (3.34).

For dead loading associated with (3.32), the condition (3.46) takes the form

$$\int_G \dot{\mathbf{S}}(\alpha^0,\dot{\mathbf{F}})\cdot\dot{\mathbf{F}}\, d\xi \geq 0 \qquad \text{for every } \tilde{\mathbf{v}},\ \mathbf{v}=\mathbf{0} \text{ on } S_u\,. \tag{3.47}$$

This inequality (but strict for $\tilde{\mathbf{v}}\neq\tilde{\mathbf{0}}$) was proposed in [25] as a condition sufficient for stability of equilibrium, however, under certain additional assumptions; cf. the discussion below. Strict inequality in (3.47) for $\tilde{\mathbf{v}}\neq\tilde{\mathbf{0}}$ excludes departure from equilibrium on a *straight* deformation path $\Delta\mathbf{x}(\xi,t)=(t-t^0)\mathbf{v}(\xi)$ unless the disturbing influences are not arbitrary small. If this is so for any kinematically admissible direction $\tilde{\mathbf{v}}$ then (3.47) is sufficient for a kind of stability which may be called *directional stability of equilibrium*.

More can be said about stability *in the first approximation*, where an auxiliary (*partially* linearized) problem is considered in place of the "true" fully nonlinear problem. That approach, to be used below, can be regarded as an extension of the so-called linear theory of elastic stability to incrementally nonlinear, rate-independent plasticity.

Stability in the first approximation

The attention is restricted to isothermal processes and rate-independent plasticity. Let a bar over a symbol denote either a quantity in the state $\mathcal{S}^0$ or a function obtained by approximation at $\mathcal{S}^0$. Consider a small neighborhood of $\mathcal{S}^0 \equiv (\tilde{\bar{\Delta}})$, $\bar{\Delta}(\boldsymbol{\xi}) \equiv (\bar{\mathbf{x}}, \bar{T}, \bar{\boldsymbol{\alpha}})(\boldsymbol{\xi})$, defined by *pointwise* inequalities $|\alpha_K - \bar{\alpha}_K| < R$ and $|\mathbf{F}-\bar{\mathbf{F}}| < R$ in G, where R is a small positive number. In that neighborhood, we wish to consider the *second-order* approximation of the left-hand expression in the stability condition (3.45), with path-dependence of the dissipation and deformation work taken to that order into account. For this purpose, we shall consider an auxiliary problem where the (isothermal) Helmholtz free energy density $\bar{\phi}$ is defined by

$$\begin{aligned}\bar{\phi}(\mathbf{F},\boldsymbol{\alpha}) \equiv\; & \phi(\bar{\Delta}) + \phi_{,\mathbf{F}}(\bar{\Delta})\cdot(\mathbf{F}-\bar{\mathbf{F}}) + \phi_{,\boldsymbol{\alpha}}(\bar{\Delta})\cdot(\boldsymbol{\alpha}-\bar{\boldsymbol{\alpha}}) \\ & + \frac{1}{2}(\mathbf{F}-\bar{\mathbf{F}})\cdot\phi_{,\mathbf{FF}}(\bar{\Delta})\cdot(\mathbf{F}-\bar{\mathbf{F}}) + (\mathbf{F}-\bar{\mathbf{F}})\cdot\phi_{,\mathbf{F}\boldsymbol{\alpha}}(\bar{\Delta})\cdot(\boldsymbol{\alpha}-\bar{\boldsymbol{\alpha}}) \\ & + \frac{1}{2}(\boldsymbol{\alpha}-\bar{\boldsymbol{\alpha}})\cdot\phi_{,\boldsymbol{\alpha\alpha}}(\bar{\Delta})\cdot(\boldsymbol{\alpha}-\bar{\boldsymbol{\alpha}}). \end{aligned} \tag{3.50}$$

The two Examples of the constitutive framework from Chapter 2 will be discussed in more detail. In Example (1) the dissipation function $\mathcal{D}$ is now linearized but only *with respect to* $\boldsymbol{\alpha}$, viz.

$$\bar{\mathcal{D}}(\boldsymbol{\alpha},\dot{\boldsymbol{\alpha}}) \equiv \mathcal{D}(\bar{\boldsymbol{\alpha}},\bar{T},\dot{\boldsymbol{\alpha}}) + (\boldsymbol{\alpha}-\bar{\boldsymbol{\alpha}})\cdot\mathcal{D}_{,\boldsymbol{\alpha}}(\bar{\boldsymbol{\alpha}},\bar{T},\dot{\boldsymbol{\alpha}}) \geq -\bar{\phi}_{,\boldsymbol{\alpha}}(\mathbf{F},\boldsymbol{\alpha})\cdot\dot{\boldsymbol{\alpha}}\;; \tag{3.51}$$

nonlinearity of $\mathcal{D}$ with respect to $\dot{\boldsymbol{\alpha}}$ has been retained. In Example (2), as an analog to (3.51) we assume (cf. (2.35) and (2.48)):

$$\bar{g}_K(\mathrm{F},\gamma_J) \equiv g_K(\bar{\Delta}) + \bar{\Lambda}_K\cdot(\mathrm{F}-\bar{\mathrm{F}}) - \sum_J g_{KJ}(\bar{\Delta})\ (\gamma_J-\bar{\gamma}_J) \le 0\ ,$$

$$\alpha_K = \bar{\alpha}_K + (\gamma_K-\bar{\gamma}_K)\bar{\mathrm{P}}_K,\quad \bar{\Lambda}_K \equiv (\partial g_K/\partial \mathrm{F})(\bar{\Delta}),\quad \bar{\mathrm{P}}_K \equiv (\partial f_K/\partial \mathrm{A}_K)(\bar{\Delta})\ . \tag{3.52}$$

The internal mechanisms of plastic deformation which are not potentially active in $\mathcal{S}^0$ are regarded as frozen in the neighborhood of $\mathcal{S}^0$, that is

$$g_K(\bar{\Delta})\ \dot{\alpha}_K(t) = 0 \tag{3.53}$$

at any t in every perturbed process under consideration. This implies

$$\bar{\mathcal{D}}(\bar{\alpha},\dot{\alpha}(t)) = \mathrm{A}(\bar{\Delta})\cdot\dot{\alpha}(t) \tag{3.54}$$

$$g_K(\bar{\Delta})\dot{\gamma}_K(t) = 0 \tag{3.55}$$

in Example (1) and (2), respectively,

For simplicity, the configuration-insensitive loading corresponding to the expression (3.32) for $\Delta\Omega$ is assumed.

Under the assumptions introduced above, $\Delta\mathbb{E}$ is represented by a quantity $\Delta\bar{\mathbb{E}}$ which is of second-order with respect to $\Delta\mathbf{F}$ and $\Delta\alpha$. For, the assumption (3.53) eliminates the first-order contribution to $\Delta\bar{\mathbb{E}}$, and the adopted linearizations ensure that a local contribution to $\Delta\bar{\mathbb{E}}$ varies proportionally to the square of path length in $(\mathbf{F},\alpha)$-space when such a path is proportionally scaled down so that its length varies while its complexity is preserved (cf. the discussion of path-dependent second-order work in [26]).

Since various internal mechanisms of plastic deformation may subsequently be unloaded and activated in a perturbed process, the dissipation $\bar{T}\Delta S^p$ and the deformation work ΔW in (3.45) are in general dependent on the transitory process leading to some given state $\mathcal{S}^{qe}$ and not only on $\mathcal{S}^{qe}$ itself, also if second-order expressions are only considered. This difficulty can be overcome if a *lower bound* to ΔW or ΔS^p is determined for given $\mathcal{S}^{qe}$. That approach has been proposed in [25], and used in [27] for particular classes of solids with $\mathcal{D}$ independent of α. We present below a generalization of previous results for the two Examples of constitutive framework from Chapter 2.

Example (1)

Suppose that the symmetry property (2.42) holds. Then, by appealing to the assumed linearity of $\bar{\mathcal{D}}$ with respect to α and convexity with respect to $\dot{\alpha}$ it can be shown [28] that among all paths in α-space satisfying (3.54) at each instant, the dissipation integral is minimized on the straight path, viz.

$$\int_{t^0}^{\tau} \bar{\mathcal{D}}(\alpha(t),\ \dot{\alpha}(t))\ dt \geq \bar{\mathcal{D}}(\bar{\alpha},\ \bar{\bar{\alpha}}-\bar{\alpha}) + \frac{1}{2}\ (\bar{\bar{\alpha}}-\bar{\alpha})\cdot\bar{\mathcal{D}}_{,\alpha}(\bar{\alpha},\ \bar{\bar{\alpha}}-\bar{\alpha})\ , \qquad \alpha(t^0) = \bar{\alpha},\quad \alpha(\tau) = \bar{\bar{\alpha}}\ . \tag{3.56}$$

This minimum property and the resulting stability condition given below provide an extension of Nguyen's [16] results.

By using (3.56) and (3.32), we obtain the (second-order) inequality

$$\Delta\bar{V} \geq \Delta\bar{\mathbb{E}} \geq \Delta^2\Phi + \bar{T}\ \Delta^2 S^P \equiv \rho_E \tag{3.57}$$

where

$$\Delta^2\Phi = \int_G \left(\frac{1}{2}\ \Delta F\cdot\phi_{,FF}(\bar{\Delta})\cdot\Delta F + \Delta F\cdot\phi_{,F\alpha}(\bar{\Delta})\cdot\Delta\alpha + \frac{1}{2}\Delta\alpha\cdot\phi_{,\alpha\alpha}(\bar{\Delta})\cdot\Delta\alpha \right) d\xi\ ,$$
$$\bar{T}\ \Delta^2 S^P = \int_G \frac{1}{2}\ \Delta\alpha\cdot\bar{\mathcal{D}}_{,\alpha}(\bar{\alpha},\ \Delta\alpha)\ d\xi\ . \tag{3.58}$$

From (3.39), (3.57) and Theorem 1 we arrive at the conclusion that if (2.42) holds then

$$\Delta^2\Phi + \bar{T}\ \Delta^2 S^P > 0 \quad \text{for every } \Delta\tilde{x},\ \Delta\tilde{\alpha} \neq \tilde{0},\quad \Delta x = 0 \ \text{ on } S_u \tag{3.59}$$

is sufficient for stability of equilibrium *in the first approximation* with respect to metrics ρ^0_E and ρ_E.

Example (2)

Suppose that the conditions of uniqueness (2.50) and of symmetry (2.51) are satisfied. On using this and (3.52) it can be proved [26] that among all deformation paths satisfying (3.55) at each instant and leading to a given small *deformation* increment, the work integral is minimized, to second order, on the straight (direct) path, viz.

$$\Delta\bar{W} \equiv \int_{t^0}^{\tau} \bar{\phi}_{,F}(t)\cdot\dot{F}(t)\, dt \geq \bar{\phi}_{,F}(\bar{\Delta})\cdot\Delta F + \frac{1}{2}\Delta F\cdot\bar{\phi}_{,FF}\cdot\Delta F - \frac{1}{2}\sum_{K,J} g_{KJ}(\bar{\Delta})\,\Delta^d\gamma_K\,\Delta^d\gamma_J$$

$$= S(\bar{\Delta})\cdot\Delta F + \frac{1}{2}\Delta^d S\cdot\Delta F; \qquad \Delta() = ()\big|_{t^0}^{\tau}, \quad \Delta(t^0)=\bar{\Delta}\,. \tag{3.60}$$

The superscript "d" denotes the increments obtained on a straight (direct) deformation path from $\bar{F}=F(t^0)$ to $\bar{F}+\Delta F$.

From (3.60), (3.32) and (2.54) we obtain the inequality

$$\Delta\bar{V} \geq \Delta\bar{E} \geq \Delta^2 W \tag{3.61}$$

where

$$\Delta^2 W = \int_G \frac{1}{2}\Delta S^d\cdot\Delta F\, d\xi = \int_G U(\bar{\Delta},\ \Delta F)\, d\xi \tag{3.62}$$

is the global expression for the second-order work on straight (direct) deformation paths. Note that if (3.32) holds and $\tilde{v}$ is continuous then

$$\Delta^2 W(\Delta\tilde{x}) = \frac{1}{2}\ddot{E}(\tilde{v})\,(\Delta t)^2, \qquad \text{for } \Delta x = v\,\Delta t, \quad v = 0 \text{ on } S_u\,. \tag{3.63}$$

Under the assumption that a finite part S_u of the body surface is rigidly constrained, define a distance from $\mathcal{S}^0$ by the metric

$$\rho_{\mathrm{F}} = \|\Delta\mathbf{F}\| \equiv \left(\int_G |\Delta\mathbf{F}|^2 d\xi \right)^{1/2} . \tag{3.64}$$

From Theorem 1, the definition (3.39) and the above relationships we arrive at the conclusion that

$$\ddot{\mathbb{E}}(\tilde{\mathbf{v}}) \geq k\, \|\dot{\mathbf{F}}\|^2 \qquad \text{for every } \tilde{\mathbf{v}},\ \mathbf{v}=0 \text{ on } S_u, \tag{3.65}$$

is sufficient for stability, *in the first approximation*, equilibrium state $\mathcal{S}^0$ with respect to metrics ρ_{E}^0, ρ_{F}.

It must be noted that fulfilment of the pointwise inequality $|\Delta\mathbf{F}|<R$ is not automatically implied by ρ_{F} or ρ_{E} tending to zero. In turn, if that inequality is violated then validity of the second-order approximation can be questioned. We are faced with a serious difficulty which is met when a rigorous justification for the second-order sufficiency condition for stability of a *continuum* is searched for. The difficulty persists also in the case of purely elastic continua (cf. [29,30]).

Status of (3.46) as a *necessary* condition of stability of equilibrium of a continuous body also is not mathematically clear in general. The fact that (3.46) is necessary for *asymptotic* stability of equilibrium is not of particular importance here since such a property cannot be required in plasticity. Nevertheless, for a *discretized* system and in the first approximation, (3.46) is necessary for isothermal stability in a dynamic sense provided the incremental constitutive relationship admits a velocity-gradient potential (2.54) [31] ; to prove this, none of the sets of constitutive assumptions from Examples (1) and (2) is needed. Extension of this result to more general circumstances requires further investigations.

4. STABILITY OF A QUASI-STATIC PROCESS

4.1. The concept of stability of a process

In the preceding Section, the loading (i.e. body forces **b** in G, surface tractions **T** on S_T, and displacements **u** on S_u) did not depend *explicitly* on time t, although in a general case **b** and **T** could be configuration-dependent and vary thus in a process induced by disturbances. Now we will consider processes which take place when the loading variables **b**, **T** and **u** depend explicitly on a loading parameter $\lambda = \lambda(t)$. We shall assume that the variation of λ in a real time is vanishingly slow so that there exists an isothermal and quasi-static response of the system, called *the fundamental process* $\mathcal{P}^0: \lambda \mapsto \mathcal{S}^0(\lambda)$, such that $\mathcal{S}^0(\lambda)$ at each λ is in equilibrium.[8] In a (sufficiently short) period of $\mathcal{P}^0$, we chose λ to be strictly increasing and adopt λ as a time-like parameter playing throughout this section the role of time. Accordingly, a dot over a symbol denotes now the right-hand material time derivative (forward rate) with respect to λ. The "quasi-static" velocities $\mathbf{v} = \dot{\mathbf{x}} = \partial\mathbf{x}(\xi,\lambda)/\partial\lambda$ have now conventional meaning and are no longer regarded as non-equilibrium state variables.

The question now arises: does stability of each equilibrium state $\mathcal{S}^0(\lambda)$ at *fixed* λ, as examined in the preceding section, guarantee stability of the *process* $\mathcal{P}^0$ when λ is varying? The answer is negative: even if the distance between $\mathcal{S}^0$ and a state $\mathcal{S}$ reached in a perturbed process is arbitrarily small if λ=const, the possibility that the distance grows with a *finite* rate *with respect to* λ has not been excluded. Hence, it can happen that after a small increment of λ the distance between $\mathcal{S}^0(\lambda)$ and $\mathcal{S}(\lambda)$ exceeds some given positive value (ε, say) no matter how small the initial disturbance is, irrespectively of the preserved stability of equilibrium. This possibility is a peculiar

---

[8] This is obviously an approximation, as discussed in Chapter 2 in the context of a material element and extended now to a system.

property of rate-independent plastic deformations, closely related to the possibility of *bifurcation*, i.e. of a non-unique incremental response to the loading program in the absence of any disturbances.

The next question is whether an approach similar to that used to examine stability of equilibrium, i.e. via an appropriate Lyapunov functional based on thermodynamics, is possible also when λ is not kept fixed. A general answer is lacking at present; we will restrict ourselves to examining this question under certain additional assumptions. In particular, rate-independent plasticity and isothermal perturbed processes will only be considered.

Moreover, the attention will be confined to *stability in the first approximation*, in the sense analogous to that adopted for stability of equilibrium in Section 3.4. Take $\mathcal{S}^0(\bar{\lambda}) = (\tilde{\bar{\mathfrak{s}}})$, $\bar{\mathfrak{s}} = (\bar{\mathbf{F}}, \bar{T}, \bar{\boldsymbol{\alpha}}) \equiv (\mathbf{F}^0, T, \boldsymbol{\alpha}^0)|_{\bar{\lambda}}$, at some fixed $\bar{\lambda}$ as the initial state of all processes under consideration which take place at non-decreasing[9] λ in a small neighborhood of $\mathcal{S}^0(\bar{\lambda})$, defined by the pointwise inequalities $|\alpha_K - \bar{\alpha}_K| < R$, $|\mathbf{F}-\bar{\mathbf{F}}| < R$ in G as well as $\lambda - \bar{\lambda} < R$. A bar over a symbol denotes now either a quantity in the state $\mathcal{S}^0(\bar{\lambda})$ or a function obtained by approximation at $\mathcal{S}^0(\bar{\lambda})$. Consider an auxiliary problem where the Helmholtz free energy density $\bar{\phi}$ is defined by the quadratic approximation of ϕ at $\bar{\mathfrak{s}}$, with the related constitutive approximations either (3.51) or (3.52) in Example (1) or (2), respectively. We also assume that disturbances are sufficiently small in order not to violate (3.53) so that either (3.54) or (3.55) holds in every perturbed process under consideration. In the present auxiliary problem, the assumed configuration-independent loading functions are additionally linearized with respect to λ at $\bar{\lambda}$, viz.

$$\begin{aligned}
\bar{\mathbf{T}}(\boldsymbol{\xi},\lambda) &= \mathbf{T}(\boldsymbol{\xi},\bar{\lambda}) + (\lambda-\bar{\lambda})\,\dot{\mathbf{T}}(\boldsymbol{\xi},\bar{\lambda}) && \text{on } S_T\,, \\
\bar{\mathbf{b}}(\boldsymbol{\xi},\lambda) &= \mathbf{b}(\boldsymbol{\xi},\bar{\lambda}) + (\lambda-\bar{\lambda})\,\dot{\mathbf{b}}(\boldsymbol{\xi},\bar{\lambda}) && \text{in } G\,, \\
\bar{\mathbf{x}}(\boldsymbol{\xi},\lambda) &= \mathbf{x}(\boldsymbol{\xi},\bar{\lambda}) + (\lambda-\bar{\lambda})\,\mathbf{v}(\boldsymbol{\xi},\bar{\lambda}) && \text{on } S_u\,.
\end{aligned} \tag{4.1}$$

---

9) *The possibility* of a dynamic transitory process at fixed λ is not excluded.

Evidently, the fundamental solution $\overline{\mathcal{P}}^0:\lambda\longrightarrow \overline{\mathcal{S}}^0(\lambda)$ to the auxiliary problem is

$$\mathbf{F}^0(\xi,\lambda) = (\lambda-\bar{\lambda})\,\dot{\mathbf{F}}^0(\xi,\bar{\lambda}), \quad \alpha^0(\xi,\lambda) = (\lambda-\bar{\lambda})\,\dot{\alpha}^0(\xi,\bar{\lambda}) \quad \text{for } \lambda>\bar{\lambda}\,, \tag{4.2}$$

where the rates at $\bar{\lambda}$ are taken from the "true" fundamental process $\mathcal{P}^0$. $\overline{\mathcal{P}}^0$ may be regarded as a linearization (with respect to λ) of $\mathcal{P}^0$ in a small interval $[\bar{\lambda},\ \bar{\lambda}+\delta\lambda)$. Stability of the solution (4.2) is now examined as stability in the first approximation of the "true" fundamental process $\mathcal{P}^0$.

The mathematical concept of stability of equilibrium with respect to two metrics, discussed in Section 3.1, can be extended to processes [20]. For our purposes, the following specifications are made. The object of stability analysis is the isothermal process $\overline{\mathcal{P}}^0$ at varying λ in the thermally insulated, compound system $\mathcal{A}$ defined in Section 3.2. Small changes of the parameters of an *equilibrium* state $\mathcal{S}(\bar{\bar{\lambda}})$ at some $\bar{\bar{\lambda}}\geq\bar{\lambda}$ with respect to those of $\mathcal{S}^0(\bar{\lambda})$ are taken as *initial* disturbances for the subsequent *quasi-static* process $\mathcal{P}$: $\lambda\longrightarrow \mathcal{S}(\lambda)$, $\lambda\geq\bar{\bar{\lambda}}$, assumed to be free of any further disturbing influences. The restriction is introduced that $\mathcal{S}(\bar{\bar{\lambda}})$ is attainable from $\mathcal{S}^0(\bar{\lambda})$ in some transitory process corresponding to some time-dependent disturbances which are small enough to preserve (3.53), in which the continuity assumptions from Section 3.2 and the constitutive relations are valid. The metric ρ^0 for the initial disturbances satisfies (3.1), where $\mathcal{S}$ and $\mathcal{S}^0$ should be understood as $\mathcal{S}(\bar{\bar{\lambda}})$ and $\overline{\mathcal{S}}^0(\bar{\bar{\lambda}})$, respectively. The metric ρ in (3.2) refers to the distance between $\mathcal{S}(\lambda)$ and $\overline{\mathcal{S}}^0(\lambda)$ at some $\lambda\geq\bar{\bar{\lambda}}$, such that $\rho(\lambda)\equiv\rho(\mathcal{S}(\lambda),\overline{\mathcal{S}}^0(\lambda))$ is continuous with respect to λ (cf. (3.3)). In analogy to (3.4), it is assumed that $\rho(\bar{\bar{\lambda}})$ is continuous at $(\overline{\mathcal{S}}^0(\bar{\bar{\lambda}}),\overline{\mathcal{S}}^0(\bar{\bar{\lambda}}))$ with respect to ρ^0.

Definition 2. The process $\overline{\mathcal{P}}^0$ (or $\mathcal{P}^0$) is called stable (or stable in the first approximation at $\bar{\lambda}$) for initial disturbances with respect to metrics ρ^0, ρ if and only if for every $\varepsilon > 0$ there is $\delta(\varepsilon) > 0$ such that $\rho^0(\bar{\bar{\lambda}}) < \delta(\varepsilon)$ implies $\rho(\lambda) < \varepsilon$ at every $\lambda > \bar{\bar{\lambda}}$.

A theorem analogous to Theorem 1 can be proved [20] which motivates the approach below.

Following the concept of stability in the energy sense proposed in [24], consider the possibility of taking the *difference* of the energy functional in the processes $\mathcal{P}$ and $\overline{\mathcal{P}}^0$, viz.

$$\Delta\mathbb{E}|_\lambda = (\Delta\mathbb{W} + \Delta\Omega)|_\lambda, \qquad \lambda \geq \bar{\lambda} \tag{4.3}$$

as the Lyapunov functional for *quasi-static* processes at varying λ.[10)] Here and in the following the prefix Δ denotes the difference of any corresponding quantities *at the same* λ in the perturbed and (linearized) fundamental processes, $\mathcal{P}$ and $\overline{\mathcal{P}}^0$; the quantities in $\overline{\mathcal{P}}^0$ are distinguished by the superscript "0". To simplify the notation, bars over quantities dependent on λ will be omitted if this causes no confusion. Of course, the actual value of $\Delta\mathbb{E}$ at λ depends on how the perturbed state $\mathcal{S}(\lambda)$ has been reached from $\mathcal{S}^0(\bar{\lambda})$.

From the assumption of configuration-independent loading, the formula (3.32) retains its validity with the changed meaning of Δ. By using $(4.1)_3$ and the conditions of mechanical equilibrium satisfied in $\overline{\mathcal{P}}^0$ (cf. (3.33)), after straightforward transformations we obtain

$$\Delta\dot{\Omega} = -\int_G (\mathbf{b}\cdot\Delta\mathbf{x})^{\boldsymbol{\cdot}}\, d\xi - \int_{S_T} (\mathbf{T}\cdot\Delta\mathbf{x})^{\boldsymbol{\cdot}}\, dS = -\int_G \dot{\mathbf{S}}^0\cdot\Delta\mathbf{F}\, d\xi - \int_G \mathbf{S}^0\cdot\Delta\dot{\mathbf{F}}\, d\xi \,.$$

Now, one easily shows that

$$\Delta\dot{\mathbb{E}} = \int_G (\Delta\mathbf{S}\cdot\dot{\mathbf{F}}^0 - \dot{\mathbf{S}}^0\cdot\Delta\mathbf{F})\, d\xi + \int_G \Delta\mathbf{S}\cdot\Delta\dot{\mathbf{F}}\, d\xi \,. \tag{4.4}$$

---

10) One can admit also dynamic perturbed processes, and then $\mathbb{E}$ has to be replaced by $\mathbb{E}+\mathbb{K}$.

Below we will examine the sign of the former integrand; the second integral vanishes for a quasi-static process $\mathcal{P}$ free of persistent disturbances.

4.2. A constitutive inequality

Suppose that no abrupt unloading takes place in the fundamental process $\mathcal{P}^0$ at $\lambda=\bar{\lambda}+0$, which is equivalent to

$$g_K^0 \equiv g_K(\Delta^0(\lambda)) = 0 \quad \text{in } \bar{\mathcal{P}}^0 \qquad \text{if } g_K(\bar{\Delta}) = 0. \tag{4.5}$$

In combination with the assumption (3.53), we obtain that

$$\bar{\mathcal{D}}(\alpha^0(\lambda), \dot{\alpha}(\lambda)) = \mathbf{A}(\Delta^0(\lambda))\cdot\dot{\alpha}(\lambda) \tag{4.6}$$

$$g_K^0\, \dot{\gamma}_K(\lambda) = 0 \tag{4.7}$$

at any λ and in any process $\mathcal{P}$, in Example (1) or (2), respectively. This implies

$$\bar{\mathcal{D}}_{,\alpha}(\bar{\alpha}, \dot{\alpha}(\lambda))\cdot\dot{\alpha}^0 = \dot{\mathbf{A}}^0\cdot\dot{\alpha}(\lambda) \ , \tag{4.8}$$

$$\dot{g}_K^0(\bar{\Delta})\ \dot{\gamma}_K(\lambda) = 0 \ . \tag{4.9}$$

These equalities are now assumed to hold as the necessary conditions for non-unloading in the fundamental process $\mathcal{P}^0$ at $\lambda=\bar{\lambda}+0$. Alternatively, (4.8) and (4.9) result from the (stronger) assumption that the fundamental process $\mathcal{P}^0$ is *smooth* at $\bar{\lambda}$.

It has been shown [32][33] that the above non-unloading condition in both Examples (1) and (2) results in the following constitutive inequality:

$$\dot{\mathbf{S}}(\bar{\Delta}, \dot{\mathbf{F}}^0)\cdot\dot{\mathbf{F}} - \dot{\mathbf{S}}(\bar{\Delta}, \dot{\mathbf{F}})\cdot\dot{\mathbf{F}}^0 \geq 0 \qquad \text{for every } \dot{\mathbf{F}} \tag{4.10}$$

provided the symmetry property (2.42) or (2.51) has been assumed. A certain consequence of (4.10) will be used later, at the moment (4.10) is quoted to indicate the possibility that the similar inequality

$$\dot{S}^0\cdot\Delta F - \Delta S\cdot\dot{F}^0 \geq 0 \qquad \text{for every } \Delta F = F - F^0, \qquad (4.11)$$
$$\Delta S = \bar{\phi}_{,F}(F,\alpha) - \bar{\phi}_{,F}(F^0,\alpha^0)$$

may be valid under the same assumptions; note that ΔF and ΔS in the auxiliary problem under consideration should be regarded as first-order quantities. This possibility is examined below.

Example (1)

The left-hand side of the inequality (4.11) is transformed as follows

$$\dot{S}^0\cdot\Delta F - \Delta S\cdot\dot{F}^0 = \Delta F\cdot\bar{\phi}_{,F\alpha}\cdot\dot{\alpha}^0 - \dot{F}^0\cdot\bar{\phi}_{,F\alpha}\cdot\Delta\alpha = -\Delta A\cdot\dot{\alpha}^0 + \dot{A}^0\cdot\Delta\alpha$$
$$= -A\cdot\dot{\alpha}^0 + \bar{\mathcal{D}}(\alpha^0,\dot{\alpha}^0) + \dot{A}^0\cdot\Delta\alpha$$
$$= -\Delta\alpha\cdot\left(\bar{\mathcal{D}}_{,\alpha}(\bar{\alpha},\dot{\alpha}^0) - \dot{A}^0\right) + \left(\bar{\mathcal{D}}(\alpha,\dot{\alpha}^0) - A\cdot\dot{\alpha}^0\right). \qquad (4.12)$$

By writing $\Delta\alpha = \int(\dot{\alpha}-\dot{\alpha}^0)d\lambda$, using the symmetry property (2.42) and substituting (4.8) we obtain that the former term after the last equality sign in (4.12) vanishes. Since the last term in (4.12) is nonnegative on account of (3.51), we arrive at the conclusion that (4.11) is valid in Example (1) provided (4.8) and (2.42) have been assumed.

Example (2)

In place of (4.12) we obtain

$$\dot{S}^0\cdot\Delta F - \Delta S\cdot\dot{F}^0 = \sum_K \bar{\Lambda}_K\cdot(\dot{F}^0\Delta\gamma_K - \Delta F\,\dot{\gamma}_K^0)$$
$$= \sum_K\Big(\dot{g}_K^0\cdot\Delta\gamma_K + \sum_J g_{KJ}(\bar{\omega})(\dot{\gamma}_J^0\,\Delta\gamma_K - \dot{\gamma}_K^0\,\Delta\gamma_J)\Big) - \sum_K g_K\dot{\gamma}_K\,. \qquad (4.13)$$

The former sum over K after the last equality sign vanishes on account of

(4.9) and (2.51). The last sum over K in (4.13) is evidently nonpositive, and validity of (4.11) has been proved also in Example (2) provided (4.9) and (2.51) are valid.

The essential point in the proof of (4.11), besides the assumed symmetry and linearizations, is that in a perturbed process only those internal mechanisms of plastic deformation are activated which are active or at least non-unloaded in the linearized fundamental process $\bar{\mathcal{P}}^0$. In other words, a perturbed process can involve unloading with respect to $\bar{\mathcal{P}}^0$ but not conversely; this is given a mathematical formulation in (4.8) or (4.9), alternatively. As already mentioned, this can be regarded as a consequence of smoothness of the fundamental process at the stage $\bar{\lambda}$ under consideration.

4.3. Conditions of stability in the first approximation

We are now ready to formulate conditions of stability (in the first approximation) of our (linearized) fundamental process. Referring to Definition 2, suppose that after some transitory, possibly dynamic process which takes place in a sufficiently small R-neighborhood of $\mathcal{S}^0(\bar{\lambda})$ as discussed above, the system is in an equilibrium state $\mathcal{S}(\bar{\bar{\lambda}}) \neq \bar{\mathcal{S}}^0(\bar{\bar{\lambda}})$, $\bar{\bar{\lambda}}\geq\bar{\lambda}$. In a subsequent quasi-static process $\mathcal{P}$ starting from $\mathcal{S}(\bar{\bar{\lambda}})$ and induced by quasi-statically varying loading in the absence of further disturbing influences, the latter integral in (4.4) vanishes by the standard transformation with the use of Green's theorem, while the former is non-positive under the assumptions leading to (4.11). Thus, if (4.11) holds then we arrive at the desired result that

$$\Delta\dot{E} \leq 0 \qquad \text{for } \lambda\geq\bar{\bar{\lambda}} \tag{4.14}$$

in any quasi-static process $\mathcal{P}$ (free of persistent disturbances). However, it must be noted that, contrary to a similar property for the functional V when λ is fixed, (4.14) is not universally valid: it has been derived only for the specified two classes of rate-independent elastic-plastic solids, under the additional assumptions: either (4.8), (2.42) or (4.9),

(2.51), and only for the auxiliary problem.

Under the conditions leading to (4.14), a natural condition of stability of the fundamental process is

$$\Delta\mathbb{E}|_{\lambda} > 0 \quad \text{for every } \mathcal{S}(\lambda) \neq \mathcal{S}^{0}(\lambda), \quad \lambda \geq \bar{\lambda}\,, \tag{4.15}$$

and for every transitory process from $\mathcal{S}^0(\bar{\lambda})$ to $\mathcal{S}(\lambda)$. Evidently, this is a stronger condition than the similar condition for stability of equilibrium discussed in Section 3.4 at λ=const. Note that $\Delta\mathbb{E}$ in (4.15) is understood as a *second-order quantity* with respect to the increments of $\mathbf{F}$, $\boldsymbol{\alpha}$ and λ from their values in $\mathcal{S}^0(\bar{\lambda})$. For, the first-order contribution to $\Delta\mathbb{E}$ vanishes on account of the equilibrium conditions and (3.53), while the higher-order terms have been eliminated by the linearizations involved in the definition of the auxiliary problem. It is possible not to discuss the auxiliary problem but rather to perform all calculations for the "true" process in a *R*-neighborhood of $\mathcal{S}^0(\bar{\lambda})$ with accuracy to the second-order terms only.

The quantity $\Delta\mathbb{E}|_{\bar{\bar{\lambda}}}$ can be interpreted as the energy supplied to the compound system $\mathcal{A}$ due to disturbing influences[11] in a transitory (isothermal) process such that $\Delta^{ext}S_{A}=0$. If (4.15) holds then the supremum of $\Delta\mathbb{E}|_{\lambda}$ in $[\bar{\lambda},\bar{\bar{\lambda}}]$ may be identified with a metric ρ^{0}_{E} generated by the energy functional $\mathbb{E}$ for the initial disturbance at $\bar{\lambda}$ (cf. 3.39). Analogously, $\Delta\mathbb{E}|_{\lambda}$ may be taken as the Lyapunov pseudo-metric ρ_E generated by $\mathbb{E}$ for the distance between $\mathcal{S}(\lambda)$ and $\bar{\mathcal{P}}^{0}(\lambda)$. From the theorem analogous to Theorem 1 it follows that *the condition* (4.15) *is sufficient for stability of the process* $\bar{\mathcal{P}}^{0}$ *with respect to these metrics* in the sense of Definition 2. Another interpretation is discussed below.

For fulfilment of the condition (4.15) it is necessary that

---

[11] Note that $\Delta\mathbb{E}|_{\lambda}$ is *not* equal in general to the work done directly by some perturbing forces since the energy flux from external sources to the body $\mathcal{B}$ through S_u, or to the loading device, is in general also influenced by a disturbance when λ is varying.

$$\Delta\ddot{E}(\tilde{v}) > 0 \quad \text{in } \mathscr{S}^0(\bar{\lambda}) \quad \text{for every } \tilde{v}\neq\tilde{v}^0,\ v = v^0 \ \text{ on } S_u\,, \tag{4.16}$$

where, for the configuration-independent loading,

$$\Delta\ddot{E}(\tilde{v}) = \Delta\left(\int_G \dot{S}(\dot{F},\bar{\Delta})\cdot\dot{F}\,d\xi - 2\int_G \dot{b}\cdot v\,d\xi - 2\int_{S_T} \dot{T}\cdot v\,dS\right) \quad \text{in } \mathscr{S}^0(\bar{\lambda}). \tag{4.17}$$

When multiplied by $\frac{1}{2}(\lambda-\bar{\lambda})^2$, the expression (4.17) represents the value of $\Delta\mathbb{E}|_\lambda$ reached on a straight path in $\tilde{F}$-space from $\mathscr{S}^0(\bar{\lambda})$ to $\mathscr{S}(\lambda)$. As discussed in Section 3.4 in the context of stability of equilibrium, if in Example (2) the conditions of symmetry (2.51) and of constitutive uniqueness (2.50) are satisfied then among all transitory processes which satisfy (3.55) at each instant and lead to the same increment of $\tilde{x}$, the second-order work integral, and consequently $\Delta\mathbb{E}$ as well, are minimized on such direct paths. Hence, under these assumptions, the second-order condition (4.15) is equivalent to (4.16). The equivalence can be demonstrated also in Example (1) provided the constitutive uniqueness condition (2.41) and the symmetry condition (2.42) are satisfied.

Now, we make one more step which was not possible in the case of examining stability of equilibrium. Namely, we may assume that $\dot{S}$ as a function of $\dot{F}$ in $\bar{\Delta}$ is differentiable at $\dot{F}^0$ almost everywhere in G. This is a natural requirement provided, of course, that $\dot{F}^0\neq 0$. Then the so-called tangent moduli $C^0(\bar{\Delta}) \equiv (\partial\dot{S}/\partial\dot{F})(\dot{F}^0,\bar{\Delta})$ are well defined almost everywhere in G. For both the sets of constitutive assumptions used to derive (4.14), we have the potential constitutive relationship (2.54) so that the tangent moduli exhibit the diagonal symmetry, $C^0_{ijkl}=C^0_{klij}$. This symmetry property is of primary importance for the possibility of (4.16) to hold and thus for validity of the whole approach discussed in this section.

The step mentioned above follows from the fact that (4.16) is, as a consequence of (4.10), implied by [31,32]

$$\mathcal{I}^0(\tilde{w}) > 0 \quad \text{in } \mathscr{S}^0(\bar{\lambda}) \quad \text{for every } \tilde{w}\neq\tilde{0},\ \ w=0 \text{ on } S_u\,, \tag{4.18}$$

where

$$\mathcal{I}^{0}(\tilde{w}) \equiv \frac{1}{2}\int_G \nabla w\cdot C^{0}(\bar{\Delta})\cdot\nabla w \, d\xi \tag{4.19}$$

is a *quadratic* functional of the single field $\tilde{w} = \Delta\tilde{v}$. Conversely, (4.16) implies nonnegativeness of (4.19) as of the second variation of $\Delta\ddot{E}$. The inequality (4.18) ensures also uniqueness of the quasi-static solution in velocities at $\bar{\lambda}$, as established originally by Hill [25] for classical elastoplasticity, extended in [17] to Example (2) and extended further in [32] to any case where (2.54) is complemented with (4.10).

To interpret (4.18) as a condition sufficient for *stability* in the first approximation of the fundamental *process*, let us rewrite it in a slightly stronger form

$$\mathcal{I}^{0}(\tilde{w}) \geq k\,\|\nabla w\|^{2} \quad \text{in } \mathcal{S}^{0}(\bar{\lambda}) \text{ for every } \tilde{w}, \quad w=0 \text{ on } S_u, \quad k>0 , \tag{4.20}$$

where k may be arbitrarily small. This is equivalent to the fulfilment of (4.18) for the moduli C^{0}_{ijkl} slightly modified by $(-k\delta_{ik}\delta_{jl})$, where δ_{ij} is the Kronecker symbol. Under the assumptions of the above mentioned theorem on the minimum of the second-order work on a straight deformation path, from (4.20) we obtain

$$\Delta E\big|_{\lambda} \geq \frac{1}{2}\,\Delta\ddot{E}(\tilde{x}(\lambda)-\tilde{x}(\bar{\lambda})) \geq \mathcal{I}^{0}(\Delta\tilde{x}(\lambda)) \geq k\,\|\Delta F(\lambda)\| \equiv k\,\rho_F(\lambda) \; ; \tag{4.21}$$

the second inequality in (4.21) can be shown to follow from (4.10) and (2.54).

For any set of the constitutive assumptions used above to derive (4.14) and under the assumed condition of constitutive uniqueness, we arrive thus at the following conclusion: The condition (4.20) is sufficient for stability {in the first approximation at $\bar{\lambda}$} of the fundamental process $\bar{\mathcal{P}}^{0}$ $\{\mathcal{P}^{0}\}$ for initial disturbances with respect to metrics ρ^{0}_{E}, ρ_F. Uniqueness of the incremental solution follows as a corollary.

For the non-linearized fundamental process $\mathcal{P}^0$, the stability has a meaning in a R-neighborhood of $\mathcal{S}^0(\bar{\lambda})$ only. As already mentioned in the context of stability of equilibrium, ρ_F vanishingly small does not guarantee that $|\Delta\mathbf{F}|<R$ at every material point; this is the source of difficulties in obtaining a rigorous proof of a second-order sufficiency condition for stability in a continuum.

The condition

$$\Delta\ddot{E} \ge 0 \qquad \text{in } \mathcal{S}^0(\bar{\lambda}) \quad \text{for every } \tilde{v}, \quad v=v^0 \text{ on } S_u , \tag{4.22}$$

slightly weaker than (4.16), can be interpreted as *necessary* for stability of the linearized fundamental process $\overline{\mathcal{P}}^0$, or for stability in the first approximation of $\mathcal{P}^0$, under weaker constitutive assumptions than those used to justify the sufficiency condition above. Namely, one can prove the following statement [31] :

If

(i) the constitutive velocity-gradient potential (2.54) exists,
(ii) an auxiliary problem defined by (3.50) and (4.1) is considered,
(iii) the condition (3.65) of stability of equilibrium holds in $\mathcal{S}^0(\bar{\lambda})$,
(iv) $\Delta\ddot{E}(\tilde{v}) < 0$ in $\mathcal{S}^0(\bar{\lambda})$ for some $\tilde{v}$, $v=v^0$ on S_u,

then a given distance ρ_F from the fundamental process $\overline{\mathcal{P}}^0$ can be reached on a straight deformation path from $\mathcal{S}^0(\bar{\lambda})$ to some $\mathcal{S}(\lambda)$ under action of arbitrarily small perturbing forces.

The condition (iv) can be replaced by

$$\bar{I}^0(\tilde{w}) < 0 \quad \text{in } \mathcal{S}^0(\bar{\lambda}) \quad \text{for some } \tilde{w}, \quad w=0 \text{ on } S_u \tag{4.23}$$

which excludes (4.22) so that (iv) is implied. By combining this *in*stability condition with the stability condition (4.20), we arrive at the conclusion that the sign of the expression (4.19) has fundamental significance for stability of the *process* $\mathcal{P}^0$, *provided* that the symmetry of the tangent moduli (along with conservativeness of the loading) is assumed.

For a *discretized* problem, the perturbing forces mentioned in the theorem above are not required and existence of bifurcation under the conditions (i)+(iv) can be proved (op. cit). A process in which (i)+(iv) are satisfied for $\bar{\lambda}$ from some interval represents then *a continuous spectrum of bifurcation points*.

5. MATERIAL STABILITY

5.1. Intrinsic stability within a material element

Consider a material element as a particular continuous system whose *unperturbed* local state is independent of spatial coordinates. Volume of the element is taken sufficiently small so that body forces can be neglected; to indicate a refined length scale, the material coordinates within the element (in the reference configuration) are no longer denoted by $\boldsymbol{\xi}$ but by another symbol $\mathbf{X}$. A constant temperature $\bar{T}$ is prescribed over the element boundary. As particular cases of stability (of equilibrium or of a quasi-static isothermal process) in a general system, we define two kinds of *intrinsic* stability within the material element:

- Stability of equilibrium within the element if displacements are fixed over the element boundary,

- Stability of the process of uniform straining $\mathbf{F}^0(t)$ of the element if placements $\mathbf{x}(\mathbf{X},t) = \mathbf{x}^0(t) + \mathbf{F}^0(t)\cdot\mathbf{X}$ are given over the element boundary.

These properties evidently do not depend on possible configuration-sensitivity of the external loads acting from outside on the element boundary. They are thus intrinsic properties of an infinitesimal material element embedded in a quasi-statically deforming (possibly inhomogeneous) continuum. Conditions for intrinsic stability will be obtained below from stability conditions for a general system as their specifications for the particular boundary value problem and also as their prerequisites. Moreover, we will additionally investigate stability with respect to internal rearrangements described by internal variables $\boldsymbol{\alpha}$ at prescribed $\mathbf{F}$ or $\mathbf{F}(t)$, the question not discussed so far here. All quantities in this Chapter correspond to some fixed regular point $\boldsymbol{\xi}$ of G.

We will distinguish two kinds of disturbances according to whether they are associated with uniform of non-uniform deformations of the element. In the former case, besides the intrinsic stability also a certain other type of material stability is briefly discussed.

5.2. Stability at uniform deformations

We have already derived the (classical) condition (3.25) for thermal stability. i.e. for the intrinsic stability of equilibrium with respect to (non-uniform) temperature fluctuations within the material element.

In this section, we will examine stability with respect to internal rearrangements within the element, described by variations of α_K at a constant temperature $\bar{T}$ for prescribed $\mathbf{F}$ or $\mathbf{F}(t)$. Contrary to other sections, the compatibility condition (3.38) is no longer assumed in advance and is allowed to be violated by disturbing influences which perturb intrinsic equilibrium within the element. Since α_K are not subject to any boundary conditions, there is no loss in generality in treating here the material element as a point.

Stability of equilibrium

Contribution $\mathrm{d}\mathfrak{e}$ (per unit reference volume) to $\mathrm{d}\mathbb{E}$ from the element in a given *equilibrium* state $\mathfrak{s}=(\mathbf{F},\bar{T},\boldsymbol{\alpha})$ at fixed $\mathbf{F}$ and $\bar{T}$ reads (cf. (3.42))

$$\mathrm{d}\mathfrak{e} = \phi_{,\boldsymbol{\alpha}}(\mathfrak{s})\cdot\mathrm{d}\boldsymbol{\alpha} + \bar{T}\sigma^{\mathrm{P}}\mathrm{d}t = -\mathbf{A}(\mathfrak{s})\cdot\mathrm{d}\boldsymbol{\alpha} + \mathcal{D}(\mathfrak{s},\mathrm{d}\boldsymbol{\alpha}) \ . \tag{5.1}$$

This expression vanishes if $\mathrm{d}\boldsymbol{\alpha}$ is consistent with the actual thermodynamic forces $\mathbf{A}(\mathfrak{s})$. However, we may consider disturbances at a microscopic level which induce a change of a thermodynamic force from the actual value A_K to some A_K^* such that the associated flux $\dot{\alpha}_K^*$ is nonzero even if $g_{K}(\triangle) < 0$. In order that $d\omega \geq 0$ also for such disturbances, it is necessary that

$$\mathcal{D}(\triangle,\ddot{\alpha}_{K}^{*}) \equiv A_{K}^{*}\cdot\ddot{\alpha}_{K}^{*} \geq A_{K}\cdot\ddot{\alpha}_{K}^{*} \qquad \text{for any admissible } \ddot{\alpha}_{K}^{*} . \qquad (5.2)$$

The problem what $\ddot{\alpha}_{K}^{*}$ may be regarded as admissible from a physical point of view is not discussed here since this would require a specification of the physical meaning of α_{K}.

Suppose, as a hypothesis, that (5.2) holds in every stable equilibrium state $\triangle$ and for any $\ddot{\alpha}_{K}^{*}$ corresponding to a small perturbation of the associated thermodynamic force A_{K}^{*}, i.e. for $|A_{K}^{*}-A_{K}|< \varepsilon$, where ε is an arbitrarily small positive number. Suppose also that points on the boundary $\partial C(\overline{T},\alpha)$ of the current elastic domain correspond to stable equilibrium states, at least in vicinity of the actual A_{K}. In rate-independent plasticity this readily implies the local convexity of $C(\overline{T},\alpha)$ in A-space at the actual A_{K} as well as the generalized normality rule for $\dot{\alpha}$, i.e. (2.38) restricted to $|A_{K}^{*}-A_{K}|< \varepsilon$. If all points on $\partial C(\overline{T},\alpha)$ correspond to stable equilibrium then we obtain also the global convexity of the (connected) domain $C(\overline{T},\alpha)$ in A-space and the principle of maximum dissipation rate (2.38). If $(2.43)_{1}$ is assumed then we additionally obtain $(2.43)_{2}$, i.e. the normality of $\dot{\alpha}_{K}$ to the hypersurface $f_{K}=0$ at any $A_{K} \in \partial C(\overline{T},\alpha)$, for each K separately. Hence, essential constitutive assumptions used in previous chapters result now from a study of material stability. It must be pointed out, however, that the necessity of the underlying condition $d\omega \geq 0$ for material stability is not clear.

In passing, we note that convexity of the elastic domain in A-space does not necessarily imply convexity of the elastic domain in *stress*-space (nor strain-space) if the dependence of A on the stress (or strain) is nonlinear.

For the rate-dependent plasticity discussed in Section 2.3, if (5.2) is assumed to hold whenever A_{K} and A_{K}^{*} are in the ε-neighborhood of the examined point on $\partial C(\overline{T},\alpha)$ then we obtain an analogous normality property but in the limit sense, viz.

$$\delta\mathbf{A}_K\cdot\dot{\alpha}_K(\mathbf{A}+\delta\mathbf{A},\bar{T},\alpha) \geq 0 \qquad \text{at } \mathbf{A}\in\partial C(\bar{T},\alpha)\ , \tag{5.3}$$

where $\delta\mathbf{A}_K$ is an infinitesimal variation of $\mathbf{A}_K$. Local or global convexity of the elastic domain $C(\bar{T},\alpha)$ in $\mathbf{A}$-space can analogously be concluded as a consequence of $d\omega\geq 0$. It is worth pointing out that $\dot{\alpha}_K$ as a continuous function of $\mathbf{A}_K$ cannot be analytic across ∂C_K, so that care is needed when (5.3) is further exploited to obtain explicit conditions in terms of partial derivatives. For instance, (5.3) can be satisfied if

$$\partial_p\dot{\alpha}_K/\partial\mathbf{A}_K = r\,\mathbf{P}_K\otimes\mathbf{P}_K \neq 0 \quad \text{at } \mathbf{A}_K\in\partial C(\bar{T},\alpha), \qquad \mathbf{P}_K = \partial f_K/\partial\mathbf{A}_K\ , \quad r>0\ , \tag{5.4}$$

where $\partial_p/\partial\mathbf{A}_K$ denotes one-sided partial derivative at the "plastic" side. For a scalar variable α_K, (5.3) means that the conjugate thermodynamic force A_K increases (at least initially) when $\dot{\alpha}_K$ becomes positive (i.e. positive rate-sensitivity at ∂C).

Consider now a departure from the equilibrium state Δ with $\dot{\alpha}\neq 0$ being initially consistent with the thermodynamic forces $\mathbf{A}(\Delta)$. Then (3.38) is satisfied (and $d^+\omega/dt=0$) *at an initial instant* of a non-equilibrium process of internal rearrangements. This is possible for initial $\dot{\alpha}\neq 0$ only in rate-independent plasticity to which our attention will now be confined. The contribution $\ddot{\omega}$ from the material element in state Δ to $\ddot{E}$ for fixed $\mathbf{F}$ and $\bar{T}$ reads

$$\ddot{\omega} = \ddot{\phi} + \dot{\mathcal{D}} = \sum_{K:\,\dot{\alpha}_K\neq 0} (-\mathbf{A}_K + \mathcal{D}_{,\dot{\alpha}_K}(\alpha,\dot{\alpha}))\cdot\ddot{\alpha}_K + (-\dot{\mathbf{A}} + \mathcal{D}_{,\alpha}(\alpha,\dot{\alpha}))\cdot\dot{\alpha} \tag{5.5}$$

or

$$\ddot{\omega} = \ddot{\phi} + \dot{\mathcal{D}} = -\sum_{K:\,\dot{\gamma}_K\neq 0} g_K(\Delta)\,\ddot{\gamma}_K + \sum_{K,J} g_{KJ}(\Delta)\,\dot{\gamma}_K\,\dot{\gamma}_J \tag{5.6}$$

for the dissipation function $\mathcal{D}$ of the form (2.36) or (2.44), respectively. To have $\ddot{\omega} > 0$ for nonzero $\dot{\alpha}$ it is necessary that these expressions are positive. In the former case with $\ddot{\alpha}$ unrestricted, this implies the potentiality condition

$$A_K = \partial\mathcal{D}/\partial\dot{\alpha}_K \qquad \text{if } \dot{\alpha}_K \neq 0 \tag{5.7}$$

and (cf. [16])

$$\dot{\alpha}\cdot\phi_{,\alpha\alpha}\cdot\dot{\alpha} + \mathcal{D}_{,\alpha}(\alpha,\dot{\alpha})\cdot\dot{\alpha} > 0 \qquad \text{for every } \dot{\alpha} \neq 0 \quad \text{such that } \mathcal{D}(\alpha,\dot{\alpha}) = A(\Delta)\cdot\dot{\alpha}\ . \tag{5.8}$$

In the latter case, we obtain $g_K(\Delta)\,\dot{\gamma}_K = 0$ and

$$\sum_{K,J} g_{KJ}\,\dot{\gamma}_K\,\dot{\gamma}_J > 0 \qquad \text{for every } \dot{\gamma}_K > 0 \quad \text{such that } \dot{\gamma}_K g_K(\Delta) = 0\ . \tag{5.9}$$

The conditions (5.8) or (5.9) ensure that no spontaneous internal rearrangements are possible at fixed T and $\mathbf{F}$ if disturbances are absent. Their sufficiency for the stability with respect to internal rearrangements at fixed T and $\mathbf{F}$ can be demonstrated under additional assumptions, analogous to those at the end of Section 3.4.

Suppose now that $\dot{\alpha}$ is uniquely defined by $\dot{\mathbf{e}}$ in any Δ, and consider stability of equilibrium of a material element under *uniform* variations of $\mathbf{F}$ within the element at fixed $\bar{T}$ and at a passive loading on the element boundary. This leads to the stability condition of the type frequently met in the literature:

$$\dot{\mathbf{t}}(\dot{\mathbf{e}})\cdot\dot{\mathbf{e}} > 0 \qquad \text{for every } \dot{\mathbf{e}} \text{ in } \Delta \tag{5.10}$$

obtained as a local version of $\Delta\ddot{E}>0$ (at λ=const) when the potential energy of the loading environment of the element is taken as $(-\bar{\mathbf{t}}\cdot\mathbf{e})$ with $\bar{\mathbf{t}}=\mathbf{t}(\Delta)$ fixed. However, that inequality is not invariant with respect to the choice of stress and strain measures [1]. This reflects the fact that (5.10) involves implicitly a definition of the loading environment and can thus hardly be regarded as a condition for intrinsic stability of the material itself.

There is one exception, namely, if the material element is in a *relaxed* state ($\mathbf{t}=0$). Then the element can be imagined to be cut off from a continuum and subject to uniform perturbations of $\mathbf{F}$ at zero tractions on its surface. The resulting condition of stability of equilibrium

$$\dot{\mathbf{t}}(\dot{\mathbf{e}})\cdot\dot{\mathbf{e}} > 0 \qquad \text{for every } \dot{\mathbf{e}} \text{ at } \mathbf{t}=0 \tag{5.11}$$

is invariant under transformation to another pair $(\mathbf{t},\mathbf{e})$ of work-conjugate measures of stress and strain. Of course, the stability is neutral with respect to rigid-body rotations of the element. If the relaxed state corresponds to an interior point of the elastic domain then (5.11) reduces to the classical condition that the fourth-order tensor of the (isothermal) elastic moduli *at a stress-free state* is positive definite in the space of symmetric second-order tensors.

Stability of a deformation process

Consider a process of macroscopically uniform deformation, $\rho^0_F:\lambda\longrightarrow \mathbf{F}^0(\lambda)$, in a homogeneous material element. In the fundamental thermodynamic process ρ^0, the deformation process ρ^0_F is accompanied by a process $\rho^0_\alpha:\lambda\longrightarrow \alpha^0(\lambda)$ of internal structural rearrangements whose stability for given ρ^0_F will be now examined. We shall combine the procedure adopted in Section 3.4 with the approach used above to study intrinsic stability of equilibrium; the attention is restricted to rate-independent isothermal processes.

An auxiliary problem being a specification of that defined in Section 4.1 is formulated at some $\bar{\lambda}$ by adopting the assumptions (3.50), (3.53) and either (3.51) or (3.52). The stability of ρ^0_α in the first approximation is examined under the restriction that the macroscopic deformation is unperturbed, $\rho_F=\rho^0_F$. The additional assumptions of symmetry and non-unloading in $\bar{\rho}^0$, (2.42), (4.8) or (2.51), (4.9), are also introduced. As an appropriate Lyapunov functional, we then take (with Δ indicating the difference of corresponding quantities in a perturbed process and in $\bar{\rho}^0_\alpha$) :

$$\Delta\upsilon|_\lambda \equiv \Delta\bar{\phi}|_\lambda + \Delta\int_{\bar{\lambda}}^{\lambda}\bar{\mathcal{D}}\,d\lambda \tag{5.12}$$

so that

$$\Delta\dot{\upsilon} = \Delta\mathbf{S}\cdot\dot{\mathbf{F}}^0 + \left(\bar{\mathcal{D}}(\alpha,\dot{\alpha}) - \mathbf{A}\cdot\dot{\alpha}\right) \tag{5.13}$$

or

$$\Delta\dot{\upsilon} = \Delta\mathbf{S}\cdot\dot{\mathbf{F}}^0 - \sum_K g_K\dot{\gamma}_K \tag{5.14}$$

in Example (1) or (2), respectively. In a process p_α free of persistent disturbances in which the internal state is always in equilibrium, the second term vanishes while $\Delta\mathbf{S}\cdot\dot{\mathbf{F}}^0$ will be non-positive as a particular case of (4.11). Consequently, $\Delta\dot{\upsilon} \le 0$ in such p_α as required. Note the essential role of the assumed symmetry and non-unloading in p^0 at $\bar{\lambda}$.

The (second-order) condition of stability of p^0_α with respect to internal rearrangements becomes

$$\Delta\upsilon|_\lambda > 0 \quad \text{at every } \alpha(\lambda),\quad \lambda>\bar{\lambda}\,,\quad |\alpha_K(\lambda)-\alpha_K(\bar{\lambda})| < R, \tag{5.15}$$

and for every transitory process leading from $\alpha^0(\bar{\lambda})$ to $\alpha(\lambda)$ (at given p^0_F) and satisfying (3.53).

In the first instance, the condition (5.15) reduces for straight paths in α-space to $\Delta\ddot{\upsilon} > 0$ in $\bar{\mathfrak{s}}=\mathfrak{s}^0(\bar{\lambda})$, viz.

$$\Delta\ddot{\upsilon} = 2\,\dot{\mathbf{F}}^0\cdot\phi_{,F\alpha}\cdot\Delta\dot{\alpha} + \Delta\left(\dot{\alpha}\cdot\phi_{,\alpha\alpha}\cdot\dot{\alpha} + \dot{\alpha}\cdot\mathcal{D}_{,\alpha}(\bar{\alpha},\dot{\alpha})\right) > 0 \tag{5.16}$$

$$\text{for every } \dot{\alpha} \ne \dot{\alpha}^0 \text{ satisfying } \mathcal{D}(\bar{\alpha},\dot{\alpha})=\bar{\mathbf{A}}\cdot\dot{\alpha}\,,$$

or (on using (2.13) and (3.52)):

$$\Delta\ddot{\upsilon} = -2\,\dot{\mathbf{F}}^0\cdot\sum_K \Lambda_K(\bar{\mathfrak{s}})\,\Delta\dot{\gamma}_K + \sum_{K,J} g_{KJ}(\bar{\mathfrak{s}})\,\Delta(\dot{\gamma}_K\dot{\gamma}_J) > 0 \tag{5.17}$$

$$\text{for every } \dot{\gamma}_K>0,\ \dot{\gamma}_K\ne\dot{\gamma}^0_K,\ g_K(\bar{\mathfrak{s}})\dot{\gamma}_K=0\,,$$

in Example (1) or (2), respectively.

Example (1)

Using (2.42), (4.8) and the expression for $\dot{\mathbf{A}}$ resulting from (3.50), the stability condition (5.16) is transformed to the form

$$\Delta\ddot{\omega} = \Delta\dot{\alpha}\cdot\phi_{,\alpha\alpha}\cdot\Delta\dot{\alpha} + \Delta\dot{\alpha}\cdot(\mathcal{D}_{,\alpha}(\bar{\alpha},\dot{\alpha}) - \mathcal{D}_{,\alpha}(\bar{\alpha},\dot{\alpha}^0)) > 0 \qquad (5.18)$$

$$\text{for every } \dot{\alpha} \neq \dot{\alpha}^0 \text{ satisfying } \mathcal{D}(\bar{\alpha},\dot{\alpha})=\bar{\mathbf{A}}\cdot\dot{\alpha}\,.$$

As already discussed, under the assumed symmetry property (2.42) and linearity of $\bar{\mathcal{D}}$ with respect to α, among all transitory paths in α-space satisfying (3.54) at each instant, the dissipation integral in (5.12) is minimized on the straight paths (cf. (3.56)). In that case (5.18) is equivalent to (5.15) and is thus sufficient for stability (in the first approximation) of the process ρ^0_α for given ρ^0_F, with $\Delta\omega$ as the measure of perturbations. In particular, (5.18) ensures that $\Delta\alpha_K(\lambda) = 0$ (to first order) at the absence of disturbances; cf. the uniqueness condition (2.41).

Example (2)

Using (3.52), (4.9) and (2.51), the condition (5.17) is rearranged as follows

$$\Delta\ddot{\omega} = \sum_{K,J} g_{KJ}(\bar{\Delta})\,\Delta\dot{\gamma}_K\,\Delta\dot{\gamma}_J > 0 \qquad \text{for every } \Delta\dot{\gamma}_K \not\equiv 0,\ g_K(\bar{\Delta})\Delta\dot{\gamma}_K = 0\,. \qquad (5.19)$$

One can show (cf. [26]) that under the assumptions introduced, (5.19) ensures that $\Delta\omega$ corresponding to given $\Delta\gamma$ and $\Delta\mathbf{F}\equiv\mathbf{0}$ is minimized on direct paths (to second order), so that (5.19) is equivalent to (5.15). It follows that in Example (2) under the assumptions (4.9) and (2.51) the condition for stability in the first approximation of the process ρ^0_α at given ρ^0_F in a neighborhood of $\Delta^0(\bar{\lambda})$ reduces to the requirement of positive definiteness of the matrix (g_{KJ}) for the indices running over the set of potentially active mechanisms in $\Delta^0(\bar{\lambda})$. In this case, the stability means that

$$k|\Delta\gamma_K(\lambda)|^2 \leq \sum_{K,J} g_{KJ}(\bar{\Delta})\, \Delta\gamma_K(\lambda)\, \Delta\gamma_J(\lambda) \leq \Delta\omega|_{\bar{\lambda}} \quad \text{for } \lambda\geq\bar{\lambda}, \quad k>0, \qquad (5.20)$$

so long as $\lambda-\bar{\lambda}$ remains sufficiently small to justify the linearizations made. In particular, it follows that $\Delta\gamma_K(\lambda)=0$ (to first order) for given p_F^0 if disturbances are absent; cf. the uniqueness condition (2.50).

It can be concluded that various constitutive inequalities, adopted in Chapter 2 simply as convenient working assumptions, can also be derived as conditions of material stability with respect to internal structural rearrangements.

5.3. Material stability at non-uniform deformations

Stability of equilibrium

Examine now intrinsic stability of an equilibrium state Δ^0 of a (continuous and homogeneous) material element with respect to *non-uniform* isothermal perturbations which preserve continuity of displacements and are compatible with zero displacements over the element boundary. Material stability with respect to internal rearrangements at prescribed deformations is taken as granted; in particular, (unlike in the preceding subsection) we again assume that $\dot{\alpha}$ in a given state is uniquely defined by $\dot{\mathbf{F}}$ (at $\dot{T}=0$) and that (3.38) holds universally.

Conditions of the intrinsic stability of equilibrium within the element are obtained by specification of, and also as necessary conditions for, the conditions formulated in Chapter 3 for a general inhomogeneous body. To begin with, let us discuss first *elastic* stability when α is frozen. Assuming that (3.25) always holds, the restriction to considering only isothermal perturbations is in that case fully justified. As a specification of (and a necessary condition for, *[34][35],)* (3.30), at an arbitrary regular point ξ of G we obtain

$$\int_{\mathcal{M}} \phi(\mathbf{F}^0+\nabla\tilde{\mathbf{u}}(\mathbf{X}),\bar{T},\alpha)\,d\mathbf{X} \geq |\mathcal{M}|\ \phi(\mathbf{F}^0,\bar{T},\alpha) \quad \text{for every } \tilde{\mathbf{u}},\ \tilde{\mathbf{u}}=0 \text{ over } \partial\mathcal{M}\ . \qquad (5.21)$$

$\tilde{\mathbf{u}}$ is a (continuous and piecewise smooth) variation of the actual placement field $\mathbf{x}^0(\mathbf{X}) = \bar{\mathbf{x}}^0 + \mathbf{F}^0\cdot\mathbf{X}$ within the element. $\mathbf{X}$ denotes a position vector, $\nabla() = \partial()/\partial\mathbf{X}$, and $\mathcal{M}$ is a domain of volume $|\mathcal{M}|$ occupied by the element, all considered in the reference configuration. Note that the $\mathbf{X}$-variable refers to a spatial scale of observation "finer" than that for a finite body: in writing (5.21), the ξ-variable has been fixed, so that $\mathbf{F}^0=\mathbf{F}^0(\Delta^0,\xi)$ and $\alpha=\alpha(\Delta^0,\xi)$ are independent of $\mathbf{X}$. If (5.21) holds then ϕ as a function of $\mathbf{F}$ at fixed $\bar{T}$ and α is called *quasi-convex at* $\mathbf{F}^0$. One easily shows that the shape and size of $\mathcal{M}$ in (5.21) is unessential.

To derive a simpler condition of material stability, we can consider the possibility that the perturbed deformation mode is effectively concentrated in a planar band of the reference orientation $\mathbf{n}$ and of the thickness h negligible in comparison with dimensions of the material element (Fig. 1). The band thickness must somehow tend to zero as the element boundary is approached, but the boundary effects can be neglected. The deformation can be assumed constant within the band and of the form $\mathbf{F} = \mathbf{F}^0+\mathbf{g}\otimes\mathbf{n}$ (in components $F_{ji} = F^0_{ji}+g_j n_i$), where $\mathbf{n}$ is the unit normal to the band in the reference configuration and $\mathbf{g}$ is some nonzero vector.

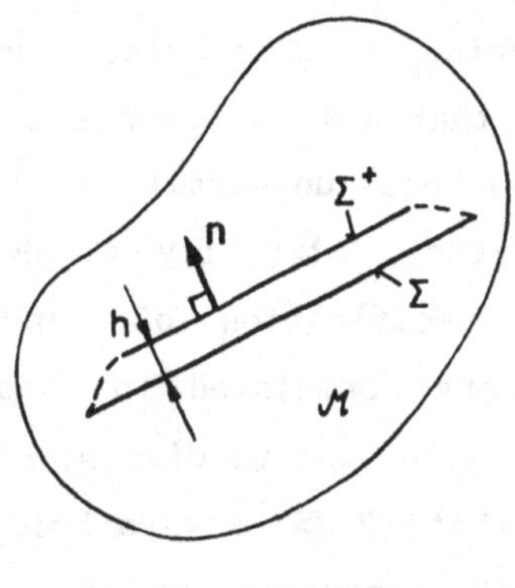

Fig. 1. A material element with a planar band of localized deformation.

From the Green theorem and the boundary condition **u**=0 over $\partial\mathcal{M}$, we readily obtain

$$\int_{\mathcal{M}} (\phi(\mathbf{F}^0+\nabla\mathbf{u}(\mathbf{X})) - \phi(\mathbf{F}^0))\,d\mathbf{X} = \int_{\mathcal{M}} \mathcal{E}_\phi(\mathbf{F}^0,\mathbf{F}^0+\nabla\mathbf{u}(\mathbf{X}))\,d\mathbf{X}\ , \tag{5.22}$$

where

$$\mathcal{E}_\phi(\mathbf{F}^0,\mathbf{F}) \equiv \phi(\mathbf{F}) - \phi(\mathbf{F}^0) - \phi_{,\mathbf{F}}(\mathbf{F}^0)\cdot(\mathbf{F}-\mathbf{F}^0), \qquad T,\alpha \text{ fixed}\ , \tag{5.23}$$

is the Weierstrass function associated with ϕ (cf. (3.28)); the frozen arguments $\overline{T}$ and α are omitted.

The displacements at the band boundaries, **u** = 0 at the plane Σ and **u** = h**g** at the surface Σ^+, say (Fig. 1), can easily be accommodated outside the band by a smooth field $\tilde{\mathbf{u}}$ such that $\nabla\mathbf{u}$ is everywhere of order h provided the band profile has a "cusp" at its perimeter. Then the right-hand integral in (5.22) is of order h^2 everywhere except within the band. For $h \longrightarrow 0$ we thus obtain

$$\int_{\mathcal{M}} \phi(\mathbf{F}^0+\nabla\mathbf{u}(\mathbf{X}))d\mathbf{X} - |\mathcal{M}|\ \phi(\mathbf{F}^0) = \mathcal{E}_\phi(\mathbf{F}^0,\mathbf{F}^0+\mathbf{g}\otimes\mathbf{n}) \int_{\Sigma} h\ d\Sigma\ + O(h^2)\ . \tag{5.24}$$

Hence, a necessary condition for (5.21), and thus also for (3.30), is

$$\mathcal{E}_\phi(\mathbf{F}^0,\mathbf{F}^0+\mathbf{g}\otimes\mathbf{n}) \ge 0 \qquad \text{for all vectors } \mathbf{g},\mathbf{n} \tag{5.25}$$

at every regular point in G. A rigorous proof of that result in the general context of the calculus of variations was given first by Graves [36]. In the usually used terminology [35], (5.25) is the condition of *rank 1 convexity* of $\phi(\mathbf{F})$ *at* $\mathbf{F}^0$, on account of the rank of the matrix $(g_j n_i)$ equal to 1. Further discussion of (5.25) and of its extension to elastic-plastic solids is deferred to the next section.

In the above considerations, *finite* perturbations of **F** (which can, however, be associated with arbitrarily small variations of **u**) are allowed for; in the calculus of variations such perturbations are called *strong* variations. If only infinitesimal perturbations of $\mathbf{F}$ are considered then, as a specification of the condition (3.34) involving the second *weak* variation, in place of the condition (5.21) we obtain

$$\int_{\mathcal{M}} \nabla \mathbf{u} \cdot \phi_{,\mathbf{FF}}(\mathbf{F}^0) \cdot \nabla \mathbf{u} \, dX \geq 0 \qquad \text{for every } \tilde{\mathbf{u}},\ \mathbf{u}=0 \text{ over } \partial\mathcal{M} \,. \tag{5.26}$$

Perhaps surprisingly at a first glance, (5.26) is *equivalent* [37] to the Legendre-Hadamard condition

$$(\mathbf{g}\otimes\mathbf{n}) \cdot \phi_{,\mathbf{FF}}(\mathbf{F}^0) \cdot (\mathbf{g}\otimes\mathbf{n}) \equiv E_{jilk}(\mathbf{F}^0)\, g_j n_i g_l n_k \geq 0 \qquad \text{for every } \mathbf{g},\mathbf{n} \,; \tag{5.27}$$

this provides another counterargument against treatment of positive definiteness of elastic moduli (at non-zero stress) in full space of strain tensors as a thermodynamic condition of stability under *arbitrary* boundary conditions. The condition (5.27) is also obtained as a second-order version of (5.25). In any case, the condition (5.27) can be regarded as necessary for material stability in a thermodynamic sense. If the inequality in (5.27) is strict for nonzero $\mathbf{g},\mathbf{n}$ then it is called the *strong ellipticity condition*, which clearly excludes a non-trivial equality in (5.26).

To find analogous conditions of intrinsic stability of equilibrium for an elastic-plastic material, we begin with the second-order energy condition (3.47). We restrict ourselves to examining the constitutive rate equations in the potential form (2.54). As a specification of and the necessary condition for (3.47), we then obtain

$$\int_{\mathcal{M}} U(\Delta^0, \nabla \mathbf{v}(X))\, dX \geq 0 \qquad \text{for every } \tilde{\mathbf{v}},\ \mathbf{v}=0 \text{ over } \partial\mathcal{M} \,. \tag{5.28}$$

By using the above construction of a deformation mode effectively concentrated in a band but with velocities in place of displacements, or directly from the Graves theorem, we obtain the following condition necessary for (5.28), and thus also for (3.46):

$$U(\Delta^0, \mathbf{g}\otimes\mathbf{n}) \geq 0 \quad \text{for all vectors } \mathbf{g},\mathbf{n}\ ; \tag{5.29}$$

note that the velocity-gradient potential U vanishes with its first derivative $\partial U/\partial(\nabla\mathbf{v})$ at $\nabla\mathbf{v}=\mathbf{0}$. Magnitude of the vectors $\mathbf{g},\mathbf{n}$ is unessential in (5.29) since U is a homogeneous function of degree two of $\dot{\mathbf{F}}$. The condition (5.29) can be regarded as a thermodynamic condition necessary for intrinsic stability of equilibrium of the homogeneous material element (under *any* boundary conditions). It represents an extension of the Legendre-Hadamard condition obtained for *linear* constitutive rate equations to the case when the constitutive rate equations are nonlinear. A mechanical interpretation of (5.29) will be discussed in the next section.

Stability of deformation process

The second-order energy condition (4.16) of stability of a quasi-static deformation process obeying (2.54) has a mathematical structure similar to (3.30), with the difference that a velocity field appears in place of a displacement field and the velocity-gradient potential replaces the free energy density. In analogy to (5.21), the quasi-convexity condition for $U(\dot{\mathbf{F}})$ at $\dot{\mathbf{F}}^0$ can be formulated as a local specification and consequence of the stability condition (4.16) for the homogeneous element deforming with uniform velocity gradient $\dot{\mathbf{F}}^0$, as defined in Section 5.1. However, of more interest will be the consequence of (4.22) in the form of the condition of *rank 1 convexity of $U(\dot{\mathbf{F}})$ at $\dot{\mathbf{F}}^0$*, viz.

$$\mathcal{E}_U(\bar{\Delta};\ \dot{\mathbf{F}}^0,\ \dot{\mathbf{F}}^0+\mathbf{g}\otimes\mathbf{n}) \geq 0 \quad \text{for all vectors } \mathbf{g},\mathbf{n} \tag{5.30}$$

at every regular material point in G at any $\bar{\lambda}$, where

$$\mathcal{E}_U(\Delta;\ \dot{\mathbf{F}}^0,\dot{\mathbf{F}}) \equiv U(\Delta;\dot{\mathbf{F}}) - U(\Delta,\dot{\mathbf{F}}^0) - \dot{\mathbf{S}}(\Delta,\dot{\mathbf{F}}^0)\cdot(\dot{\mathbf{F}}-\dot{\mathbf{F}}^0) \tag{5.31}$$

is the Weierstrass function associated with the velocity-gradient potential U. The condition (5.30) is obtained in an analogous way as

(5.25). Since (4.22) can be interpreted as necessary for stability of the fundamental process for an arbitrary constitutive potential U, (5.30) may be interpreted as necessary for intrinsic stability of a process of uniform deformation with velocity gradient $\dot{\mathbf{F}}^0$ within the material element embedded in a *deforming* continuum.

If the global stability condition is taken in the form (4.18) then (5.30) reduces to Legendre-Hadamard condition formulated for *the tangent moduli*,

$$(\mathbf{g}\otimes\mathbf{n})\cdot\mathbf{C}^0(\bar{\Delta})\cdot(\mathbf{g}\otimes\mathbf{n}) \geq 0 \qquad \text{for every } \mathbf{g},\mathbf{n}\ . \tag{5.32}$$

If the constitutive inequality (4.10) holds then (5.32) is equivalent to (5.30). However, (5.29) can still hold even if (5.32) is violated; this reflects at the level of a material element the basic fact that in plasticity the concepts and thus also the conditions of stability of equilibrium and of stability of a quasi-static deformation process are distinct.

5.4. Acceleration waves and shear bands

To give another interpretation to the conditions of material stability obtained in the preceding section, consider the problem of propagation of *acceleration waves* in a rate-independent material. An acceleration wave is, by definition, a geometric surface Σ which moves relatively to the material and across which the acceleration $\dot{\mathbf{v}}$, velocity gradient $\dot{\mathbf{F}}=\nabla\mathbf{v}$ and stress-rate $\dot{\mathbf{S}}$ are discontinuous while the displacement, velocity, strain and stress are continuous. Only isothermal deformations are considered and, for simplicity, the reference configuration is chosen to coincide momentarily with the actual configuration ($\mathbf{F}=\mathbf{1}$). The well-known kinematic compatibility and wave propagation conditions read

$$[[\ \dot{\mathbf{v}}\ ]] = -\,v\mathbf{g}\ , \qquad [[\ \dot{\mathbf{F}}\ ]] = \mathbf{g}\otimes\mathbf{n}, \qquad [[\ \dot{\mathbf{S}}\ ]]\cdot\mathbf{n} = \bar{\rho}\ v^2\mathbf{g}\ , \tag{5.33}$$

where $[[\cdot]]$ denotes a jump across Σ, $\mathbf{n}$ is the unit normal to Σ, $\mathbf{g}$ is a nonzero vector, $\bar{\rho}$ is the mass density and v is the normal speed of propagation of Σ relative to the material.

It is well known that (5.27) fulfilled at every $\mathbf{F}^0$ is necessary and sufficient for the speed of propagation of all acceleration waves in *elastic* materials to be real. The speed is always real and nonzero if and only if the inequality in (5.27) is strict for nonzero $\mathbf{g},\mathbf{n}$, i.e. if the elastic moduli are strongly elliptic.

Consider now rate-independent inelastic materials obeying the constitutive rate equations which admit an arbitrarily nonlinear potential (2.54). The results given below are taken from the recent paper [38]. It has been proved that for such inelastic materials the condition (5.29) is necessary and sufficient for the speed v of propagation of all acceleration waves into *non-deforming* material to be real. If

$$U(\mathbf{g}\otimes\mathbf{n}) > 0 \qquad \text{for all } \mathbf{g},\mathbf{n} \neq \mathbf{0} \tag{5.34}$$

then the speed is real and nonzero. If (5.29) holds with equality for some nonzero $\mathbf{g}^*,\mathbf{n}^*$ then it can be shown that such a pair $(\mathbf{g}^*,\mathbf{n}^*)$ constitutes a solution to the equation (5.33) with v=0, that is, the condition for a *stationary discontinuity* is met. This corresponds to a nonzero quasi-static incremental solution $\dot{\mathbf{F}}^*=\mathbf{g}^*\otimes\mathbf{n}^*$ in a band of the normal $\mathbf{n}^*$, compatible with constant nominal tractions on the band boundaries moving as rigid planes. If (5.29) is violated then it has been argued that a vanishingly small disturbance can induce a dynamic process of localized deformation in a sufficiently thin band of the orientation $\mathbf{n}$. If $\mathbf{g}^*$ is orthogonal to $\mathbf{n}^*$ then the localization has the form of a *shear band*.

Suppose now that at one side of an acceleration wave the material is *deforming* with a velocity gradient $\dot{\mathbf{F}}^0$. Such a distinction was immaterial for purely elastic deformations, but is important for a non-quadratic potential U. It has been proved that the condition (5.30) is necessary for the speed of propagation of all such acceleration waves to be real. In particular, (5.32) is also necessary for this, and with the interpretation of the left-hand expression in (5.32) as a second-order approximation of that in (5.30), it becomes clear that (5.32) is relevant to the waves of an amplitude sufficiently small for the tangent moduli $\mathbf{C}^0$ to apply on *both* sides of Σ.

If (5.30) holds with equality for some non-zero $\mathbf{g}^*, \mathbf{n}^*$ then a quasi-static *bifurcation* in a band of orientation $\mathbf{n}^*$ becomes possible, in the sense that the quasi-static incremental solution $\dot{\mathbf{F}}^* = \dot{\mathbf{F}}^0 + \mathbf{g} \otimes \mathbf{n}$ in the band is kinematically and statically compatible with the uniform deformation rate $\dot{\mathbf{F}}^0$ outside the band. Moreover, existence of such a bifurcation in a band of the orientation $\mathbf{n}$ has also been proved when the condition (5.30) of intrinsic stability of the process is violated while the condition (5.34) of stability of equilibrium still holds. This is again a demonstration of the distinction between stability of equilibrium and stability of a quasi-static process in plastic solids.

Let us proceed to the condition (5.21). If the elastic moduli $\mathbf{E}(\Delta^0)$ are strongly elliptic then it might seem, in view of the resulting strict inequality in (5.26) for $\tilde{\mathbf{u}} \neq \tilde{\mathbf{0}}$, that an energy barrier exists in a vicinity of the uniform equilibrium state Δ^0 which excludes a departure from equilibrium unless a positive amount of work is supplied to the element. However, this is not so if (5.25) is violated and *moving strong discontinuities in strain* are allowed for.

Namely, suppose that the band illustrated schematically in Fig. 1 is (artificially) created by the movement of one (Σ^+) of its planar boundaries with an initial normal speed (in the reference configuration) $\dot{h} > 0$, *starting from zero initial thickness* at $t=0$, say. This is kinematically admissible if the jump in $\mathbf{F}$ has the form $[[\ \mathbf{F}\ ]] = \mathbf{g} \otimes \mathbf{n}$ and is associated with initial velocity at the moving discontinuity surface Σ^+ being equal to zero from inside and to $\dot{h}\mathbf{g}$ from outside of the formed band. The terms "speed" and "velocity" are conventional since t need not be here a natural time. If α is frozen then from (5.24) one obtains the formula

$$\frac{\mathrm{d}}{\mathrm{d}t} \int_{\mathcal{M}} \phi(\mathbf{F}^0 + \nabla \mathbf{u}(\mathbf{X}))\, \mathrm{d}\mathbf{X} \Bigg|_{\Delta^0} = \mathcal{E}_{\phi}(\mathbf{F}^0, \mathbf{F}^0 + \mathbf{g} \otimes \mathbf{n}) \int_{\Sigma} \dot{h}\, \mathrm{d}\Sigma \,. \qquad (5.35)$$

It follows that fulfilment of (5.25) at every regular point of the body $\mathcal{B}$ being in equilibrium is necessary for V_e to be nondecreasing for any *kinematically* admissible mode of quasi-static departure from the equilibrium state if moving strong discontinuities in strain are not excluded. We do not discuss here the question of dynamic admissibility of the moving discontinuities (cf. e.g. [39]).

Equilibrium states of an elastic solid body which are stable with respect to small variations of **F** but such that (5.25) is violated in some part of G can be called *metastable*. Metastable states can be treated as stable or not depending on the strength of actual local disturbances or inhomogeneities which are neglected or smoothed out in the macroscopic description but are always present in reality; it should not be forgotten that we are discussing here only *models* of real materials.

Finally, we formulate a condition analogous to (5.25) but for a rate-independent elastic-plastic material where α varies with **F** and when the energy condition of isothermal stability takes the form (3.45). Specification of (3.45) for the material element, when transformed in analogy to (5.22), reads

$$\int_{\mathcal{M}} \Delta w(\nabla \mathbf{u})\, dX = \int_{\mathcal{M}} (\Delta w(\nabla \mathbf{u}) - \mathbf{S}(\Delta^0)\cdot\nabla \mathbf{u}\,)dX \ \geq 0 \quad \text{for every } \tilde{\mathbf{u}},\ \mathbf{u}=\mathbf{0} \text{ over } \partial\mathcal{M}\,. \tag{5.36}$$

The value of the deformation work Δw in a transition from $\mathbf{F}^0$ to $\mathbf{F}^0+\nabla\mathbf{u}$ depends now on the deformation path. We restrict ourselves to examining *straight* deformation paths. Consider a deformation mode concentrated in a thin planar band as in the derivation of (5.24). Proceeding exactly as before, we arrive at

$$\int_{\mathcal{M}} \Delta w(\nabla \mathbf{u})\, dX = \mathcal{E}_W(\Delta^0, \mathbf{g}\otimes\mathbf{n}) \int_{\Sigma} h\, d\Sigma \ + O(h^2)\ , \tag{5.37}$$

where

$$\mathcal{E}_W(\Delta^0, \mathbf{g}\otimes\mathbf{n}) = \int_0^1 \Big(\, \mathbf{S}(\Delta(z)) - \mathbf{S}(\Delta^0)\ \Big)\cdot\dot{\mathbf{F}}\ dz$$

$$= \phi(\Delta(1)) - \phi(\Delta^0) - S(\Delta^0)\cdot(g\otimes n) + \bar{T}\int_0^1 \sigma^p(\Delta(z),\dot{\alpha}(z))\,dz\,,$$

$$\Delta(z) = (F(\Delta^0)+zg\otimes n,\ \bar{T},\ \alpha(z)),\qquad \dot{\alpha}(z) = \dot{\alpha}(\Delta(z),\dot{F})\,,\qquad \dot{F}=g\otimes n\,. \tag{5.38}$$

$\mathcal{E}_w$ represents the difference between the actual work of deformation on the straight path from $F(\Delta^0)$ to $F(\Delta^0)+g\otimes n$ and the corresponding work done by equilibrium stresses $S(\Delta^0)$. For elastic solids $\mathcal{E}_w$ reduces to $\mathcal{E}_\phi$.

From (5.37) as a necessary condition for (5.36), and thus also necessary for the global stability condition (3.45) to hold, we obtain

$$\mathcal{E}_w(\Delta^0, g\otimes n) \geq 0 \qquad \text{for all vectors } g, n\,. \tag{5.39}$$

Note that (5.39) must be satisfied if (3.45) is to hold in a neighborhood of $\mathcal{G}^0$ relative to the distance $\|\Delta F\|$. Equilibrium states which are stable with respect to small variations of F and α but such that (5.39) is violated in some part of G can be called *metastable*; cf. the discussion above on metastability in elastic solids.

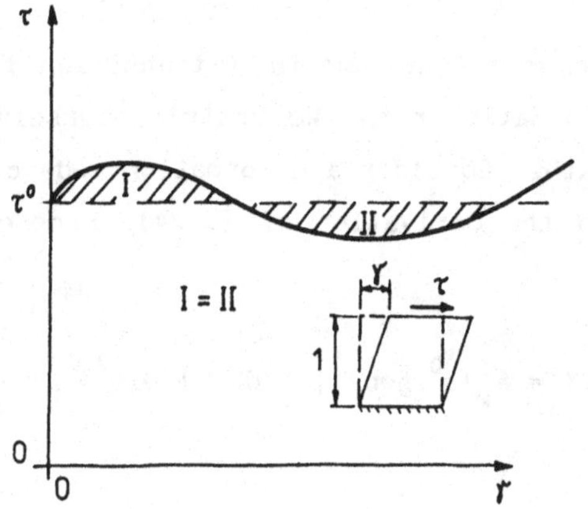

Fig. 2. Schematic stress-strain diagram for simple shear superimposed on the critical state defined by condition (5.39).

It is an open question in which cases the metastability associated with violation of (5.39) can be treated as *sufficient* for formation of shear bands, for instance, in plastically deformed metals. Suppose that existing strong nonuniformities in polycrystals justify that hypothesis, introduced in [40]. The critical stage of the deformation process beyond which the onset of formation of shear bands might be expected is then found by seeking the first Maxwell line for the shear stress - shear strain curves drawn for simple shear (γ) superimposed in various directions on the traversed states (Fig. 2). Preliminary calculations performed with the use of simple models of plasticity have given quite realistic results [40], although full justification of that procedure is yet to be found. This question deserves further study.

Acknowledgement. I am indebted to Prof. B. Raniecki for his helpful comments on an earlier version of the manuscript.

REFERENCES

1. Hill, R.: Aspects of invariance in solids mechanics, *Advances in Applied Mechanics, vol.* **18**, Acad. Press, New York 1978, 1-75.

2. Mandel, J.: Plasticité Classique et Viscoplasticité, CISM Course, Udine, Springer 1971.

3. Rice, J.R.: Inelastic constitutive relations for solids: an internal- variable theory and its application to metal plasticity, *J. Mech. Phys. Solids*, **19** (1971) 433-455.

4. de Groot, S.R. and Mazur, P.: *Non-equilibrium Thermodynamics*, North-Holland, Amsterdam 1962.

5. Muschik, W.: Internal variables in non-equilibrium thermodynamics, *J. Non-Equilib. Thermodyn.*, **15** (1990) 127-137.

6. Kestin, J.: A note on the relation between the hypothesis of local equilibrium and the Clausius-Duhem inequality, *J. Non-Equilib. Thermodyn.*, **15** (1990) 193-212.

7. Rice, J.R.: Continuum mechanics and thermodynamics of plasticity in relation to microscale deformation mechanisms, in: *Constitutive Equations in Plasticity* (Ed. A.S. Argon), MIT Press, Cambridge Mass. 1975, 23-79.

8. Lehmann, T. (Ed.): *The Constitutive Law in Thermoplasticity*, CISM Course, Udine, Springer 1984.

9. Kleiber, M. and Raniecki, B.: Elastic-plastic materials at finite strains, in: *Plasticity Today: Modelling, Methods and Applications* (Ed. A. Sawczuk and G. Bianchi), Elsevier, London 1985, 3-46.

10. Neuhäuser, H.: Physical manifestation of instabilities in plastic flow, in: *Mechanical Properties of Solids: Plastic Instabilities* (Ed. V. Balakrishnan and E.C. Bottani), World Scientific, Singapore 1986, 209-252.

11. Hill, R. and Rice, J.R.: Elastic potentials and the structure of inelastic constitutive laws, *SIAM J. Appl. Math.*, **25** (1973) 448-461.

12. Kestin, J. and Rice, J.R.: Paradoxes in the application of thermodynamics to strained solids, in: *A Critical Review of Thermodynamics* (Ed. E.B. Stuart *et. al.*), Mono Book Corp., Baltimore 1970, 275-298.

13. Morreau, J.J.: Sur les lois de frottement, de plasticité et de viscosité, *C. R. Acad. Sci. Paris, t.* **271**, *Serie A* (1970) 608-611.

14. Morreau, J.J.: On Unilateral Constraints, Friction and Plasticity, CIME Course, Bressansone, in: *New Variational Techniques in Mathematical Physics,* Edizioni Cremonese, 1974, 175-322.

15. Nguyen, Q.S.: Stabilité et bifurcation des systèmes dissipatifs standards à comportement indépendant du temps physique, *C. R. Acad. Sci. Paris, t.* **310**, *Serie II* (1990) 1375-1380.

16. Nguyen, Q.S.: Bifurcation and stability of time-independent standard dissipative systems, CISM Course, Udine 1991.

17. M.J. Sewell: A survey of plastic buckling, in: *Stability* (H.H.E. Leipholz, ed.), Univ. of Waterloo Press, Ontario 1972, 85-197.

18. Hill, R. and Rice, J.R.: Constitutive analysis of elastic-plastic crystals at arbitrary strain, *J. Mech. Phys. Solids*, **20** (1972) 401-413.

19. Hill, R.: Some basic principles in the mechanics of solids without a natural time, *J. Mech. Phys. Solids,* **7** (1959) 209-225.

20. Movchan, A.A.: Stability of processes with respect to two metrics (in Russian), *Prikl. Math. Mekh.*, **24** (1960) 988-1001.

21. Bazant, Z.P.: Stable states and stable paths of propagation of damage zones and interactive fractures, in: *Cracking and Damage* (Ed. J. Mazars and Z.P. Bazant), Elsevier, London 1989, 183-206.

22. Ericksen, J.L.: A thermo-kinetic view of elastic stability theory, *Int. J. Solids Structures,* **2** (1966) 573-580.

23. Gurtin, M.E.: Thermodynamics and Stability, *Arch. Rational Mech. Anal.*, **59** (1975) 63-96.

24. Petryk, H.: A consistent energy approach to defining stability of plastic deformation processes, in: *Stability in the Mechanics of Continua, Proc. IUTAM Symp.* (Ed. F.H. Schroeder), Springer, Berlin 1982, 262-272.

25. Hill, R.: A general theory of uniqueness and stability in elastic-plastic solids, *J. Mech. Phys. Solids*, **6** (1958) 236-249.

26. Petryk, H.: On the second-order work in plasticity, *Arch. Mech.*, **43** (1991) 377-397.

27. Nguyen, Q.S. and Radenkovic, D.: Stability of equilibrium in elastic plastic solids, *Springer Lecture Notes in Mathematics, Vol.* **503** (1976) 403-414.

28. Petryk, H.: *(to be published).*

29. Knops, R.J. and Wilkes, E.W.: Theory of Elastic Stability, *Handbuch der Physik* VI a/3, Springer, Berlin 1973.

30. Koiter, W.T.: A basic open problem in the theory of elastic stability, *Springer Lecture Notes in Mathematics,* Vol. **503** (1976) 366-373.

31. Petryk, H.: The energy criteria of instability in time-independent inelastic solids, *Arch. Mech.* **43** (1991) 519-545.

32. Petryk, H.: On constitutive inequalities and bifurcation in elastic-plastic solids with a yield-surface vertex, *J. Mech. Phys. Solids,* **37** (1989) 265-291.

33. Nguyen, Q.S. and Petryk, H.: A constitutive inequality for time-independent dissipative solids, *C. R. Acad. Sci. Paris, t.* **312,** *Serie II,* (1991) 7-12.

34. Morrey, C.B.: Quasi-convexity and the lower semicontinuity of multiple integrals, *Pacific J. Math.*, **2** (1952) 25-53.

35. Ball, J.M.: Convexity conditions and existence theorems in nonlinear elasticity, *Arch. Rational Mech. Anal.*, **63** (1977) 337-403.

36. Graves, L.M.: The Weierstrass condition for multiple integral variation problems, *Duke Math. J.* **5** (1939) 656-660.

37. van Hove, L.: Sur l'extension de la condition de Legendre du calcul des variations aux intégrales multiples à plusieurs fonctions inconnues, *Proc. Kon. Ned. Acad. Wet.*, **50** (1947) 18-23.

38. Petryk, H.: Material instability and strain-rate discontinuities in incrementally nonlinear solids, *J. Mech. Phys. Solids,* **40** (1992) 1227-1250.

39. Abeyratne, R. and Knowles, J.K.: On the driving traction acting on a surface of strain discontinuity in a continuum, *J. Mech. Phys. Solids,* **38** (1990) 345-360.

40. Petryk, H.: The energy criteria of instability of plastic flow, *XVII IUTAM Congress*, Grenoble (1988).

Printed in the United States
By Bookmasters